INDUSTRIAL WATER POLLUTION CONTROL

McGraw-Hill Series in Water Resources and Environmental Engineering

Rolf Eliasen, Paul H. King, and Ray K. Linsley – Consulting Editors

Bailey and Ollis: *Biochemical Engineering Fundamentals*
Bishop: *Marine Pollution and Its Control*
Biswas: *Models for Water Quality Management*
Bouwer: *Groundwater Hydrology*
Canter: *Environmental Impact Assessment*
Chanlett: *Environmental Protection*
Chow, Maidment and Mays: *Applied Hydrology*
Eckenfelder: *Industrial Water Pollution Control*
Linsley and Franzini: *Water Resources Engineering*
Metcalf and Eddy, Inc.: *Wastewater Engineering Collection and Pumping of Wastewater*
Metcalf and Eddy, Inc.: *Wastewater Engineering: Treatment, Disposal, Reuse*
Peavy, Rowe, and Tchobanoglous: *Environmental Engineering*
Rich: *Low-Maintenance, Mechanically-Simple Wastewater Treatment Systems*
Sawyer and McCarty: *Chemistry for Environmental Engineering*
Steel and McGhee: *Water Supply and Sewerage*
Tchobanoglous, Theisen, and Eliassen: *Solid Wastes: Engineering Principles and Management Issues*

INDUSTRIAL WATER POLLUTION CONTROL

Second Edition

W. Wesley Eckenfelder, Jr.
Distinguished Professor
Vanderbilt University
Board Chairman
Aware, Inc.

McGraw-Hill Book Company

New York St. Louis San Francisco Auckland Bogotá Caracas
Colorado Springs Hamburg Lisbon London Madrid Mexico Milan
Montreal New Delhi Oklahoma City Panama Paris San Juan
São Paulo Singapore Sydney Tokyo Toronto

This book was set in Times Roman.
The editors were B. J. Clark and John M. Morriss;
the production supervisor was Louise Karam;
the cover was designed by Carla Bauer.
Project supervision was done by Santype International Limited.
R. R. Donnelley & Sons Company was printer and binder.

Industrial Water Pollution Control

1234567890DOCDOC89321098
ISBN 0-07-018903-X

Library of Congress Cataloging-in-Publication Data
Eckenfelder, W. Wesley (William Wesley), (date).
Industrial water pollution control.
(McGraw-Hill series in water resources and environmental engineering)
Includes index.
1. Sewage—Purification. 2. Factory and trade waste.
I. Title. II. Series.
TD745.E23 1989 628.3 87-36209
ISBN 0-07-018903-X
ISBN 0-07-018904-8 (solutions manual)

ABOUT THE AUTHOR

W. Wesley Eckenfelder, Jr., received degrees in Civil Engineering from Manhattan College and Environmental Engineering from Pennsylvania State University and New York University. He was Associate Professor at Manhattan College, Professor at the University of Texas, and is presently Distinguished Professor of Environmental Engineering at Vanderbilt University. He was a founder of the consulting firm of Hydroscience, Inc., a partner in Weston Eckenfelder & Assoc. (now Roy F. Weston, Inc.), and has served as Board Chairman of AWARE Incorporated since 1979.

He has developed continuing education courses in Water Pollution Control for Manhattan College, the University of Texas, Vanderbilt University, the University of Queensland, Australia, the American Institute of Chemical Engineers, and the Chemical Manufacturers Association.

He has served as consultant to over 150 industries, municipalities, consulting firms, and governmental agencies.

He is the author or editor of 18 books and over 200 technical papers in water pollution control including *Industrial Water Pollution Control* (1967), McGraw Hill; *Principles of Water Quality Management* (1980), Canners Books; and *Water Pollution Control* (1978), Jenkins Publishing Co. These books have been translated into Japanese and one into French. Professor Eckenfelder is on numerous editorial advisory boards including Water Technology Letters (U.K.) and Hazardous Waste and Hazardous Materials.

He is a member of numerous technical societies including the Water Pollution Control Federation, the American Chemical Society, the American Institute of Chemical Engineers, and the American Society of Civil Engineers. He is an honorary member of the International Association of Water Pollution Research and Control.

In 1974 he was awarded the SOCMA gold medal for excellence in environmental chemistry, in 1975 the Industrial Wastes Medal, and in 1981 the Thomas Camp Medal from the Water Pollution Control Federation.

He is a registered Professional Engineer in the State of Texas.

CONTENTS

PREFACE

It has been twenty years since this book was first published. During that period not only have regulations undergone a vast change, but both conventional technologies have been further refined and new technologies have been developed to meet increasingly more stringent water quality criteria. Effluent limitation on specific priority pollutants and toxicity to aquatic organisms have rendered many of the older conventional treatment facilities obsolete. The challenge today is to meet these new requirements in both an environmentally acceptable and cost effective manner.

In order to address these new challenges the present volume reviews the existing theory and addresses the application of state of the art technology to the solution to todays' problems in industrial water pollution control.

Of necessity, this book does not develop the detailed principles or the theory of processes applicable to specific areas of water pollution control. Rather it stresses the application of these theories and its application to specific industrial problems. Publications and texts are referenced for the reader who wishes a more detailed development of the theory.

Where applicable, case histories are used to illustrate the application of technology to specific industrial applications.

Problems drawn from field experience are used to illustrate the application of technology to a range of problem areas.

This book is intended as a text for the student in courses related to industrial water pollution control and as a guide for the engineer in industry, governmental agencies, and consulting engineering firms involved in developing state of the art solutions to industrial water pollution control problems.

While all specific problems cannot be answered in one text, it is hoped that this volume will provide guidance and direction to those faced with the increasingly more complex solutions of industrial water pollution control.

I would like to express my thanks for the many useful comments and suggestions provided by colleagues who reviewed this text during the course of its development, especially to Anthony Collins, Clarkson University; Benjamin Ewing, University of Illinois—Urbana; Houston Huckabay, Louisiana Technical University; Paul King, Northeastern University; and Anthony Wachinski, University of Colorado. Special recognition is due to Dee Biggert and Jackie Thomas of Aware, Inc., for manuscript preparation and to Dr. Y. Argamon of Technion University and Victor Boero of Vanderbilt University for technical reviews.

W. Wesley Eckenfelder

INDUSTRIAL WATER POLLUTION CONTROL

CHAPTER 1

SOURCE AND CHARACTERISTICS OF INDUSTRIAL WASTES

1.1 UNDESIRABLE WASTE CHARACTERISTICS

Depending on the nature of the industry and the projected uses of the waters of the receiving stream, various waste constituents may have to be removed before discharge. These may be summarized as follows:

1. Soluble organics causing depletion of dissolved oxygen. Since most receiving waters require maintenance of minimum dissolved oxygen, the quantity of soluble organics is correspondingly restricted to the capacity of the receiving waters for assimilation or by specified effluent limitations.
2. Suspended solids. Deposition of solids in quiescent stretches of a stream will impair the normal aquatic life of the stream. Sludge blankets containing organic solids will undergo progressive decomposition resulting in oxygen depletion and the production of noxious gases.
3. Trace organics. When a receiving water is to be used as a potable water supply, phenol and other organics discharged in industrial wastes will cause tastes and odors in the water. If these contaminants are not removed before discharge, additional water treatment will be required.
4. Heavy metals, cyanide, and toxic organics. The EPA has defined a list of toxic organic and inorganic chemicals that now appear as specific limitations in most permits. The identified priority pollutants are listed in Table 1.1.

TABLE 1.1
EPA list of organic priority pollutants

Compound name	Compound name
1. Acenaphthene†	Dichlorobenzidine†
2. Acrolein†	28. 3,3′-Dichlorobenzidine
3. Acrylonitrile†	
4. Benzene†	Dichloroethylenes† (1,1-dichloroethylene and 1,2-dichloroethylene)
5. Benzidine†	
6. Carbon tetrachloride† (tetrachloromethane)	29. 1,1-Dichloroethylene
	30. 1,2-*trans*-Dichloroethylene
	31. 2,4-Dichlorophenol†
Chlorinated benzenes (other than dichlorobenzenes)	Dichloropropane and dichloropropene†
7. Chlorobenzene	32. 1,2-Dichloropropane
8. 1,2,4-Trichlorobenzene	33. 1,2-Dichloropropylene (1,2-dichloropropene)
9. Hexachlorobenzene	
Chlorinated ethanes† (including 1,2-dichloroethane, 1,1,1-trichloroethane, and hexachloroethane)	34. 2,4-Dimethylphenol†
	Dinitrotoluene†
10. 1,2-Dichloroethane	35. 2,4-Dinitrotoluene
11. 1,1,1-Trichloroethane	36. 2,6-Dinitrotoluene
12. Hexachloroethane	37. 1,2-Diphenylhydrazine†
13. 1,1-Dichloroethane	38. Ethylbenzene†
14. 1,1,2-Trichloroethane	39. Fluoranthene†
15. 1,1,2,2-Tetrachloroethane	
16. Chloroethane (ethyl chloride)	Haloethers† (other than those listed elsewhere)
Chloroalkyl ethers† (chloromethyl, chloroethyl, and mixed ethers)	40. 4-Chlorophenyl phenyl ether
	41. 4-Bromophenyl phenyl ether
17. Bis(chloromethyl) ether	42. Bis(2-chloroisopropyl) ether
18. Bis(2-chloroethyl) ether	43. Bis(2-chloroethoxy) methane
19. 2-Chloroethyl vinyl ether (mixed)	
Chlorinated napthalene†	Halomethanes† (other than those listed elsewhere)
20. 2-Chloronapthalene	
	44. Methylene chloride (dichloromethane)
Chlorinated phenols† (other than those listed elsewhere; includes trichlorophenols and chlorinated cresols)	45. Methyl chloride (chloromethane)
	46. Methyl bromide (bromomethane)
	47. Bromoform (tribromomethane)
21. 2,4,6-Trichlorophenol	48. Dichlorobromomethane
22. *para*-Chloro-*meta*-cresol	49. Trichlorofluoromethane
23. Chloroform (trichloromethane)†	50. Dichlorodifluoromethane
24. 2-Chlorophenol†	51. Chlorodibromomethane
	52. Hexachlorobutadiene†
Dichlorobenzenes†	53. Hexachlorocyclopentadiene†
25. 1,2-Dichlorobenzene	54. Isophorone†
26. 1,3-Dichlorobenzene	55. Naphthalene†
27. 1,4-Dichlorobenzene	56. Nitrobenzene†

TABLE 1.1—(*continued*)

Compound name	Compound name
Nitrophenols† (including 2,4-dinitrophenol and dinitrocresol)	87. Trichloroethylene†
57. 2-Nitrophenol	88. Vinyl chloride† (chloroethylene)
58. 4-Nitrophenol	
59. 2,4-Dinitrophenol†	Pesticides and metabolites
60. 4,6-Dinitro-*o*-cresol	89. Aldrin†
	90. Dieldrin†
Nitrosamines†	91. Chlordane† (technical mixture and metabolites)
61. *N*-Nitrosodimethylamine	
62. *N*-Nitrosodiphenylamine	DDT and metabolites†
63. *N*-Nitrosodi-*n*-propylamine	92. 4-4′-DDT
64. Pentachlorophenol†	93. 4,4′-DDE (*p,p*′-DDX)
65. Phenol†	94. 4,4′-DDD (*p,p*′-TDE)
Phthalate esters†	Endosulfan and metabolites†
66. Bis(2-ethylhexyl) phthalate	95. α-Endosulfan-alpha
67. Butyl benzyl phthalate	96. β-Endosulfan-beta
68. Di-*n*-butyl phthalate	97. Endosulfan sulfate
69. Di-*n*-octyl phthalate	
70. Diethyl phthalate	Endrin and metabolites†
71. Dimethyl phthalate	98. Endrin
	99. Endrin aldehyde
Polynuclear aromatic hydrocarbons (PAH)†	
72. Benzo(a)anthracene (1,2-benzanthracene)	Heptachlor and metabolites†
73. Benzo(a)pyrene (3,4-benzopyrene)	100. Heptachlor
74. 3,4-Benzofluoranthene	101. Heptachlor epoxide
75. Benzo(k)fluoranthene (11,12-benzofluoranthene)	
76. Chrysene	Hexachlorocyclohexane (all isomers)†
77. Acenaphthylene	102. α-BHC-alpha
78. Anthracene	103. β-BHC-beta
79. Benzo(ghi)perylene (1,12-benzoperylene)	104. γ-BHC (lindane)-gamma
80. Fluorene	105. δ-BHC-delta
81. Phenanthrene	
82. Dibenzo(a,h)anthracene (1,2,5,6-dibenzanthracene)	Polychlorinated biphenyls (PCB)†
83. Indeno (1,2,3-cd) pyrene (2,3-*o*-phenylenepyrene)	106. PCB-1242 (Arochlor 1242)
84. Pyrene	107. PCB-1254 (Arochlor 1254)
85. Tetrachloroethylene†	108. PCB-1221 (Arochlor 1221)
86. Toluene†	109. PCB-1232 (Arochlor 1232)
	110. PCB-1248 (Arochlor 1248)
	111. PCB-1260 (Arochlor 1260)
	112. PCB-1016 (Arochlor 1016)
	113. Toxaphene†
	114. 2,3,7,8-Tetrachlorodibenzo-*p*-dioxin (TCDD)†

† Specific compounds and chemical classes as listed in the consent degree.

5. Color and turbidity. These present aesthetic problems even though they may not be particularly deleterious for most water uses. In some industries, such as pulp and paper, economic methods are not presently available for color removal.
6. Nitrogen and phosphorus. When effluents are discharged to lakes, ponds, and other recreational areas, the presence of nitrogen and phosphorus is particularly undesirable since it enhances eutrophication and stimulates undesirable algae growth.
7. Refractory substances resistant to biodegradation. These may be undesirable for certain water-quality requirements. ABS (alkyl benzene sulfonate) from detergents is substantially nondegradable and frequently leads to a persistence of foam in a watercourse. Some refractory organics are toxic to aquatic life.
8. Oil and floating material. These produce unsightly conditions and in most cases are restricted by regulations.
9. Volatile materials. Hydrogen sulfide and other volatile organics will create air-pollution problems and are usually restricted by regulation.

1.2 SOURCES AND CHARACTERISTICS OF WASTEWATERS

The volume and strength of industrial wastewaters are usually defined in terms of units of production (e.g., gallons per ton of pulp or cubic meters per tonne of pulp and pounds of BOD per ton of pulp or kilograms of BOD per tonne of pulp for a pulp-and-paper-mill waste) and the variation in characteristic by a statistical distribution. In any one plant there will be a statistical variation in waste-flow characteristic. The magnitude of this variation will depend on the diversity of products manufactured and of process operations contributing waste, and on whether the operations are batch or continuous. Good housekeeping procedures to minimize dumps and spills will reduce the statistical variation. Typical plots of a sequence of batch processes are shown in Fig. 1-1. Variation in waste flow and characteristics within a single plant are shown in Fig. 1-2.

Wide variation in waste flow and characteristics will also appear among similar industries, e.g., the paperboard industry. This is due to differences in housekeeping and water reuse as well as to variations in the production processes. Very few industries are identical in their sequence of process operations; as a result, an industrial waste survey is usually required to establish waste loadings and their variations. Variations for several industries are shown in Table 1.2. The variation in suspended solids and BOD discharge from 11 paperboard mills is shown in Fig. 1-3.

1.3 INDUSTRIAL WASTE SURVEY

The industrial waste survey involves a definite procedure designed to develop a flow and material balance of all processes using water and producing wastes and to establish the variation in waste characteristics from specific process operations

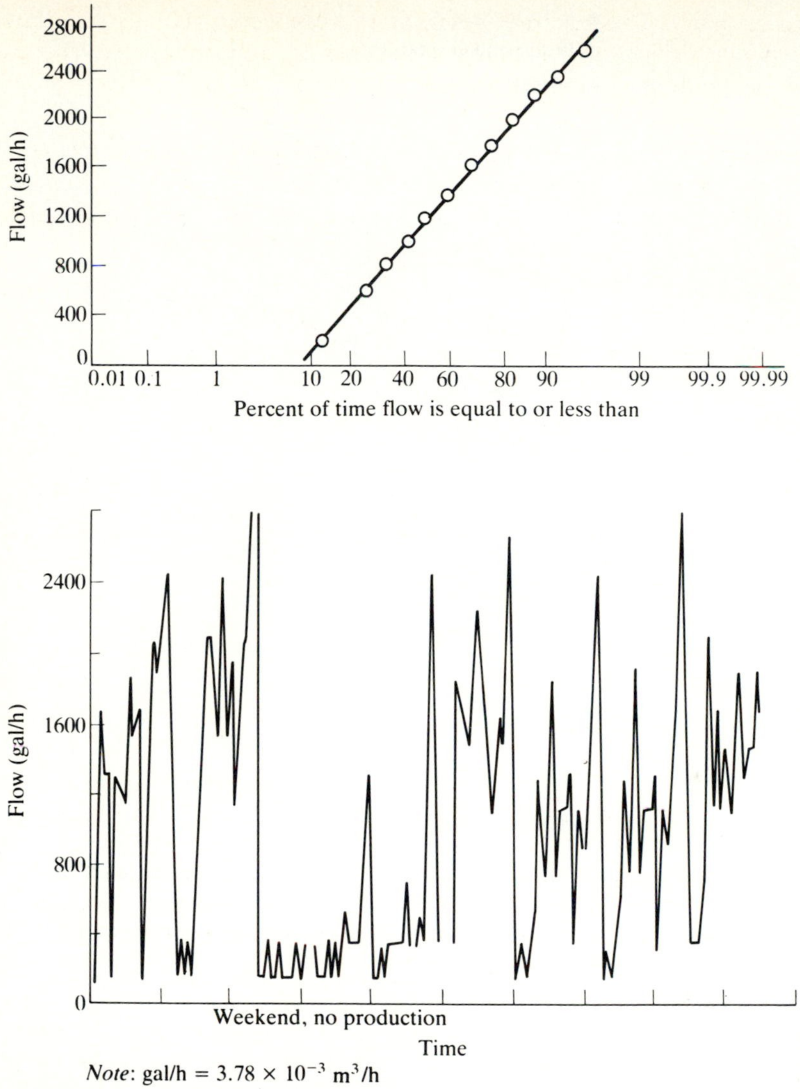

FIGURE 1-1
Variation in flow from a batch operation.

as well as from the plant as a whole. The results of the survey should establish possibilities for water conservation and reuse and finally the variation in flow and strength to undergo wastewater treatment.

The general procedure to be followed in developing the necessary information with a minimum of effort can be summarized in four steps:

1. Develop a sewer map from consultation with the plant engineer and an inspection of the various process operations. This map should indicate possible sampling stations and a rough order of magnitude of the anticipated flow.

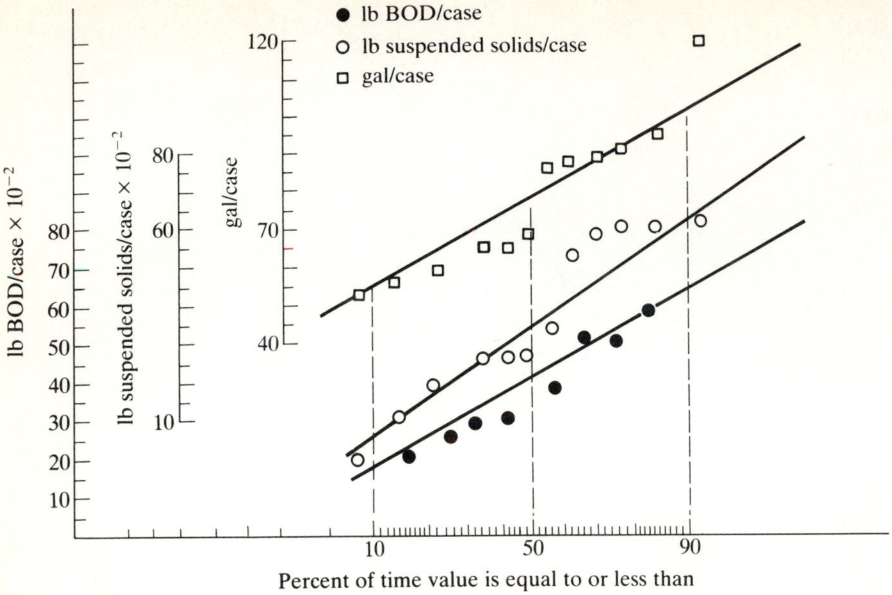

Note:
lb = 0.45 kg
gal = 3.78×10^{-3} m^3

FIGURE 1-2
Daily variation in flow and characteristics: tomato waste.

TABLE 1.2
Variation in flow and waste characteristics for some representative industrial wastes

Waste	Flow, gal/production unit % frequency			BOD, lb/production unit % frequency			Suspended solids, lb/ production unit % frequency		
	10	50	90	10	50	90	10	50	90
Pulp and paper†	11,000	43,000	74,000	17.0	58.0	110.0	26.0	105.0	400.0
Paperboard†	7,500	11,000	27,500	10	28	46	25	48	66
Slaughterhouse‡	165	800	4,300	3.8	13.0	44	3.0	9.8	31.0
Brewery§	130	370	600	0.8	2.0	44	0.25	1.2	2.45
Tannery¶	4.2	9.0	13.6	575††	975	1400	600††	1900	3200

† Tons paper production.

‡ 1000 lb live weight kill.

§ bbl beer.

¶ Pounds of hides; sulfides as S vary from 260 mg/l (10%) to 1230 mg/l (90%).

†† As mg/l.

Note:
gal = 3.78×10^{-3} m^3
lb = 0.45 kg
ton = 907 kg
bbl beer = 0.164 m^3

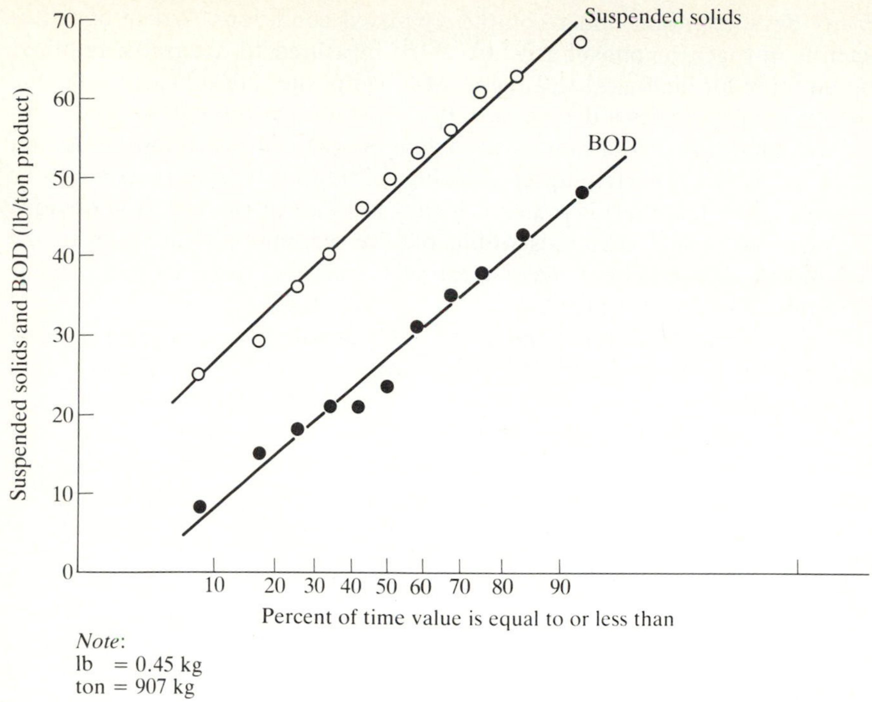

FIGURE 1-3
Variations in suspended solids and BOD from eleven paperboard mills.

2. Establish sampling and analysis schedules. Continuous samples with composites weighted according to flow are the most desirable, but these either are not always possible or do not lend themselves to the physical sampling location. The period of sample composite and the frequency of sampling must be established according to the nature of the process being investigated. Some continuous processes can be sampled hourly and composited on an 8-, 12-, or even 24-h basis, but those that exhibit a high degree of fluctuation may require a 1- or 2-h composite and analysis. More frequent samples are rarely required, since most industrial waste-treatment processes have a degree of built-in equalization and storage capacity. Batch processes should be composited during the course of the batch dump.

The analyses to be run on the samples depend both on the characteristics and on the ultimate purpose of the analysis. For example, pH must be run on grab samples since it is possible in some cases for compositing to result in neutralization of highly acidic and basic wastes, and this would yield highly misleading information for subsequent design. Variations in BOD loading may require 8-h or shorter composites for certain biological-treatment designs involving short detention periods, while 24-h composites will usually suffice for aerated lagoons with many days' retention under completely mixed conditions. Where constituents such as nitrogen or phosphorus are to be measured to determine required nutrient addition for biological treatment, 24-h composites are sufficient since the biological system possesses a degree of buffer capacity. One exception is the presence of toxic discharges to a biological system. Since a one-shot dose of certain toxic materials can completely upset a biological-treatment process, continuous monitoring of such materials is required if they are known to exist. It is obvious that the presence of such materials would require separate consideration in the waste-treatment design. Other waste-treatment processes may require similar considerations in sampling schedules.

The selected method of flow measurement will usually be contingent on the physical location to be sampled. When the waste flows through a sewer, it is frequently possible to measure the velocity of flow and the depth of water in the sewer and calculate the flow from the continuity equation. Since $Q = AV$, the area in a partially filled sewer can be determined given the depth from Fig. 1-4. This method applies only for partially filled sewers of constant cross section. The

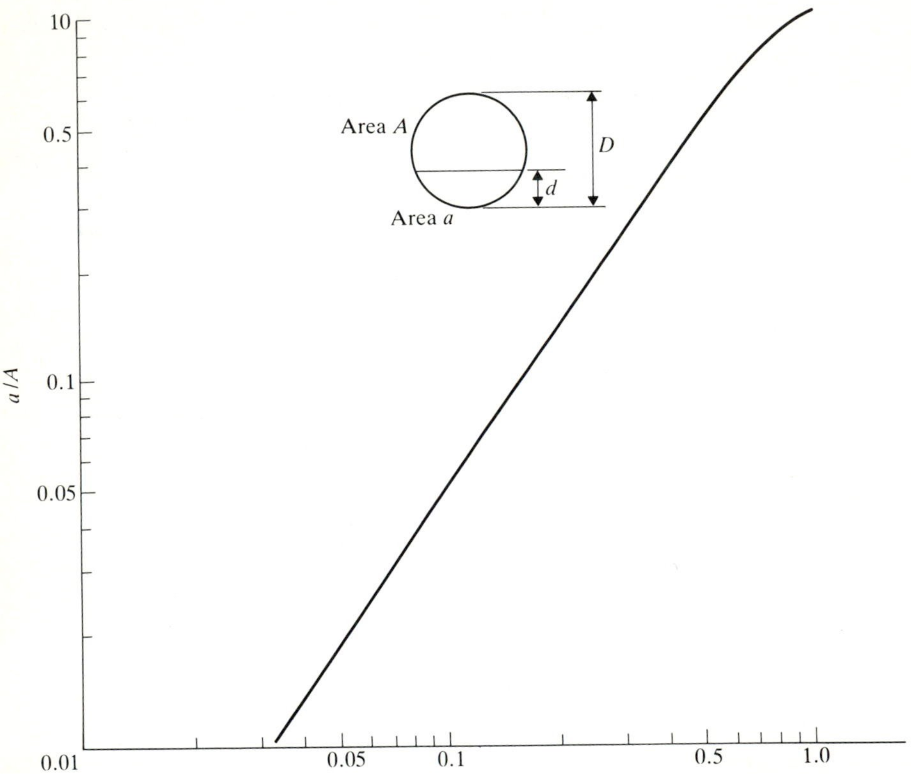

FIGURE 1-4
Determination of waste flow in partially filled sewers.

average velocity of flow can be estimated as 0.8 of the surface velocity timed from a floating object between manholes. More accurate measurements can be obtained by the use of a current meter. In gutters or channels, either a small weir can be constructed or the flow can be estimated as above by measuring the velocity and depth of flow in the channel. In some cases the flow can be obtained from the pumping rate and the duration of pumping of a wastestream. Total waste flow from an industrial plant can be measured by use of a weir or other suitable measuring device. In certain instances the daily waste flow can be estimated from water consumption records.

3. Develop a flow-and-material-balance diagram. After the survey data are collected and the samples analyzed, a flow-and-material-balance diagram should be developed that considers all significant sources of waste discharge. How closely the summation of the individual sources checks the measured total effluent provides a check on the accuracy of the survey. A typical flow-and-material-balance diagram for a corn-processing plant in shown in Fig. 1-5.
4. Establish statistical variation in significant waste characteristics. As was previously shown, the variability of certain waste characteristics is significant for waste-treatment plant design. These data should be prepared as a probability plot showing frequency of occurrence.

Data from industrial waste surveys are highly variable and are usually susceptible to statistical analysis. Statistical analysis of variable data provides the

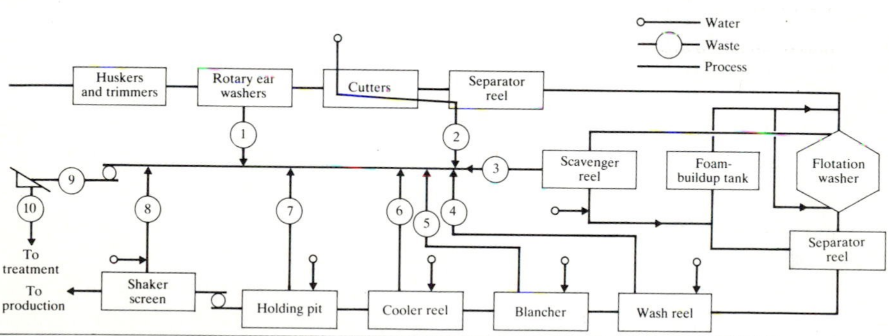

Line from→	1 Washer	2 Cutters	3 Reel	4 Reel	5 Blanch	6 Cool	7 Hold	8 Shaker	9 Sewer	10 Screen
to→				Sewer					Screen	Treatment
Flow, g/min	21.7	27.0	10.4	18.0	4.5	24.5	16.9	2.1	125.1	121
BOD, lb/d	2,500	2,300	390	973	610	1630	186		8,600	6250
COD, lb/d	3,640	4,640	555	1030	870	2140	192		13,000	9980
SS, lb/d	1,820	2,480	184	281	144	530	50		5,500	1700
VSS, lb/d	1,740	2,360	95	91	92	266	38		4,700	1900
Analysis:										
BOD, mg/l	9,830	7,112	3130	4600	11,300	5630	918		5,730	6200
COD, mg/l	14,000	14,400	4450	4780	16,100	7280	950		3,670	6030
SS, mg/l	6,950	7,660	1460	1300	2,670	1830	250		3,670	1170
VSS, mg/l	6,690	7,290	760	420	1,710	910	190		3,140	1030

Note:
gal/min = 3.78×10^{-3} m^3/min
lb/d = 0.45 kg/d

FIGURE 1-5
Waste flow diagram and material balance at a corn plant.

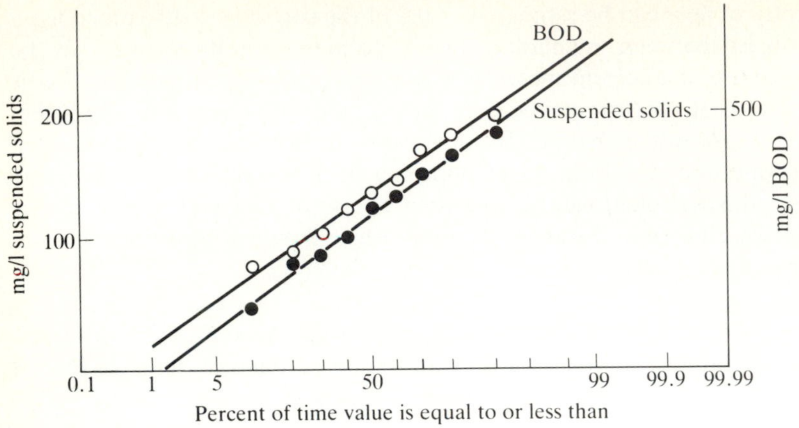

FIGURE 1-6
Probability of occurrence of BOD and suspended solids in raw waste.

basis for process design. The data are reported in terms of frequency of occurrence of a particular characteristic, which is that value of the characteristic that may be expected to be equaled or not exceeded 10, 50, or 90 percent of the time. The 50 percent chance value is approximately equal to the average. Correlation in this manner will linearize variable data, as shown in Fig. 1-6. The probability of occurrence of any value, such as flow, BOD, or suspended solids, may be determined as shown in the probability plot.

The suspended solids and BOD values are each arranged in order of increasing magnitude. n is the total number of solids or BOD values and m is the assigned serial number for 1 to n. The plotting positions $m/(n + 1)$ are equivalent to the percent occurrence of the value. The actual values are then plotted against the percent occurrences on probability paper, as indicated in Fig. 1-6. A line of best fit may usually be drawn by eye or, if desirable, it may be calculated by standard statistical procedures. The probability of occurrence of any values can now be obtained. The statistical calculation is shown in Example 1.1. In order to extrapolate the results of an industrial waste survey to future production, it is desirable to relate waste flow and loading to production schedules. Since some effluent-producing operations do not vary directly with production increase or decrease, the scale-up is not always linear. This is true of the cannery operation shown in Fig. 1-7, in which six process operations were independent of the number of washing and cleaning rigs in operation.

Example 1.1. For small amounts of industrial waste survey data (i.e., less than 20 datum points), the statistical correlation procedure is as follows:

1. Arrange the data in increasing order of magnitude (first column of Table 1.3).
2. In the second column of Table 1.3, m is the assigned serial number from 1 to n where n is the total number of values.

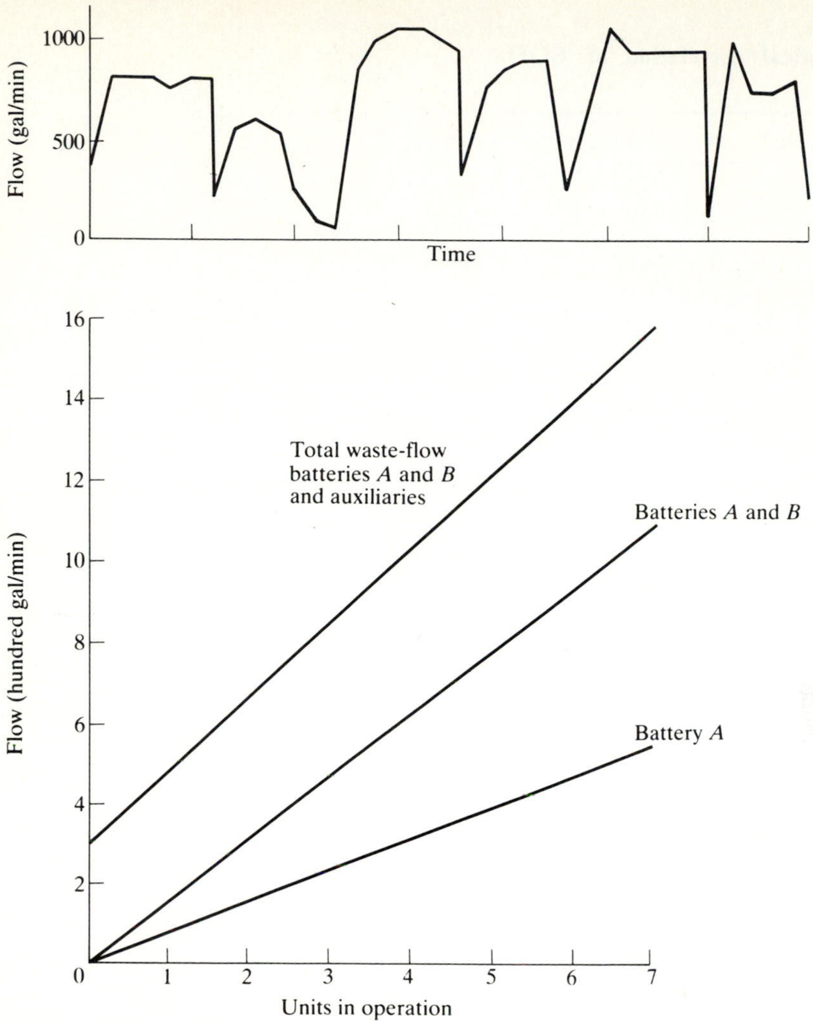

Note: gal/min = 3.78×10^{-3} m^3/min

FIGURE 1-7
Variation in plant waste flow from unit operations.

3. The plotting position is determined by dividing the total number of samples into 100 and assigning the first value as one-half this number (third column of Table 1.3):

$$\text{Plotting position} = \frac{100}{n} + \text{previous probability}$$

For $m = 5$:

$$\text{Plotting position} = \frac{100}{9} + 38.85$$

$$= 49.95$$

TABLE 1.3
Statistical correlation of BOD data

BOD, mg/l	m	Plotting position
200	1	5.55
225	2	16.65
260	3	27.75
315	4	38.85
350	5	49.95
365	6	61.05
430	7	72.15
460	8	83.75
490	9	94.35

4. These data are illustrated in Fig. 1-8. The standard deviation of these data (S) is calculated by

$$S = \frac{X_{84.1\%} - X_{15.9\%}}{2}$$

From Fig. 1-8:

$$S = \frac{436 - 254}{2}$$

$$= 91$$

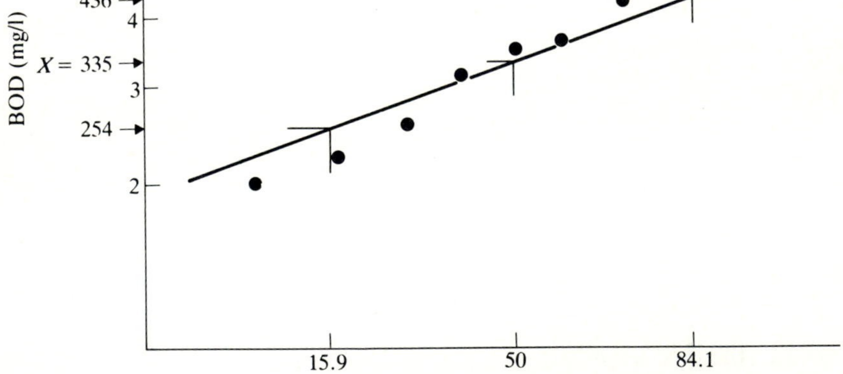

FIGURE 1-8

and mean:

$$\bar{X} = X_{50.0\%}$$
$$= 335$$

When large numbers of data are to be analyzed it is convenient to group the data for plotting, for example, 0 to 50, 51 to 100, 101 to 150, etc. The plotting position is determined as $m/(n + 1)$, where m is the cumulative number of points and n is the total number of observations. The statistical distribution of data serves several important functions in developing the industrial waste management program.

1.4 WASTE CHARACTERISTICS—ESTIMATING THE ORGANIC CONTENT

Although the interpretation of most of the waste characteristics is straightforward and definitive, special consideration must be given to the organic content.

The organic content of the waste can be estimated by each of four tests, although considerable caution should be exercised in interpreting the results:

1. The BOD test measures the biodegradable organic carbon and, under certain conditions, the oxidizable nitrogen present in the waste.
2. The COD test measures the total organic carbon with the exception of certain aromatics, such as benzene, which are not completely oxidized in the reaction. The COD test is an oxidation-reduction reaction, so other reduced substances, such as sulfides, sulfites, and ferrous iron, will also be oxidized and reported as COD.
3. The TOC test measures all carbon as CO_2, and hence the inorganic carbon (CO_2, HCO_3^-, and so on) present in the wastewater must be removed prior to the analysis or corrected for in the calculation.
4. The TOD test measures organic carbon and unoxidized nitrogen and sulfur.

Remember to exercise considerable caution in interpreting the test results and in correlating the results of one test with another.

The BOD by definition is the quantity of oxygen required for the stabilization of the oxidizable organic matter present after five days of incubation at 20°C. The BOD is conventionally formulated as a first-order reaction:

$$\frac{dL}{dt} = kL \tag{1.1}$$

which integrates to

$$L = L_o e^{-kt} \tag{1.2}$$

Since L, the amount of oxygen demand remaining at any time, is not known, Eq. (1.2) is reexpressed as

$$y = L_o(1 - e^{-kt})$$

where

$$y = L_o - L$$

or

$$y = L_o(1 - 10^{-k_{10}t}) \qquad (1.3)$$

By definition, L_o is the oxygen required to stabilize the total quantity of biologically oxidizable organic matter present; if k is known, L_5 is a fixed percentage of L_o. In order to interpret the BOD_5 obtained on industrial wastes, certain important factors must be considered.

It must be recognized that the oxygen consumed in the BOD test is the sum of (1) oxygen used for synthesis of the organic matter present and (2) endogenous respiration of the microbial cells as shown in Fig. 1-9. (The basis for this is discussed in detail on page 146.) The rate of oxygen utilization during phase 1 is 10 to 20 times that during phase 2. In most readily degradable substrates, phase 1 is complete in 24 to 36 hours.

In wastes containing readily oxidizable substrates, e.g., sugars, there will be a high oxygen demand for the first day, as the substrate is rapidly utilized, fol-

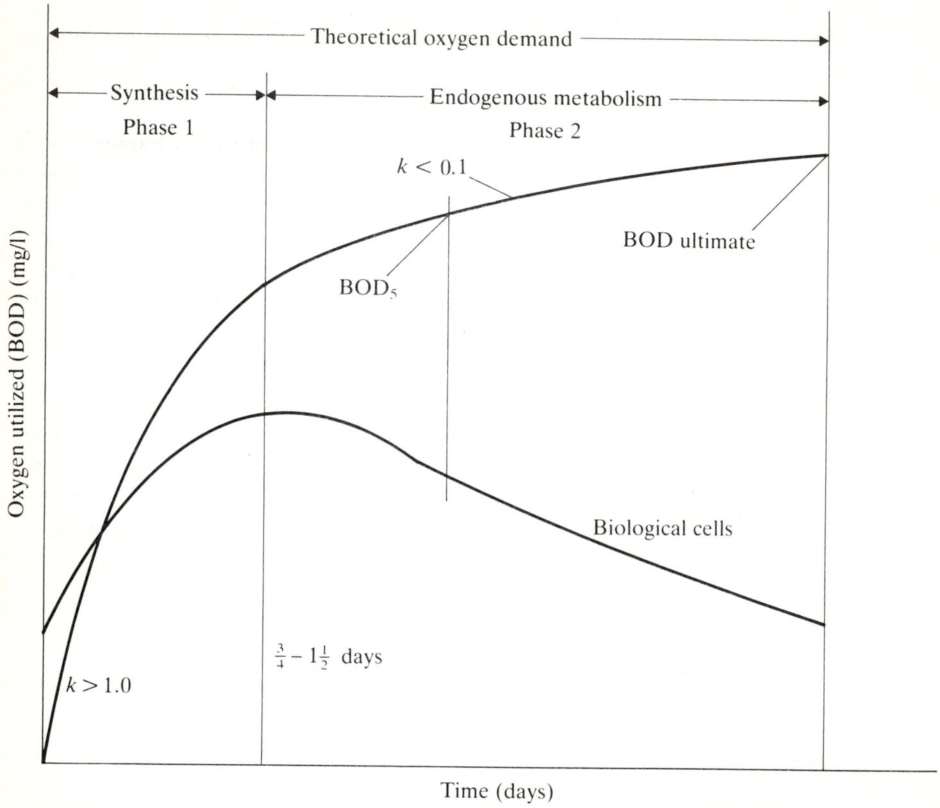

FIGURE 1-9
Reactions occurring in the BOD bottle.

TABLE 1.4
Average BOD rate constants at 20°C

Substance	k_{10}, day^{-1}
Untreated wastewater	0.15–0.28
High-rate filters and anaerobic contact	0.12–0.22
High-degree biotreatment effluent	0.06–0.10
Rivers with low pollution	0.04–0.08

lowed by a slower endogenous rate over the subsequent days of incubation. When these data are fitted to a first-order curve over a five-day period, a high K value will result because of the high initial slope of the curve. Conversely, a well-oxidized effluent will contain very little available substrate and for the most part only endogenous respiration will occur over the five-day incubation period. Since this rate of oxygen utilization is only a fraction of the rate obtaining in the presence of available substrate, the resulting K rate will be corresponding lower. Schroepfer[1] showed this by comparing the k_{10} rates of a well-treated sewage effluent and raw sewage containing a large quantity of available substrate. The average rate was 0.17 per day for the sewage and 0.10 per day for the effluent. It is obvious that under these conditions a direct comparison of five-day BODs is not valid. Typical rate constants are shown in Table 1.4.

Many industrial wastes are difficult to oxidize; they require a bacterial seed acclimated to the specific waste, or a lag period may occur which yields an erroneous interpretation of the five-day BOD values. Stack[2] showed that the five-day BOD of synthetic organic chemicals varied markedly depending on the type of acclimation of seed used. Some typical BOD curves are shown in Fig. 1-10. Curve A is normal exertion of BOD. Curve B is representative of what might be expected from sewage which slowly adjusted to the waste. Curves C and D are characteristic of nonacclimated seed or a toxic wastewater.

In the pulp and paper industry, both one- and five-day BODs have been run. The results are shown in Fig. 1-11. Note that for the untreated wastewater 70 percent of the five-day BOD is exerted in one day, while for the effluent, only 50 percent is exerted. Referring to the discussion on page 14, it can be interpreted that the one-day BOD reflects the soluble organics remaining in the sample. In many cases, the one-day BOD may provide a good control test for treatment plant performance.

Although modifications of the BOD procedure such as the short-term test proposed by Busch[3] may eliminate some of the errors resulting from the first-order assumption and the variation in k_{10} due to substrate level, these procedures have not found broad application in industry. It is essential, therefore, that the following factors be considered in the interpretation of the BOD for an industrial waste:

1. That the seed is acclimated to the waste and all lag periods are eliminated.

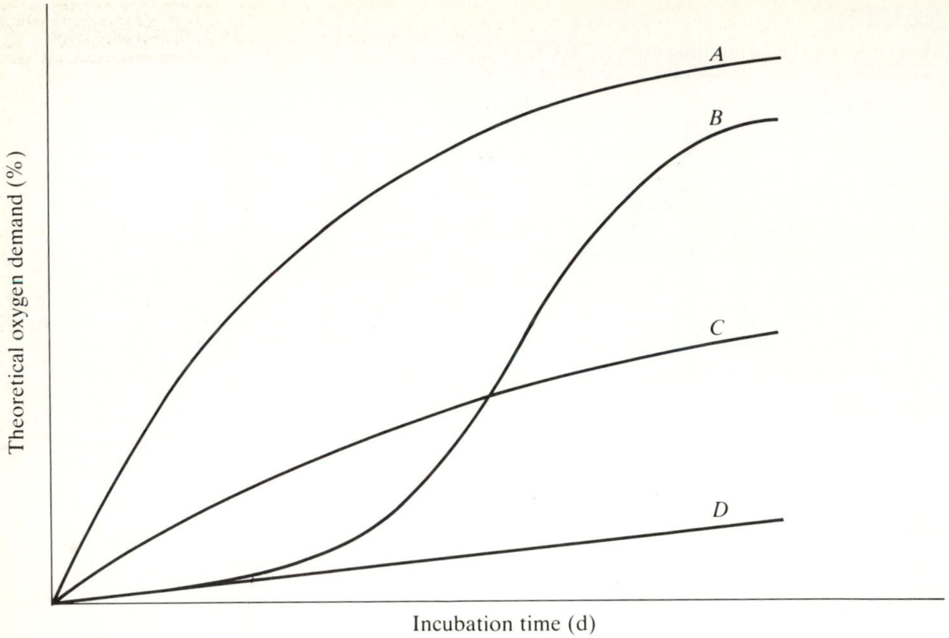

FIGURE 1-10
Characteristic BOD curves.

2. That long-term BOD tests establish the magnitude of k_{10} on both the waste and treated effluent. In the case of acidic wastes, all samples should be neutralized before incubation.

Toxicity in a waste is usually evidenced by so-called sliding BOD values, i.e., an increasing calculated BOD with increasing dilution. If this situation exists, it is necessary to determine the dilution value below which the computed BODs are consistent.

The COD test measures the total organic content of a waste which is oxidized by dichromate in acid solution. When a silver sulfate catalyst is used, the recovery for most organic compounds is greater than 92 percent. Since the COD will report virtually all organic compounds, many of which are either partially biodegradable or nonbiodegradable, it is proportional to the BOD only for readily assimilable substances, e.g., sugars. Such a case is shown in Fig. 1-12 for a readily assimilable pharmaceutical waste.

Because the five-day BOD will represent a different proportion of the total oxygen demand for raw wastes than for effluents, the BOD/COD ratio will frequently vary for effluents as compared with untreated wastes. There will be no correlation between BOD and COD when organic suspended solids are present in the waste that are only slowly biodegradable in the BOD bottle, and, therefore, filtered or soluble samples should always be used. Pulp and fiber in a paper-

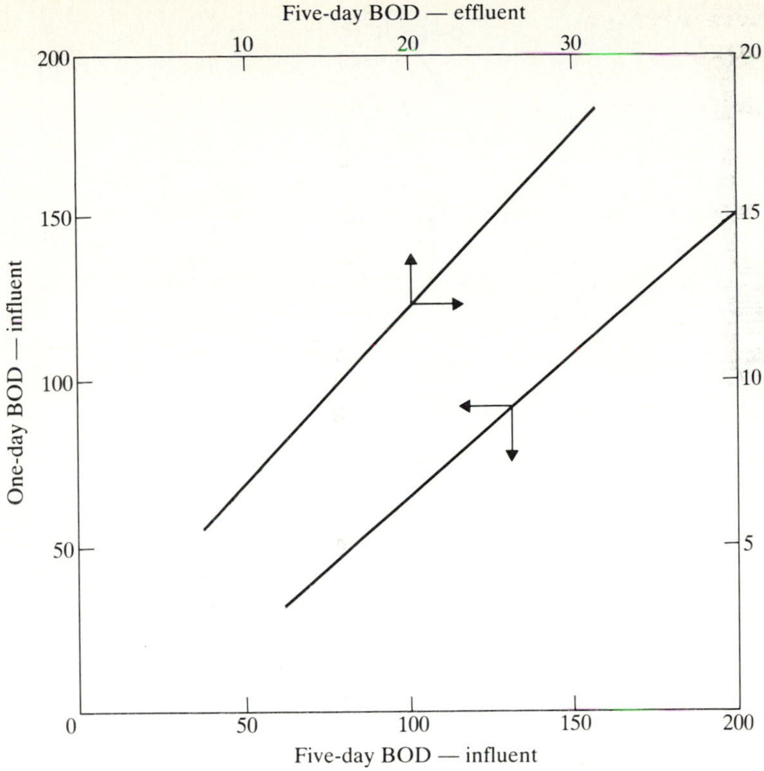

FIGURE 1-11
Relationship between one-day and five-day BOD for a pulp and paper-mill wastewater.

mill waste are an example. There will also be no correlation between BOD and COD in complex waste effluents containing refractory substances such as ABS. For this reason, treated effluents may exert virtually no BOD and yet exhibit a substantial COD.

Total organic carbon (TOC) has become a common and popular method of analysis due to its simplicity of measurement. There are presently several carbon analyzers on the market.

When considering routine plant control or investigational programs, the BOD is not a useful test due to the long incubation time. It is therefore important to develop correlations between BOD and COD or TOC.

The theoretical oxygen demand (THOD) of a wastewater is calculated as the oxygen required to oxidize the organics to end products:

$$C_6H_{12}O_6 + 6O_2 \rightarrow 6CO_2 + 6H_2O$$

$$\text{THOD} = \frac{6M_{O_2}}{M_{C_6H_{12}O_6}} = 1.07$$

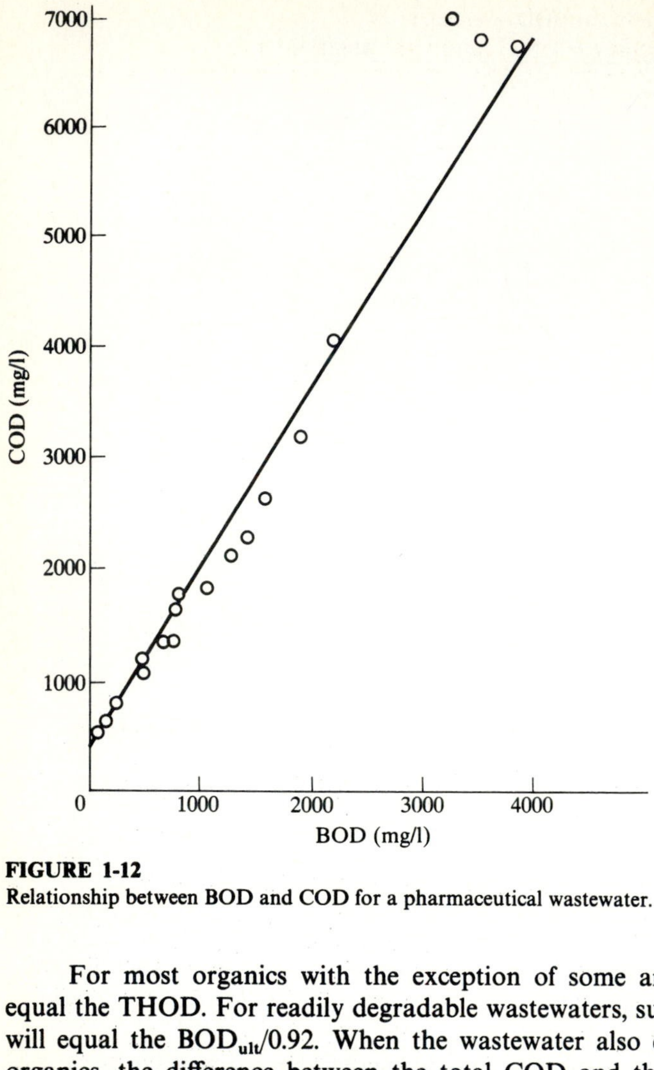

FIGURE 1-12
Relationship between BOD and COD for a pharmaceutical wastewater.

For most organics with the exception of some aromatics the COD will equal the THOD. For readily degradable wastewaters, such as a dairy, the COD will equal the $BOD_{ult}/0.92$. When the wastewater also contains nondegradable organics, the difference between the total COD and the $BOD_{ult}/0.92$ will represent the nondegradable content.

It has been found that some nondegradable organics will accumulate during biooxidation due to oxidation byproducts of the organics in the wastewater and byproducts of endogenous metabolism. Hence the effluent COD through biological treatment will increase over the influent nondegradable COD.

TOC is related to COD through a carbon-oxygen balance:

$$C_6H_{12}O_6 + 6O_2 \rightarrow 6CO_2 + 6H_2O$$

$$\frac{\text{COD}}{\text{TOC}} = \frac{6M_{O_2}}{6M_C} = 2.66$$

TABLE 1.5
Oxygen demand and organic carbon of industrial wastewaters[4]

Waste	BOD_5, mg/l	COD, mg/l	TOC, mg/l	BOD/TOC	COD/TOC
Chemical†	—	4,260	640	—	6.65
Chemical†	—	2,410	370	—	6.60
Chemical†	—	2,690	420	—	6.40
Chemical	—	576	122	—	4.72
Chemical	24,000	41,300	9,500	2.53	4.35
Chemical—refinery	—	580	160	—	3.62
Petrochemical	—	3,340	900	—	3.32
Chemical	850	1,900	580	1.47	3.28
Chemical	700	1,400	450	1.55	3.12
Chemical	8,000	17,500	5,800	1.38	3.02
Chemical	60,700	78,000	26,000	2.34	3.00
Chemical	62,000	143,000	48,140	1.28	2.96
Chemical	—	165,000	58,000	—	2.84
Chemical	9,700	15,000	5,500	1.76	2.72
Nylon polymer	—	23,400	8,800	—	2.70
Petrochemical	—	—	—	—	2.70
Nylon polymer	—	112,600	44,000	—	2.50
Olefin processing	—	321	133	—	2.40
Butadiene processing	—	359	156	—	2.30
Chemical	—	350,000	160,000	—	2.19
Synthetic rubber	—	192	110	—	1.75

† High concentration of sulfides and thiosulfates.

Depending on the organic in question, the COD/TOC ratio may vary from zero when the organic material is resistant to dichromate oxidation to 5.33 for methane. Since the organic content undergoes change during biological oxidation, it can be expected that the COD/TOC ratio will also change. This same rationale also applies to the BOD/TOC ratio. Values of BOD, COD, and TOC for a variety of organics and wastewaters is shown in Table 1.5. The calculations relative to BOD, COD, and TOC are illustrated in Example 1.2.

The BOD, COD, and TOC tests are gross measures of organic content and as such do not reflect the response of the wastewater to various types of biological treatment technologies. It is therefore desirable to partition the wastewater into several categories, as shown in Fig. 1-13. The distinction between sorbable and nonsorbable BOD is important in selecting process configuration for the control of sludge quality, as discussed on page 147.

Example 1.2. A wastewater contains the following:

150 mg/l ethylene glycol
100 mg/l phenol
40 mg/l sulfide (S^{2-})

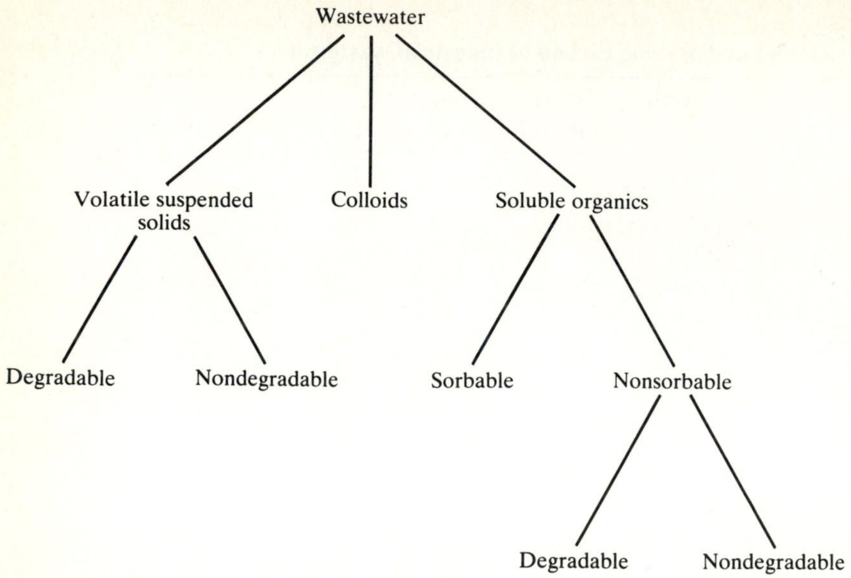

FIGURE 1-13
Partition of organic constituents of a wastewater.

125 mg/l ethylene diamine hydrate (ethylene diamine is essentially nonbiodegradable)

(*a*) Compute the COD and TOC.
(*b*) Compute the BOD_5 if the k_{10} is 0.2/day.
(*c*) After treatment, the BOD_5 is 25 mg/l. Estimate the COD ($k_{10} = 0.1$/day).

Solution

(*a*) The COD is computed:

Ethylene glycol

$$C_2H_6O_2 + 2.5O_2 \rightarrow 2CO_2 + 3H_2O$$

$$COD = \frac{2.5(32)}{62} \times 150 \text{ mg/l} = 194 \text{ mg/l}$$

Phenol

$$C_6H_6O + 7O_2 \rightarrow 6CO_2 + 3H_2O$$

$$COD = \frac{7(32)}{94} \times 100 \text{ mg/l} = 238 \text{ mg/l}$$

Ethylene diamine hydrate

$$C_2H_{10}N_2O + 2.5O_2 \rightarrow 2CO_2 + 2H_2O + 2NH_3$$

$$COD = \frac{2.5(32)}{78} \times 125 \text{ mg/l} = 128 \text{ mg/l}$$

Sulfide

$$S^{2-} + 2O_2 \rightarrow SO_4^{2-}$$

$$\text{COD} = \frac{2(32)}{32} \times 40 \text{ mg/l} = 80 \text{ mg/l}$$

The total COD is 640 mg/l.

The TOC is computed:

Ethylene glycol

$$\frac{24}{62} \times 150 \text{ mg/l} = 58 \text{ mg/l}$$

Phenol

$$\frac{72}{94} \times 100 \text{ mg/l} = 77 \text{ mg/l}$$

Ethylene diamine

$$\frac{24}{78} \times 125 \text{ mg/l} = 39 \text{ mg/l}$$

The total TOC is 174 mg/l.

(*b*) The ultimate BOD can be estimated:

$$(194 \text{ mg/l} + 238 \text{ mg/l} + 80 \text{ mg/l}) \times 0.92 = 471 \text{ mg/l}$$

Thus

$$\frac{\text{BOD}_5}{\text{BOD}_{\text{ult}}} = (1 - 10^{-(5 \times 0.2)}) = 0.9$$

BOD_5 is 471 mg/l × 0.9 = 424 mg/l.

(*c*) The BOD_{ult} of the effluent is

$$\frac{25 \text{ mg/l}}{1 - 10^{-(5 \times 0.1)}} = \frac{25 \text{ mg/l}}{0.7} = 36 \text{ mg/l}$$

The COD is 36/0.92 = 39 mg/l. Therefore the COD will be 128 mg/l + 39 mg/l + residual byproducts.

1.5 WATER REUSE AND IN-PLANT WASTE CONTROL

Although it is theoretically possible to completely close up many industrial process systems through water reuse, an upper limit on reuse is provided by product quality control. For example, a closed system in a paper mill will result in a continuous buildup of dissolved organic solids. This can increase the costs of slime control, cause more downtime on the paper machines, and under some conditions cause discoloration of the paper stock. Obviously, the maximum degree of reuse is reached before these problems occur.

Water requirements are also significant when reuse is considered. Water makeup for hydropulpers in a paper mill need not be treated for removal of suspended solids; on the other hand, solids must be removed from shower water on the paper machines to avoid clogging of the shower nozzles. Wash water for produce such as tomatoes does not have to be pure but usually requires chlorination to ensure freedom from microbial contamination.

Byproduct recovery frequently accompanies water reuse. The installation of save-alls in paper mills to recover fiber permits the reuse of treated water on cylinder showers. The treatment of plating-plant rinsewater by ion exchange yields a reusable chromic acid. There are many other similar instances in industry.

The best way to illustrate water reuse in industry is to cite several examples. Figure 1-14 shows simplified schematics of two water-reuse systems in a paperboard mill. When a disk filter is employed, the cloudy filtrate is returned to the

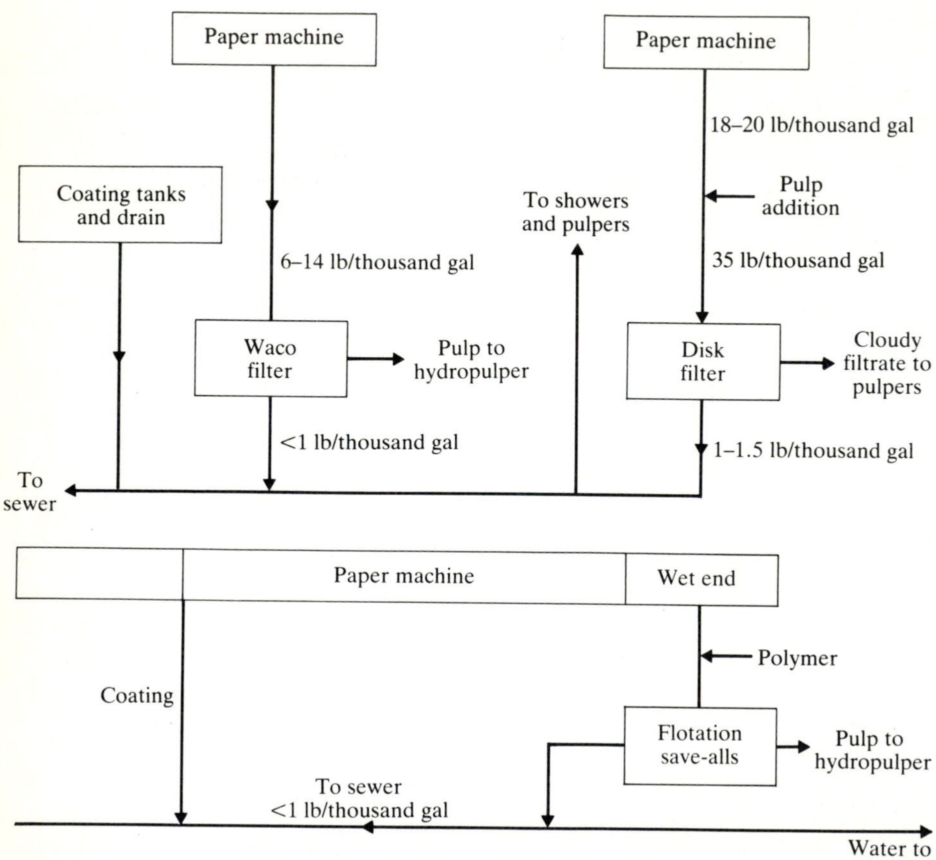

Note: lb/thousand gal = 0.120 kg/m^3

FIGURE 1-14
Pulp and fiber recovery in a board mill.

pulpers as makeup water and the clear filtrate is used for showers and other miscellaneous uses. The overflow is continuously discharged to the sewer. The use of a flotation save-all results in a similar operation. Waste from the clay-coating operations is usually discharged directly to the sewer.

Water conservation in brewery operations has been effected by reuse of the third rinse in kettle and vat cleaning for the next rinse, by reuse of cooling water for cleaning purposes, and by counterflow in pasteurizing and bottling operations. The total waste loading is reduced by removing the spent grains in a semidry state, by removing the yeast from the fermenters for filtration and drying, and by removing the sediment from the chill storage tanks as a slurry for separate disposal. In many cases these operations have yielded a salable byproduct which has more than offset the increased costs of recovery.

Reuse and makeup of sulfide dehairing solutions from tanning liquors have reduced both the volume and strength of wastes from this industry.

Water reuse in a corn-processing plant is shown in Table 1.6.

TABLE 1.6
Unit waste flows for corn-processing operation and possible changes to reduce flow

Unit	Line	Flow, gal/min	Water use	Possible change	Estimated new flow
Rotary ear	1	21.7	Loosens and removes sand, etc.	Screen 7 and reuse in washer	21.7
Cutting machines	2	27.0	Flumes kernels to flotation washer	Screen and recirculate	0.0
Scavenger reel	3	10.4	Flumes waste separated by flotation washer	Use smaller flow or remove as solid	5.2
Wash reel	4	18.0	Rewashes corn (not necessary)	Remove	8.0
Blancher	5	4.5	Overflow from makeup waste	No change	4.5
Cooling reel	6	24.5	Cool corn after blanching	Use smaller flow	10.0
Holding tank	7	16.9	Overflow from makeup water	Screen and reuse in rotary washer	0.0
Shaker screen	8	2.1	Flumes waste separated by reel	Remove as solid or reuse screen	0.0
Total		125.1			49.4

Note: gal/min = 3.78×10^{-3} m^3/min.

Spent caustic in a refinery is an alkaline waste, containing high concentrations of sulfides, mercaptans, and phenolates. Removal of spent caustic for separate treatment may markedly reduce wastewater-treatment costs and can in some cases result in a marketable product.

In the organic chemicals industry, an example might be drawn from the Union Carbide plant at South Charleston, West Virginia. The plant is reasonably representative of what might be expected from a major multiproduct chemicals complex. A detailed study of waste load reduction by in-plant changes showed that the present plant flow of 11.1 million gallons/day (42,000 m^3/d) and 55,700 pounds of BOD/day (25,300 kg/d) could be reduced to a flow of 8.3 million gal/day (31,400 m^3/d) and 37,100 lb BOD/day (16,800 kg/d). The ways of achieving these reductions are shown in Table 1.7. Equipment revision and additions and unit shutdowns refer to those units that would be replaced. Incineration involves the option of taking the more concentrated wastewater streams and, rather than discharging to the main sewer, segregating them for incineration. Reprocessing refers to taking the tank bottoms that still have product present and reprocessing them for further product recovery.

Frequently, wastestreams can be eliminated or reduced by process modifications or improvements. A notable example of this is the use of save-rinse and

TABLE 1.7
General analysis of in-plant modifications

Type of modifications	Description	Total RWL reduction, %
Equipment revision and additions	Self-explanatory	25
Unit shutdown	Shutdowns due to the age of the unit or the product. These shutdowns are not a direct result of pollution considerations, but they are somewhat hastened by these considerations	10
Scrubber replacement	Replacement of scrubbers associated with amine production by burning of the off-vapors	3
Segregation, collection, and incineration	Of specific concentrated wastewater streams	35
Raw material substitutions	Self-explanatory	3
Reprocessing	Collecting tail streams from specific processes, then putting streams through an additional processing unit to recover more product and concentrate the final wastestream	
Miscellaneous small projects	A variety of modifications which individually do not represent a large reduction in RWL	21

spray-rinse tanks in plating lines. This measure will bring about a substantial reduction in wastewater flow and strength. In dairies, equipment modifications to collect drippings will usually substantially reduce the BOD loading to the sewer. Substitution of sizing agents in textile plants may result in a lower net pollutional load for treatment.

There are many other examples from various industries. A critical evaluation of water reuse and byproduct recovery possibilities should be undertaken before the development of waste-treatment needs.

Segregation of incompatible wastestreams should be considered before treatment process design criteria are developed. In some older plants, however, this may be economically unfeasible or in some cases impossible. Segregation becomes a necessity when combination of wastestreams will result in potential safety hazards. In a plating plant, for example, a mixture of acidic metal rinses with cyanide streams could produce toxic HCN.

It is not uncommon for a portion of a waste flow to contain the majority of the suspended solids loading. Only this part of the flow need receive treatment for solids removal. This is true, for instance, in a tannery where 90 percent of the suspended solids emanates from the beam house in approximately 60 percent of the flow.

Frequently cooling water or other noncontaminated streams can be segregated and discharged directly to the receiving water.

Consideration of water reuse, byproduct recovery, and waste segregation will enable a revised material-balance-and-flow diagram to be made which will establish the basis for process design of waste-treatment facilities. This is illustrated in Example 1.3.

Example 1.3. An industrial waste survey has been conducted on a plant producing soap, detergent, and miscellaneous toilet articles. A sewer map of the plant is shown in Fig. 1-15, with sampling stations denoted. The mean analyses found during the survey are summarized in Table 1.8.

TABLE 1.8

Waste source	Sampling station	COD, mg/l	BOD, mg/l	SS, mg/l	ABS, mg/l	Flow, gal/min
Liquid soap	*D*	1,100	565	195	28	300
Toilet articles	*E*	2,680	1,540	810	69	50
Soap production	*R*	29	16	39	2	30
ABS production	*S*	1,440	380	309	600	110
Powerhouse	*P*	66	10	50	0	550
Condenser	*C*	59	21	24	0	1100
Spent caustic	*B*	30,000	10,000	563	5	2
Tank bottoms	*A*	120,000	150,000	426	20	1.5
Fly ash	*F*			6750		10
Main sewer		450	260	120	37	2150

Note: gal/min = 3.78×10^{-3} m^3/min.

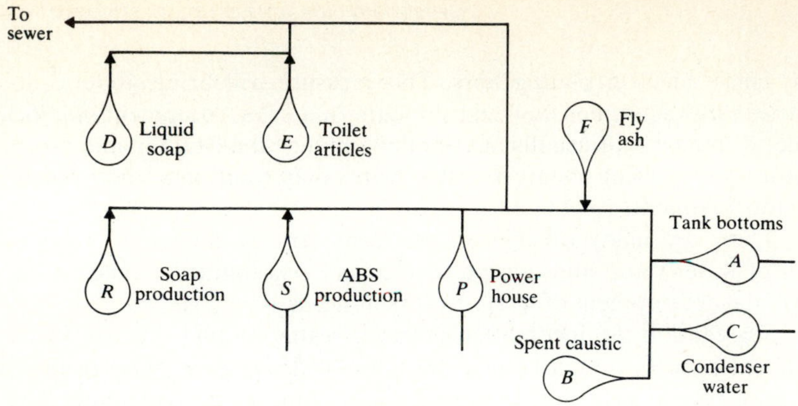

	Line									
	D	*E*	*R*	*S*	*P*	*F*	*B*	*C*	*A*	Main sewer
COD, lb/d	3950	1600	10	1900	440		720	780	2150	11,550
BOD, lb/d	2030	920	6	500	65		240	280	2700	6,740
SS, lb/d	700	485	14	410	330	800	14	320	8	3,080
ABS, lb/d	100	41		800						941
Flow, gal/min	300	50	30	110	550	10	2	1100	1.5	2,150

Note:
lb/d = 0.45 kg/d
mg/l × gal/min = 1.20 × 10^{-2} lb/d

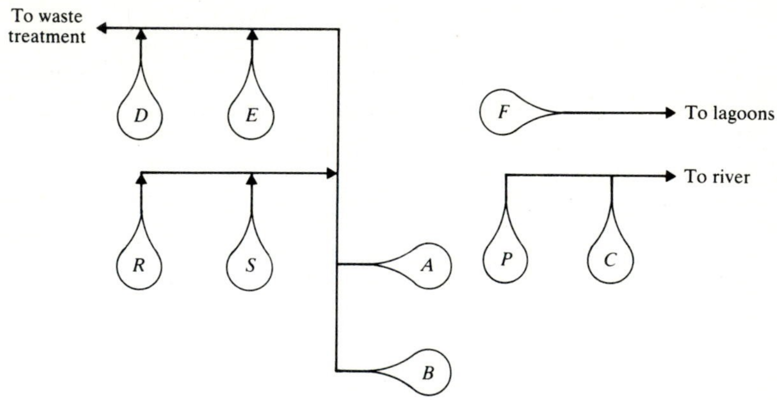

	Line						
	D	*E*	*R*	*S*	*A*	*B*	Total
COD, lb/d	3950	1600	10	1900	2150	720	10,330
BOD, lb/d	2030	920	6	500	2700	240	6,396
SS, lb/d	700	485	14	410	8	14	1,631
ABS, lb/d	100	41		800			941
Flow, gal/min	300	50	30	110	1.5	2	493.5

Note:
lb/d = 0.45 kg/d
mg/l × gal/min = 1.20 × 10^{-2} lb/d
Resulting concentrations to treatment:

COD, mg/l	1720	SS, mg/l	270
BOD, mg/l	1060	ABS, mg/l	157

FIGURE 1-15

(*a*) Develop a flow-and-material-balance diagram for this plant.

(*b*) The fly ash, tank bottoms, and spent caustic are discharged from batch-holding tanks. Neglecting internal sewer changes, indicate waste-segregation procedures you could institute and prepare a revised flow-and-material-balance diagram for this plant.

This is shown in Fig. 1-15.

1.6 STORMWATER CONTROL

In most industrial plants, it is now necessary to contain and control pollutional discharges from stormwater. Pollutional discharges can be minimized by providing adequate diking around process areas, storage tanks, and liquid transfer points with drainage into the process sewer. Contaminated stormwater is usually collected based on a storm frequency for the area in question (e.g., a ten-year storm) in a holding basin. The collected water is then passed through the wastewater-treatment plant at a controlled rate. A total storm-runoff flow and contaminant loading of a refinery petrochemical installation is shown in Fig. 1-16. An example of stormwater control is shown in Example 1.4.

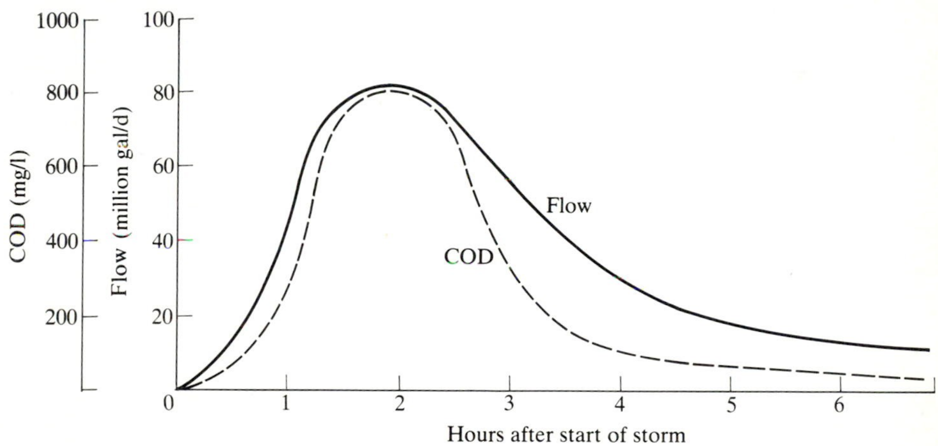

Note: million gal/d = 3.78 × 10^3 m^3/d

FIGURE 1-16
Total storm runoff flow and contaminant loading for a refinery installation.

Example 1.4. If runoff is to be stored or surged prior to eventual treatment, anticipated rainfall volumes must be estimated. The probable rainfalls for periods of one day to one year are given in Table 1.9 and Figure 1-17 for the two-year storm.

The areas and runoff coefficients associated with a refinery complex are shown in Table 1.10.

Determine:

(*a*) The cumulative runoff (million gal) as a function of time (days).

(*b*) The required minimum return rate for treatment (million gal/d or m^3/d) for storage periods of 10, 30, 60, and 120 days.

(*c*) The required storage capacity (million gal or m^3) for storage periods of 10, 30, 60, and 120 days.

TABLE 1.9

Time period, d	Cumulative rainfall, in
1	2.5
5	3.7
10	4.5
30	6.6
60	10.7
120	18.0
240	30.0
360	41.0

Note: in = 0.0254 m = 2.54 cm.

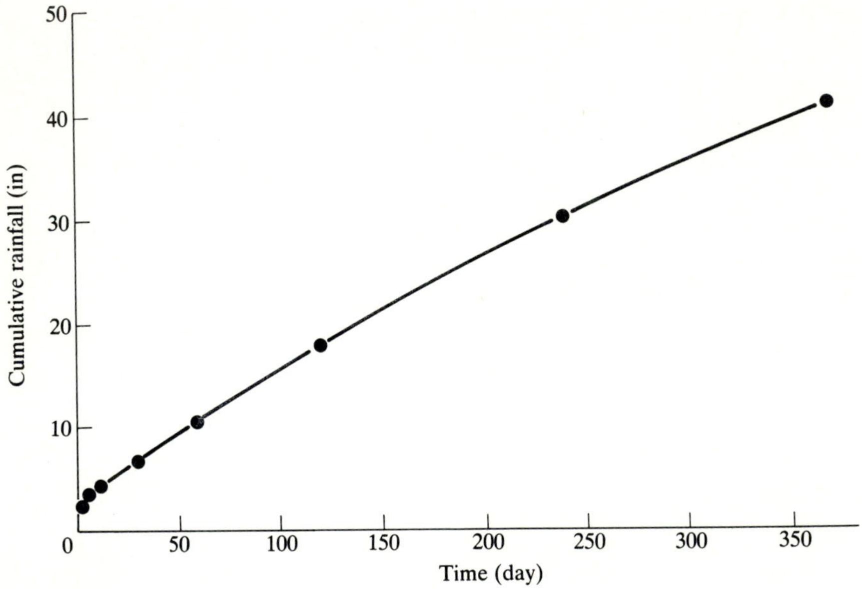

Note: in = 0.0254 m

FIGURE 1-17
Two-year recurrence level probability rainfalls.

Solution

(*a*) Total effective areas

$$= 21.8 \times 1.0 + 115.4 \times 0.5 + 166.9 \times 0.7 + 176.7 \times 0.5 + 55.6 \times 0.9$$

$$= 334.7 \text{ acre } (1.36 \times 10^6 \text{ m}^2)$$

TABLE 1.10

Type of surface	Area, acres	Runoff coefficient, C
Water surface	21.8	1.0
Dirt or gravel	115.4	0.5
Bermed tank farm	166.9	0.7
Grass	176.7	0.5
Asphalt or concrete	55.6	0.9

Note: acre = 4.05×10^3 m^2.

Based on the table of cumulative rainfall with various time periods the cumulative runoff as a function of time can be estimated by

$$q = i(CA)$$

in which CA is effective area and i is the rainfall intensity. An example calculation for a 10-day period at 4.5 in (0.114 m) rainfall is

$$q = 4.5 \text{ in} \times 334.7 \text{ acre} \times \frac{43{,}560 \text{ ft}^2}{\text{acre}} \times \frac{0.0833 \text{ ft}}{\text{in}} \times \frac{7.48 \text{ gal}}{\text{ft}^3}$$

$$= 4.088 \times 10^7 \text{ gal or } 40.9 \text{ million gal } (1.55 \times 10^5 \text{ m}^3)$$

The results are tabulated in Table 1.11 and plotted (see Fig. 1-18).

TABLE 1.11

Time, d	Runoff, million gal
1	22.7
5	33.6
10	40.9
30	60.0
60	97.2
120	163.5
240	272.5
360	372.4

Note: million gal = 3.78×10^3 m^3.

(*b*) The minimum allowable return rate for treatment from a storage basin is defined by the slope of a straight line connecting the origin and a point on the cumulative runoff curve representing the maximum practical storage period. A sample calculation for 10 days is

$$\text{Return rate} = \frac{40.9 - 0}{10 - 0} \text{ million gal/d}$$

$$= 4.1 \text{ million gal/d } (1.55 \times 10^4 \text{ m}^3\text{/d})$$

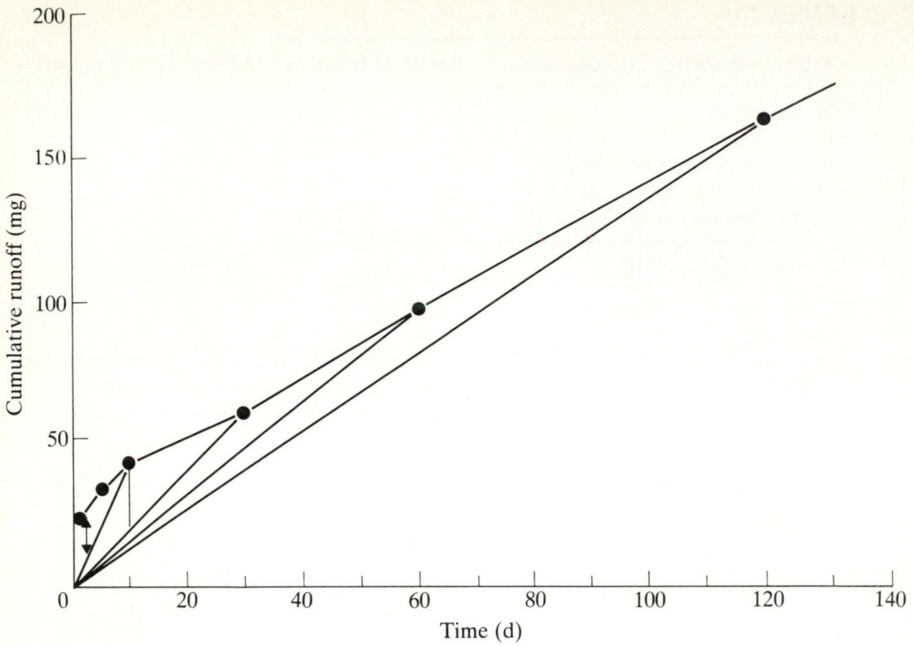

Note: million gal = 3.78×10^3 m^3

FIGURE 1-18
Cumulative runoff curve.

(*c*) The required storage is defined by the maximum vertical distance between the cumulative runoff curve and the straight line depicting the rate of removal. An example for 10 days is

$$\text{Required storage} = 18 \text{ million gal } (6.8 \times 10^4 \text{ m}^3)$$

The results are given in Table 1.12.

TABLE 1.12

Storage period, d	Return rate, million gal/d	Required storage, million gal
10	4.09	18.0
30	2.00	20.9
60	1.62	24.0
120	1.36	28.0

Note:
million gal/d = 3.78×10^3 m^3/d
million gal = 3.78×10^3m^3

PROBLEMS

1.1. The 4-h composite COD data from a brewery effluent over a 7-d period is as follows:

980			
2800	3200	6933	3325
1380	3175	1240	6000
1250	3850	580	3100
720	2870	710	2500
8650	2600	3410	1830
7200	2743	2910	3225
2800	3600	8300	2370
2570	4066	2950	1380
1780	1550	2230	2600

Develop a statistical plot of these data in order to define the 50 and the 90 percent values.

1.2. A tomato-processing plant is shown in Fig. P1-2. The industrial waste survey data are

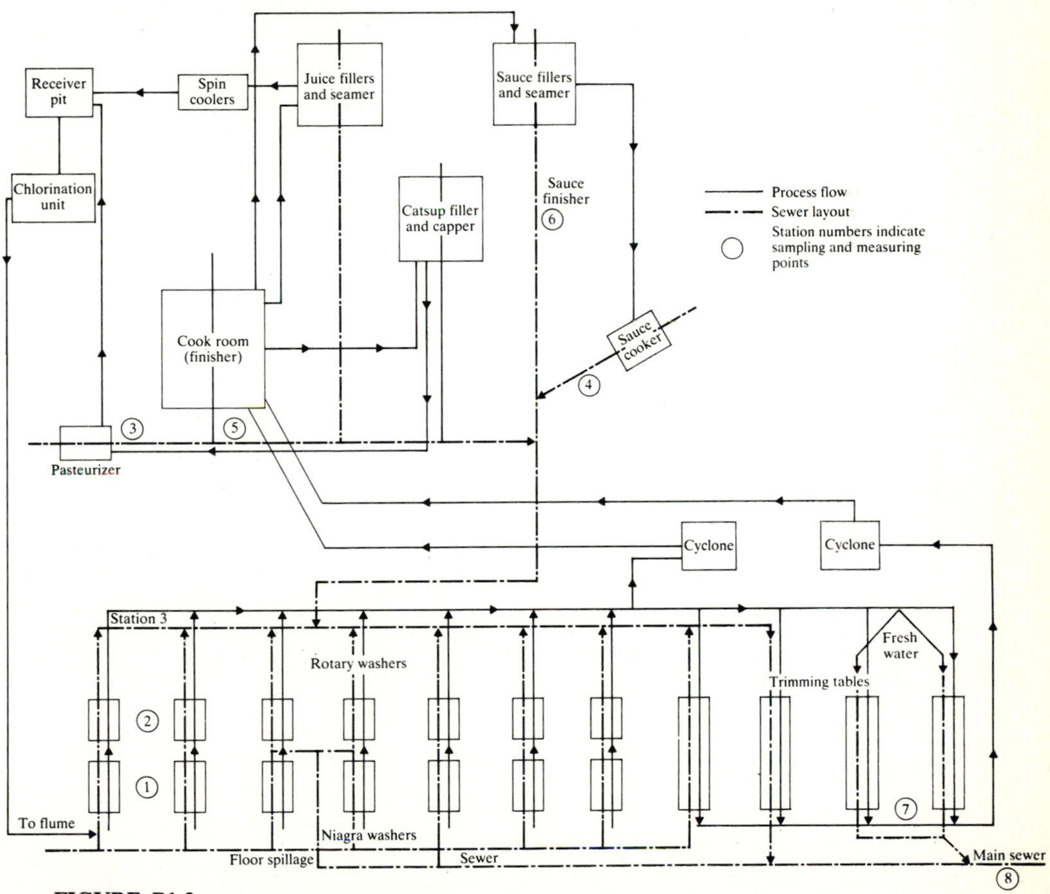

FIGURE P1-2
Flow diagram of a tomato-processing plant.

shown below. Develop a flow-and-material-balance diagram for the plant sewer system. Show possible changes to reduce the flow and loading. The waste from the trimming tables is presently flumed to the sewer.

Sampling station	Process unit	Flow, gal/min	BOD, mg/l	SS, mg/l
1	Niagra washers	500	75	180
2	Rotary washers	—	90	340
3	Pasteurizer	300	30	20
4	Source cooker	10	3520	7575
5	Cook room	20	4410	8890
6	Source finisher	150	230	170
7	Trimming tables	100	450	540
8	Main outfall	1080	240	470

Note: gal/min = 3.78×10^{-3} m^3/min.

1.3. A wastewater contains the following constituents:

40 mg/l phenol
350 mg/l glucose
3 mg/l S^{2-}
50 mg/l methyl alcohol, CH_3OH
100 mg/l isophorene, $C_9H_{14}O$ (nondegradable)

(*a*) Compute the THOD, the COD, the TOC, and the BOD_5 assuming the k_{10} for the mixed wastewater is 0.25/d.

(*b*) After treatment the soluble BOD_5 is 10 mg/l with a k_{10} of 0.1/d. Compute the residual COD and TOC.

REFERENCES

1. Schroepfer, G. J.: *Advances in Water Pollution Control*, vol. 1, Pergamon Press, New York, 1964.
2. Stack, V. T.: *Proc. 8th Ind. Waste Conf.*, 1953, p. 492, Purdue University.
3. Busch, A. W.: *Proc. 15th Ind. Waste Conf.*, 1961, p. 67, Purdue University.
4. Ford, D. L.: *Proc. 23rd Ind. Waste Conf.*, 1968, p. 94, Purdue University.

CHAPTER 2

WASTEWATER-TREATMENT PROCESSES

The schematic diagram in Fig. 2-1 illustrates an integrated system capable of treating a variety of plant wastewaters. The scheme centers on the conventional series of primary- and secondary-treatment processes, but also includes tertiary treatment and individual treatment of certain streams.

Primary- and secondary-treatment processes handle most of the nontoxic wastewaters; other waters have to be pretreated before being added to this flow. These processes are basically the same in an industrial plant as in a POTW.

Primary treatment prepares the wastewaters for biological treatment. Large solids are removed by screening, and grit is allowed to settle out. Equalization, in a mixing basin, levels out the hour-to-hour variations in flows and concentrations. There should be a spill pond, to retain slugs of concentrated wastes that could upset the downstream processes. Neutralization, where required, follows equalization because streams of different pH partly neutralize each other when mixed. Oils, greases, and suspended solids are removed by flotation, sedimentation, or filtration.

Secondary treatment is the biological degradation of soluble organic compounds—from input levels of 50 to 1000 mg/l BOD and even greater to effluent levels typically under 15 mg/l. This is usually done aerobically, in an open, aerated vessel or lagoon, but wastewaters may be pretreated anaerobically, in a pond or a closed vessel. After biotreatment, the microorganisms and other carried-over solids are allowed to settle. A fraction of this sludge is recycled in certain processes, but ultimately the excess sludge, along with the sedimented solids, has to be disposed of.

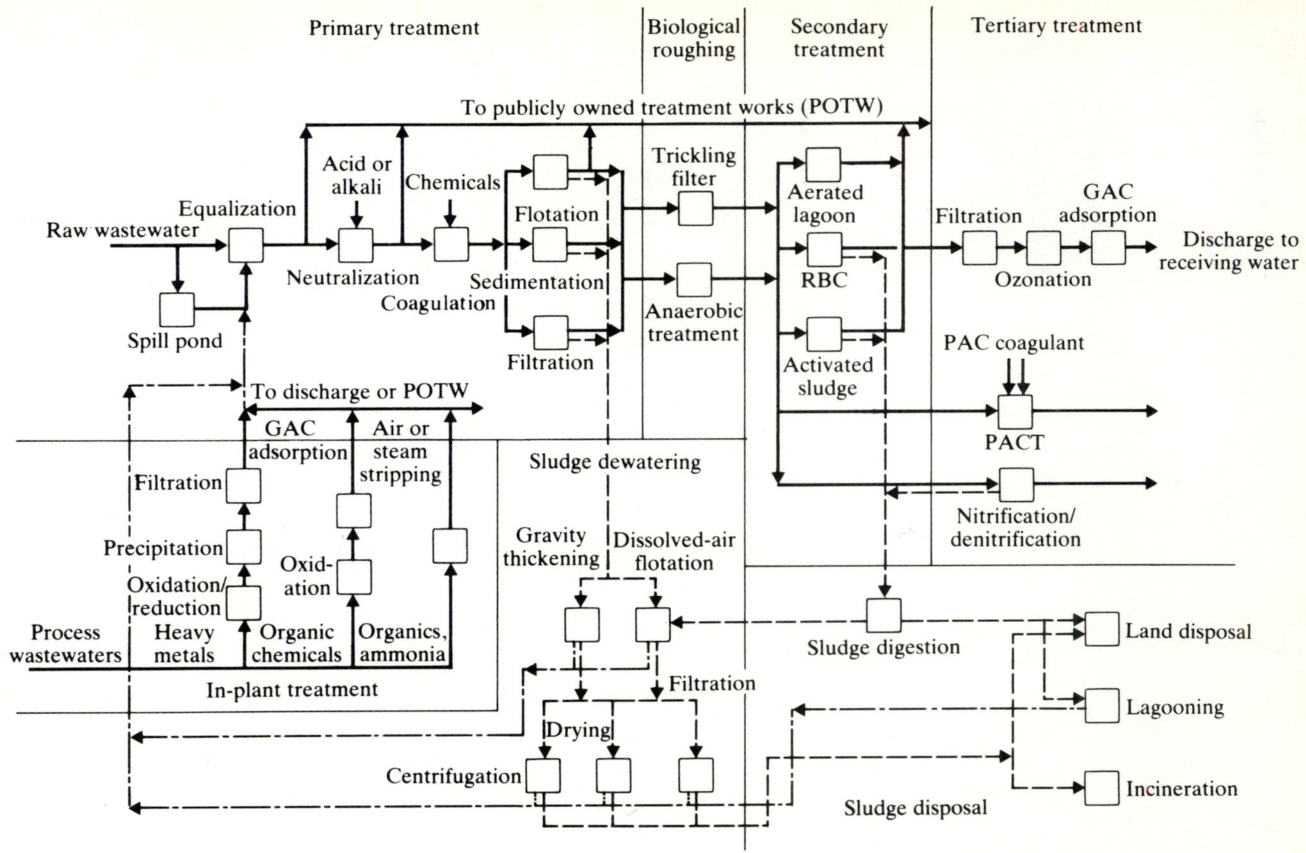

FIGURE 2-1
Alternative technologies for the treatment of industrial wastewaters.

Many existing wastewater-treatment systems were built just for primary and secondary treatment, though a plant might also have systems for removing materials that would be toxic to microorganisms. Until recently, this was adequate, but now it is not, and so new facilities have to be designed and old facilities retrofitted to include additional capabilities to remove priority pollutants and residuals toxic to aquatic life.

Tertiary-treatment processes are added on after biological treatment in order to remove specific types of residuals. Filtration removes suspended or colloidal solids; adsorption by granular activated carbon (GAC) removes organics; chemical oxidation also removes organics. Unfortunately, tertiary systems have to treat a large volume of wastewater, and so they are expensive. They can also be inefficient because the processes are not pollutant-specific. For example, dichlorophenol can be removed by ozonation or GAC adsorption, but those processes will remove most other organics as well.

In-plant treatment is necessary for streams rich in heavy metals, pesticides, and other substances that would pass through primary treatment and inhibit biological treatment. In-plant treatment also makes sense for low-volume streams rich in nondegradable materials, because it is easier and much less costly to remove a specific pollutant from a small, concentrated stream than from a large, dilute one. Processes used for in-plant treatment include precipitation, activated carbon adsorption, chemical oxidation, air or steam stripping, ion exchange, reverse osmosis, and electrodialysis and wet air oxidation.

Existing treatment systems can also be modified so as to broaden their capabilities and improve their performance; this is more widely practiced than the above options. One example is adding powdered activated carbon (PAC) to the biological-treatment process, to adsorb organics that the microorganisms cannot degrade; this is marketed as the PACT process. Another example is adding coagulants at the end of the biological-treatment basin so as to remove phosphorus and residual suspended solids.

All these processes have their place in the overall wastewater-treatment scheme.

The selection of a wastewater-treatment process or a combination of processes depends upon:

1. The characteristics of the wastewaters. This should consider the form of the pollutant, i.e., suspended, colloidal, or dissolved, the biodegradability, and the toxicity of the organic and inorganic components.
2. The required effluent quality. Consideration should also be given to possible future restrictions such as an effluent bioassay aquatic toxicity limitation.
3. The costs and availability of land for any given wastewater-treatment problem. One or more treatment combinations can produce the desired effluent. Only one of these alternatives, however, is the most cost-effective. A detailed cost analysis should therefore be made prior to final process design selection.

TABLE 2.1
Chemical waste treatment

Treatment method	Type of waste	Mode of operation	Degree of treatment	Remarks
Ion exchange	Plating, nuclear	Continuous filtration with resin regeneration	Demineralized water recovery; product recovery	May require neutralization and solids removal from spent regenerant
Reduction and precipitation	Plating, heavy metals	Batch or continuous treatment	Complete removal of chromium and heavy metals	One day's capacity for batch treatment; 3-h retention for continuous treatment; sludge disposal or dewatering required
Coagulation	Paperboard, refinery, rubber, paint, textile	Batch or continuous treatment	Complete removal of suspended and colloidal matter	Flocculation and settling tank or sludge blanket unit; pH control required
Adsorption	Toxic or organics, refractory	Granular columns of powdered carbon	Complete removal of most organics	Powdered carbon (PAC) used with activated sludge process
Chemical oxidation	Toxic and refractory organics	Batch or continuous ozone or catalyzed hydrogen peroxide	Partial or complete oxidation	Partial oxidation to render organics more biodegradable

TABLE 2.2
Biological waste treatment

Treatment method	Mode of operation	Degree of treatment	Land requirements	Equipment	Remarks
Lagoons	Intermittent or continuous discharge; facultative or anaerobic	Intermediate	Earth dug; 10–60 days' retention		Odor control frequently required
Aerated lagoons	Completely mixed or facultative continuous basins	High in summer; less in winter	Earth basin, 8–16 ft (2.44–4.88 m) deep; 8–16 acres/(million gal · d) [8.55–17.1 $m^2/(m^3 \cdot d)$]	Pier-mounted or floating surface aerators or subsurface diffusers	Solids separation in lagoon; periodic dewatering and sludge removal
Activated sludge	Completely mixed or plug flow; sludge recycle	>90% removal of organics	Earth or concrete basin; 12–20 ft (3.66–6.10 m) deep; 75,000–350,000 ft^3/(million gal · d) [0.561–2.62 $m^3/(m^3 \cdot d)$]	Diffused or mechanical aerators; clarifier for sludge separation and recycle	Excess sludge dewatered and disposed of
Trickling filter	Continuous application; may employ effluent recycle	Intermediate or high, depending on loading	225–1400 ft^2/(million gal · d) [5.52–34.4 $m^2/(10^3\ m^3 \cdot d)$]	Plastic packing 20–40 ft deep (6.10–12.19 m)	Pretreatment before POTW or activated sludge plant
RBC	Multistage continuous	Intermediate or high		Plastic disks	Solids separation required
Anaerobic	Complete mix with recycle; upflow or downflow filter, fluidized bed; upflow sludge blanket	Intermediate		Gas collection required; pretreatment before POTW or activated sludge plant	
Spray irrigation	Intermittent application of waste	Complete; water percolation into groundwater and runoff to stream	40–300 gal/(min · acre) [$6.24 \times 10^{-7} - 4.68 \times 10^{-6}\ m^3/(s \cdot m^2)$]	Aluminum irrigation pipe and spray nozzles; movable for relocation	Solids separation required; salt content in waste limited

Note:
ft = 0.305 m
acre/(million gal · d) = 1.07 $m^2/(m^3 \cdot d)$
ft^3/(million gal · d) = $7.48 \times 10^{-3}\ m^3$/(thousand $m^3 \cdot d$)
ft^2/(million gal · d) = $2.45 \times 10^{-2}\ m^2$/(thousand $m^3 \cdot d$)
gal/(min · acre) = $1.56 \times 10^{-8}\ m^3/(s \cdot m^2)$

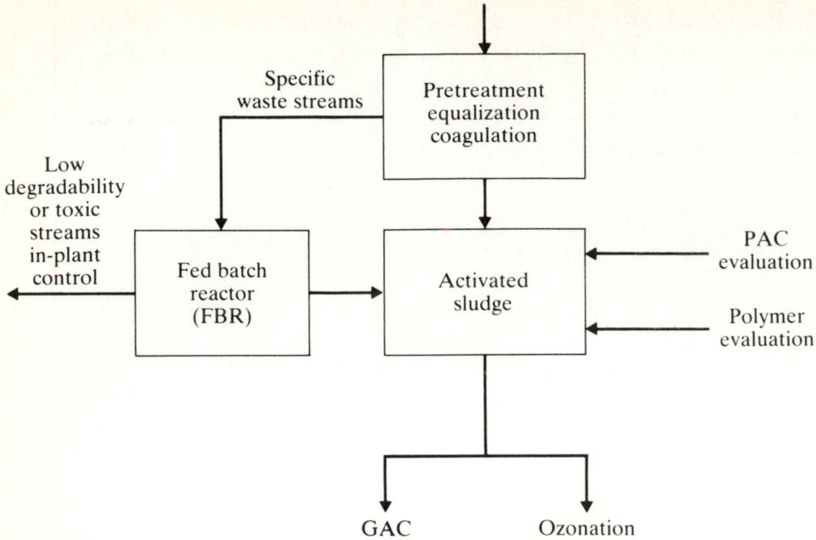

FIGURE 2-2
Evaluation procedure for industrial-waste treatment technology selection.

In many cases the quality of the wastewater can be defined and process design criteria generated from defined parameters or from a laboratory or pilot plant program. Examples are pulp and paper-mill wastewaters and food-processing wastewaters. In the case of complex chemical wastewaters containing toxic and nontoxic organics and inorganics a more defined screening procedure is

TABLE 2.3
Maximum quality attainable from waste-treatment processes

Process	BOD	COD	SS	N	P	TDS
Sedimentation, % removal	10–30	—	59–90	—	—	—
Flotation, †% removal	10–50	—	70–95	—	—	—
Activated sludge, mg/l	<25	‡	<20	§	§	—
Aerated lagoons, mg/l	<50	—	>50	—	—	—
Anaerobic ponds, mg/l	>100	—	<100	—	—	—
Deep-well disposal	Total disposal of waste					
Carbon adsorption, mg/l	<2	<10	<1	—	—	—
Denitrification and nitrification, mg/l	<10	—	—	<5	—	—
Chemical precipitation, mg/l	—	—	<10	—	<1	—
Ion exchange, mg/l	—	—	<1	¶	¶	¶

† Higher removals are attained when coagulating chemicals are used.
‡ COD_{inf} − [BOD_{ult}(removed)/0.9].
§ N_{inf} − 0.12 (excess biological sludge), lb; P_{inf} − 0.026 (excess biological sludge), lb.
¶ Depends on resin used, molecular state, and efficiency desired.
Note: lb = 0.45 kg.

needed to select candidate processes for treatment, as shown in Fig. 2-2. It is significant to note that heavy metals should be removed in pretreatment prior to any biological treatment. Heavy metals may be toxic to the biological process and will accumulate on the biological sludge, resulting in constraints for sludge-disposal options. The details of the FBR test are discussed on page 182.

The alternatives for chemical waste treatment is summarized in Table 2.1. The alternatives for biological wastewater treatment are summarized in Table 2.2. The maximum attainable effluent quality for the conventional wastewater-treatment processes are shown in Table 2.3.

CHAPTER 3

PRE- AND PRIMARY TREATMENT

3.1 EQUALIZATION

The objective of equalization is to minimize or control fluctuations in wastewater characteristics in order to provide optimum conditions for subsequent treatment processes. The size and type of equalization basin provided varies with the quantity of waste and the variability of the wastewater stream. The basin should be of a sufficient size to adequately absorb waste fluctuations caused by variations in plant-production scheduling and to dampen the concentrated batches periodically dumped or spilled to the sewer.

The purpose of equalization for industrial treatment facilities are:

1. To provide adequate dampening of organic fluctuations in order to prevent shock loading of biological systems.
2. To provide adequate pH control or to minimize the chemical requirements necessary for neutralization.
3. To minimize flow surges to physical-chemical treatment systems and permit chemical feed rates compatible with feeding equipment.
4. To provide continuous feed to biological systems over periods when the manufacturing plant is not operating.
5. To provide capacity for controlled discharge of wastes to municipal systems in order to distribute waste loads more evenly.
6. To prevent high concentrations of toxic materials from entering the biological-treatment plant.

Mixing is usually provided to ensure adequate equalization and to prevent settleable solids from depositing in the basin. In addition, the oxidation of reduced compounds present in the wastestream or the reduction of BOD by air stripping may be achieved through mixing and aeration. Methods that have been used for mixing include:

1. Distribution of inlet flow and baffling
2. Turbine mixing
3. Diffused air aeration
4. Mechanical aeration

The most common method is to provide submerged mixers or, in the case of a readily degradable wastewater such as a brewery, to use surface aerators employing a power level of approximately 15 to 20 hp/million gal (0.003 to 0.004

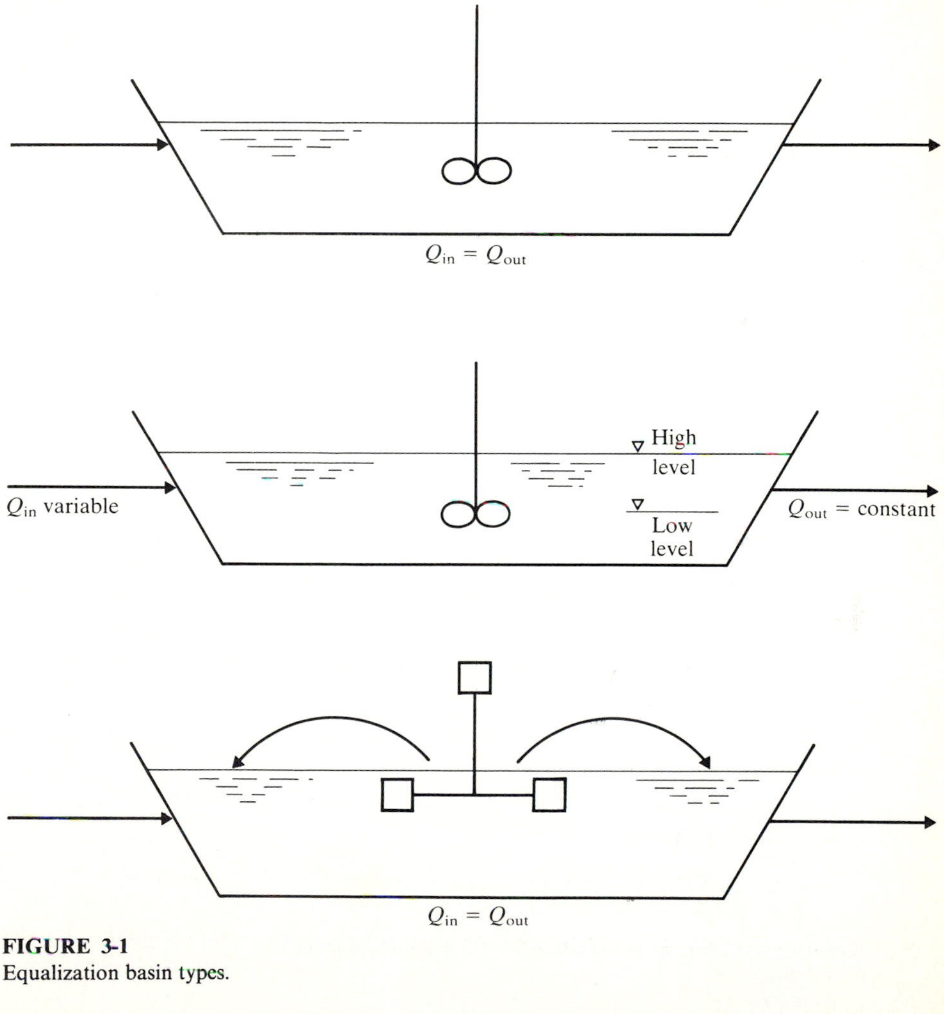

FIGURE 3-1
Equalization basin types.

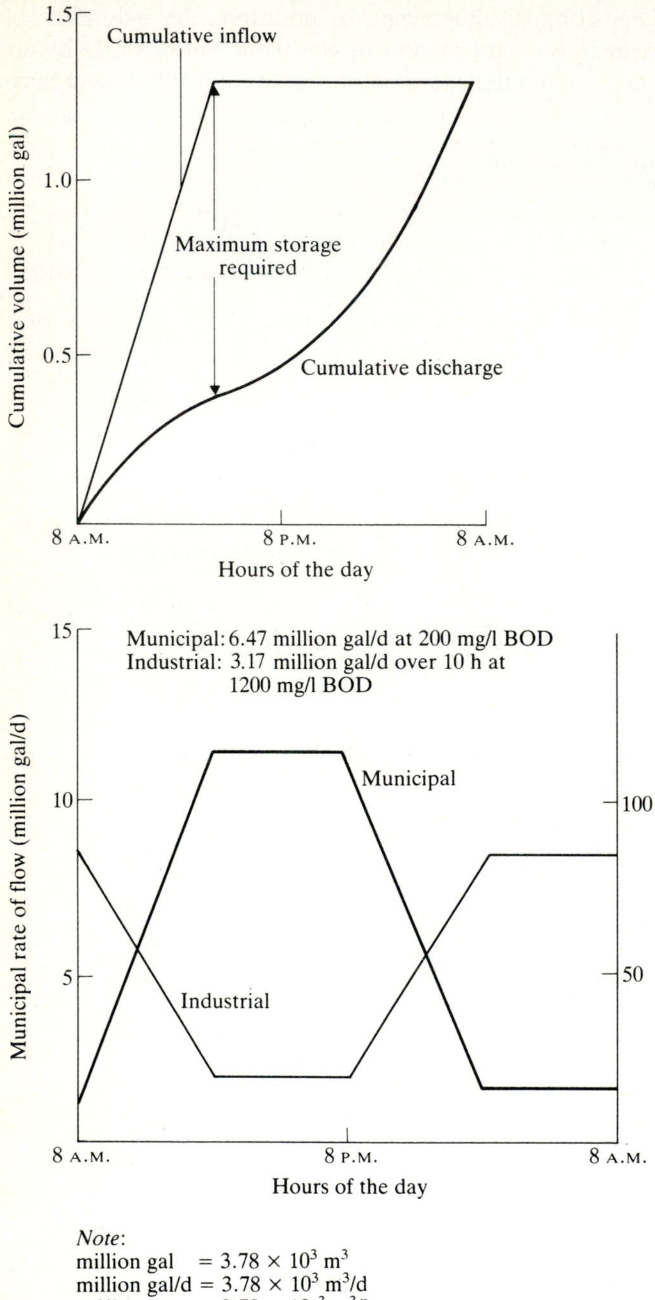

Note:
million gal $= 3.78 \times 10^3\ m^3$
million gal/d $= 3.78 \times 10^3\ m^3/d$
gal/h $= 3.78 \times 10^{-3}\ m^3/h$

FIGURE 3-2
Controlled discharge of an industrial wastewater to a municipal plant.

kW/m^3). Air requirements for diffused air aeration are approximately 0.5 ft^3 air/gal waste (3.74 m^3/m^3). Equalization basin types are shown in Fig. 3-1.

The equalization basin may be designed with a variable volume to provide a constant effluent flow or with a constant volume and an effluent flow which varies with the influent. The variable-volume basin is particularly applicable to the chemical treatment of wastes having a low daily volume. This type of basin may also be used for discharge of wastes to municipal sewers. It may be desirable to program the effluent pumping rate to discharge the maximum quantity of waste during periods of normally low flow to the municipal treatment facility as shown in Fig. 3-2. Ideally, the organic loading to the treatment plant is maintained constant over a 24-h period.

Equalization basins may be designed to equalize flow, concentration, or both. For flow equalization, the cumulative flow is plotted versus time over the equalization period (that is, 24 h). The maximum volume with respect to the constant discharge line is the equalization volume required. The required calculations are shown in Example 3.1.

Example 3.1. Given the data in Table 3.1, design an equalization basin for a constant outflow from the basin. The data are plotted as the summation of inflow

TABLE 3.1

Time	gal/min	gal	Σ gal × 10^{-3}
8	50	3,000	3.0
9	92	5,520	8.5
10	230	13,800	22.3
11	310	18,600	40.9
12	270	16,200	57.1
1	140	8,400	65.5
2	90	5,400	70.9
3	110	6,600	77.5
4	80	4,800	82.3
5	150	9,000	91.3
6	230	13,800	105.1
7	305	18,300	123.4
8	380	22,800	146.2
9	200	12,000	158.2
10	80	4,800	163.0
11	60	3,600	166.6
12	70	4,200	170.8
1	55	3,300	174.1
2	40	2,400	176.5
3	70	4,200	180.7
4	75	4,500	185.2
5	45	2,700	187.9
6	55	3,300	191.2
7	35	2,100	193.3

Note:
gal/min = 3.78 × 10^{-3} m^3/min
gal = 3.78 × 10^{-3} m^3

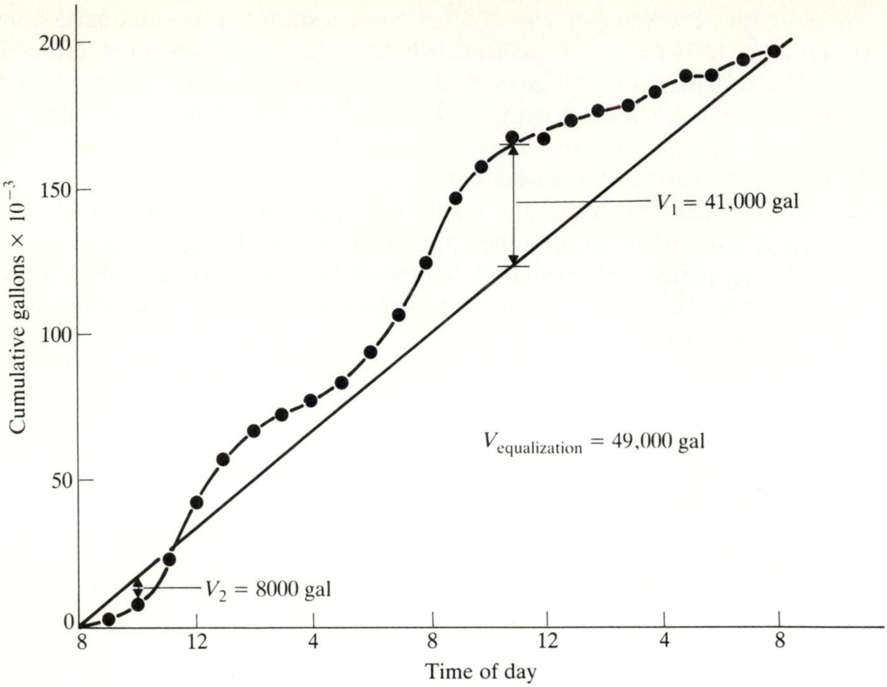

FIGURE 3-3

versus time as shown in Fig. 3-3. The treatment rate is (193,300 gal/day)/(1440 min/day) = 134 gal/min (507 l/min). The required storage volume is 41,000 gal + 8000 gal residual or 49,000 gal (186 m^3).

The equalization basin may be sized to restrict the discharge to a maximum concentration commensurate with the maximum permissible discharge from subsequent treatment units. For example, if the maximum effluent from an activated sludge unit is 50 mg/l BOD_5, the maximum effluent from the equalization basin may be computed and thereby provide a basis for sizing the unit.

For the case of near-constant wastewater flow and a normal statistical distribution of wastewater composite analyses the required equalization retention time is[1]

$$t = \frac{\Delta t(S_i^2)}{2(S_e^2)} \tag{3.1}$$

where Δt = time interval over which samples were composited, h
t = equalization detention time, h
S_i^2 = variance of the influent wastewater concentration (the square of the standard deviation)
S_e^2 = variance of the effluent concentration at a specified probability (for example, 99 percent)

Where a completely mixed basin is to be used for treatment, such as in an activated sludge basin or an aerated lagoon, this volume can be considered as part of the equalization volume. For example, if the completely mixed aeration basin retention time is 8 h and the total required equalization retention time is 16 h, then the equalization basin only needs to have a retention time of 8 h. Example 3.2 illustrates this calculation.

Example 3.2. A waste with a total flow of 5 million gal/d (0.22 m^3/s or 19,000 m^3/d) was characterized as shown in Fig. 3-4. Extensive data were collected every 4 h for 17 d. The average BOD was 690 mg/l and the maximum value was 1185 mg/l.

Design calculations with activated sludge systems have indicated that the effluent from the equalization basin must not exceed 896 mg/l in order to meet the effluent quality criteria of an average BOD of 15 mg/l and a maximum concentration of 25 mg/l from the activated sludge system.

Design an equalization basin to meet the desired effluent requirements. Base calculations on a 95 percent probability that the equalized effluent will be equal to or less than 896 mg/l.

(*a*) Calculate the mean, standard deviation, and variance of the influent. These parameters may be calculated graphically from Fig. 3-4.

(*b*) From this plot obtain the 50 percent value:

$$50 \text{ percent value} \approx \bar{X} \approx 690 \text{ mg/l}$$

(*c*) Calculate the standard deviation, S_i, as half the difference in the values that occur at the 15.9 (50.0 minus 34.1) and 84.1 (50.0 plus 34.1) percentile levels from Fig. 3-4:

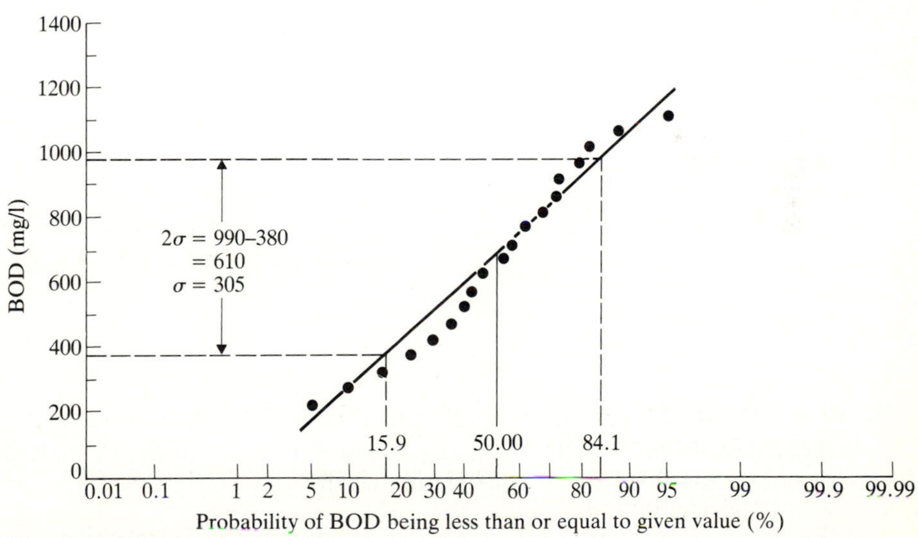

FIGURE 3-4

$$S_i = \frac{\text{value at 84.1 percent} - \text{value at 15.9 percent}}{2}$$

$$= \frac{990 - 380}{2}$$

$$= 305 \text{ mg/l}$$

(*d*) Calculate the variance as the square of the standard deviation:

$$S_i^2 = (305)^2$$

$$= 93{,}025 \text{ mg}^2/\text{l}^2$$

(*e*) Calculate the standard deviation of the effluent:

$$\bar{X} = 690 \text{ mg/l}$$

$$X_{\text{max}} = 896 \text{ mg/l}$$

For a 95 percent confidence level, $Z = 1.65$ from the normal probability tables, so

$$S_e = \frac{X_{\text{max}} - \bar{X}}{Z}$$

$$= \frac{896 - 690}{1.65}$$

$$= 125 \text{ mg/l}$$

(*f*) Calculate the effluent variance

$$S_e^2 = (125)^2$$

$$= 15{,}625 \text{ mg}^2/\text{l}^2$$

(*g*) Calculate the required detention time:

$$t = \frac{\Delta t(S_i^2)}{2(S_e^2)}$$

$$= \frac{4(93{,}025)}{2(15{,}625)}$$

$$= 11.9 \text{ h}$$

$$\approx 0.5 \text{ d}$$

Patterson and Menez[2] have developed a method to define equalization requirements when both the flow and the strength vary randomly. A material balance can be established for the equalization basin:

$$C_i QT + C_0 V = C_2 QT + C_2 V \tag{3.2}$$

in which C_i = concentration entering the equalization basin over the sampling interval T

T = sampling interval; i.e., one hour

Q = average flow rate over the sampling interval

C_0 = concentration in the equalization basin at the start of the sampling interval

V = volume of the equalization basin

C_2 = concentration leaving the equalization basin at the end of the sampling interval

It is assumed that the effluent concentration is almost constant during one time interval. This is valid assuming the time intervals are appropriately spaced.

Equation (3.2) can be rearranged to compute the effluent concentration after each time interval:

$$C_2 = \frac{C_i T + C_0 V/Q}{T + V/Q} \tag{3.2a}$$

The range of effluent concentrations can then be calculated for a range of equalization volumes, V. A peaking factor, PF, is computed for the influent strength and flow. The effluent PF for design purposes is the ratio of the maximum concentration to the average concentration. An equalization basin design is shown in Example 3.3.

Example 3.3. The following data were obtained at an industrial plant over an 8-h production cycle. Each time interval represents a sampling period of 1 h. The effluent concentrations of an equalization volume for retention times of 8 and 4 h are shown in Table 3.2. For an 8-h retention time, the volume of the equalization basin

TABLE 3.2

Time interval	Flow rate, gal/min × 10^{-3}	Influent concentration, mg/l	Effluent concentration at V_r† of 1.0	Effluent concentration at V_r† of 0.5
1	1.6	245	187	198
2	0.2	64	185	193
3	1.0	54	173	169
4	1.2	167	172	169
5	1.6	329	194	208
6	2.0	48	169	162
7	1.2	55	157	141
8	1.0	395	179	181
Average	1.225	178	178	178
PF	10.0	8.2	1.09	1.17

† $V_r = \dfrac{\text{volume of equalization basin}}{\text{total volume of daily wastewater flow}}$

Note: gal/min = 3.78×10^{-3} m^3/min.

can be calculated:

$$1.225 \times 10^3 \text{ gal/min} \times 60 \text{ min/h} \times 8 \text{ h} = 588 \times 10^3 \text{ gal } (2.23 \times 10^3 \text{ m}^3)$$

The effluent concentration after the first time interval is

$$C_2 = \frac{C_i T + C_0 V/Q}{T + V/Q}$$

$$= \frac{245 \times 1 + 178 \times 588/(1.6 \times 60)}{1 + 588/(1.6 \times 60)}$$

$$= 187 \text{ mg/l}$$

In like manner, all subsequent effluent concentrations can be calculated as shown in Table 3.2.

The PF for the influent concentration can be computed:

$$\text{PF} = \frac{395}{48} = 8.2$$

The PF for the effluent with 8 h retention is

$$\text{PF} = \frac{194}{178} = 1.09$$

The same approach can be employed for simultaneous flow and waste strength equalization, but in this case the volume of the equalization basin is a variable.

When occasional dumps or spills are anticipated, for example, 1 percent of the time, a spill basin with an automatic bypass activated by a monitor should be used, as shown in Fig. 3-5. The spill basin may be required for surges in organics, that is, TOC, TDS, temperature, or specific toxic compounds.

3.2 NEUTRALIZATION

Many industrial wastes contain acidic or alkaline materials that require neutralization prior to discharge to receiving waters or prior to chemical or biological treatment. For biological treatment, a pH in the biological system should be maintained between 6.5 and 8.5 to ensure optimum biological activity. The biological process itself provides a neutralization and a buffer capacity as a result of the production of CO_2, which reacts with caustic and acidic materials. The degree of preneutralization required depends, therefore, on the ratio of BOD removed and the causticity or acidity present in the waste. These requirements are discussed in Chap. 6.

Types of Processes

MIXING ACIDIC AND ALKALINE WASTESTREAMS. This process requires sufficient equalization capacity to effect the desired neutralization.

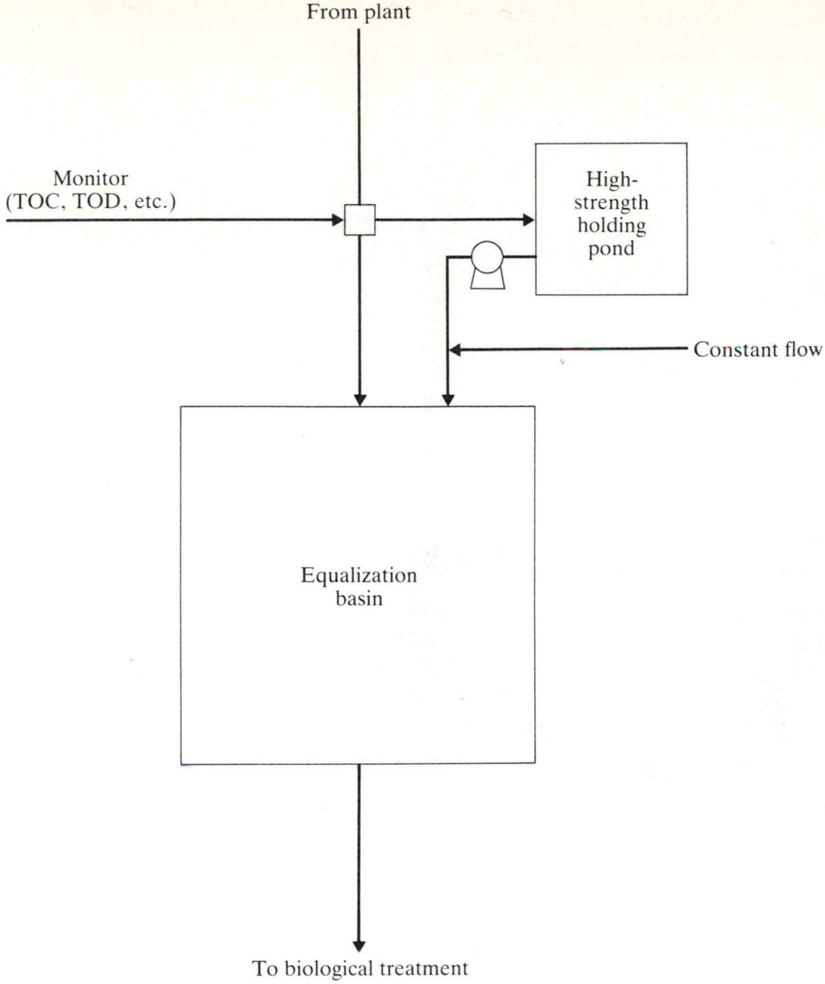

FIGURE 3-5
Use of a high-strength holding pond for spills.

ACID WASTES NEUTRALIZATION THROUGH LIMESTONE BEDS. These may be downflow or upflow systems. The maximum hydraulic rate for downflow systems is 1 gal/(min · ft^2) [4.07×10^{-2} m^3/(min · m^2)] to ensure sufficient retention time. The acid concentration should be limited to 0.6 percent H_2SO_4 if H_2SO_4 is present to avoid coating of limestone with nonreactive $CaSO_4$ and excessive CO_2 evolution, which limits complete neutralization. High dilution or dolomitic limestone requires longer detention periods to effect neutralization. Hydraulic loading rates can be increased with upflow beds, because the products of reaction are swept out before precipitation. A limestone bed system is shown in Fig. 3-6.

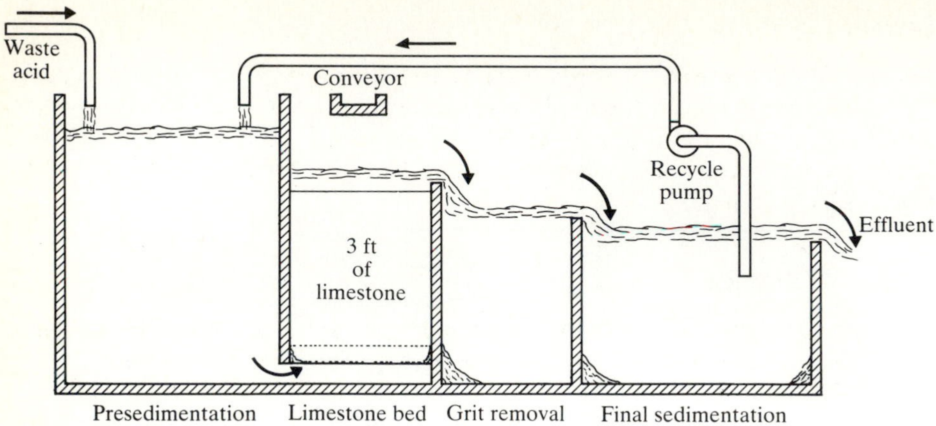

FIGURE 3-6
Simplified flow diagram of limestone neutralization.[3]

MIXING ACID WASTES WITH LIME SLURRIES. This neutralization depends on the type of lime used. The magnesium fraction of lime is most reactive in strongly acid solutions and is useful below pH 4.2. Neutralization with lime can be defined by a basicity factor obtained by titration of a one-gram sample with an excess of HCl, boiling 15 min, followed by backtitration with 0.5 *N* NaOH to the phenolphthalein endpoint.

In lime slaking, the reaction is accelerated by heat and physical agitation. For high reactivity, the lime reaction is complete in 10 min. Storage of lime slurry for a few hours before neutralization may be beneficial. Dolomitic quicklime (only the CaO portion) hydrates except at elevated temperature. Slaked quicklime is used as an 8 to 15 percent lime slurry. Neutralization can also be accomplished using NaOH, Na_2CO_3, or NH_4OH.

BASIC (ALKALINE) WASTES. Any strong acid can be used effectively to neutralize alkaline wastes, but cost considerations usually limit the choice to sulfuric or possibly hydrochloric acid. The reaction rates are practically instantaneous, as with strong bases.

Flue gases that can contain 14 percent CO_2 can be used for neutralization. When bubbled through the waste, the CO_2 forms carbonic acid, which then reacts with the base. The reaction rate is somewhat slower but is sufficient if the pH need not be adjusted to below 7 to 8. Another approach is to use a spray tower in which the stack gases are passed countercurrent to the waste liquid droplets.

All of the above processes usually work better with the stepwise addition of reagents, i.e., a staged operation. Two stages, with possibly a third tank to even out any remaining fluctuations, are generally optimum.

System

Batch treatment is used for waste flows to 100,000 gal/d (380 m^3/d). Continuous treatment employs automated pH control—where air is used for mixing, the minimum air rate is 1 to 3 $ft^3/(min \cdot ft^2)$ [0.3 to 0.9 $m^3/(min \cdot m^2)$] at 9 ft (2.7 m) liquid depth. If mechanical mixers are used, 0.2 to 0.4 hp/thousand gal (0.04 to 0.08 kW/m^3) is required.

Control of Process

The automatic control of pH for wastestreams is one of the most troublesome, for the following reasons:

1. The relation between pH and concentration or reagent flow is highly nonlinear for strong acid–strong base neutralization, particularly when close to neutral (pH 7.0). The nature of the titration curve as shown in Fig. 3-7 favors multistaging in order to insure close control of the pH.
2. The influent pH can vary at a rate as fast as 1 pH unit per minute.
3. The wastestream flow rates can double in a few minutes.
4. A relatively small amount of reagent must be thoroughly mixed with a large liquid volume in a short time interval.

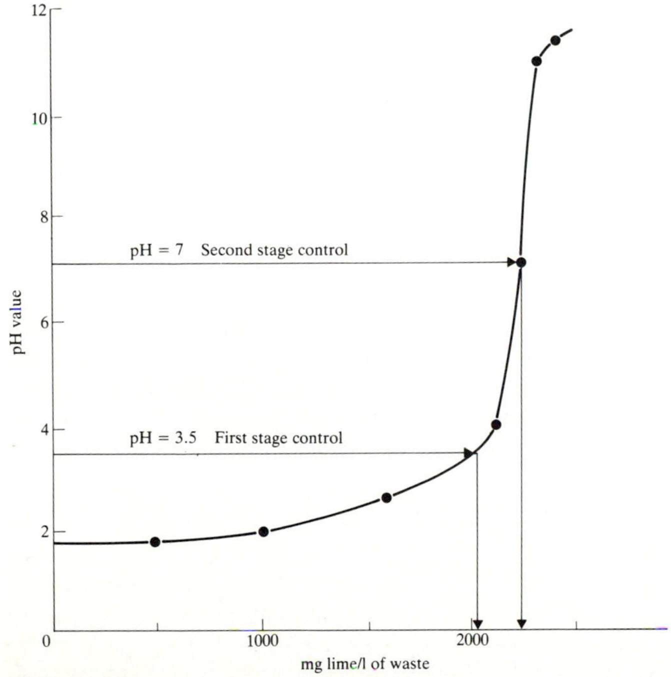

FIGURE 3-7
Lime-waste titration curve for strong acid.

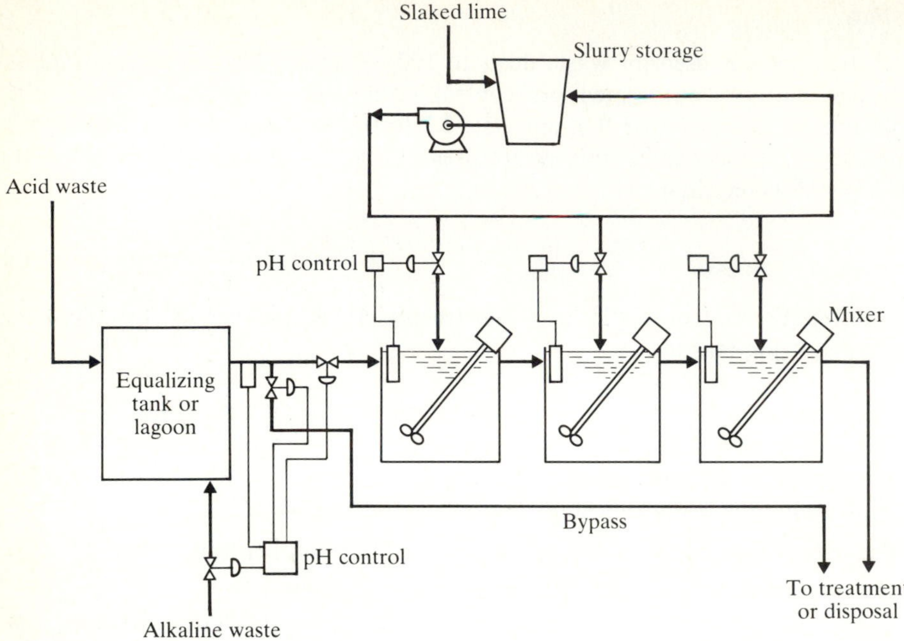

FIGURE 3-8
Multistage neutralization process.

Advantage is usually gained by the stepwise addition of chemicals (see Fig. 3-8). In reaction tank 1, the pH may be raised to 3 to 4. Reaction tank 2 raises the pH to 5 to 6 (or any other desired endpoint). If the wastestream is subject to slugs or spills, a third reaction tank may be desirable to effect complete neutralization. Acid waste neutralization is shown in Example 3.4.

Example 3.4. A wastewater (100 gal/min) (0.38 m^3/min) is highly acidic and requires neutralization prior to secondary treatment. This flow is to be neutralized to a pH of 7.0 using lime. The titration curve for the wastewater is shown in Fig. 3-7, from which a two-stage control neutralization system will be used with a total lime consumption of 2250 mg/l. The first stage requires 2000 mg/l and the second stage 250 mg/l.

The average lime dosage in the first stage is

$$(100 \text{ gal/min})(1440 \text{ min/d})(8.34 \text{ lb/million gal/mg/l})(2000 \text{ mg/l}) \times 10^{-6} \text{ (million gal/gal)} = 2400 \text{ lb/d}(1090 \text{ kg/d})$$

The average lime dosage in the second stage is

$$100 \times 1440 \times 8.34 \times 250 \times 10^{-6} = 300 \text{ lb/d}(140 \text{ kg/d})$$

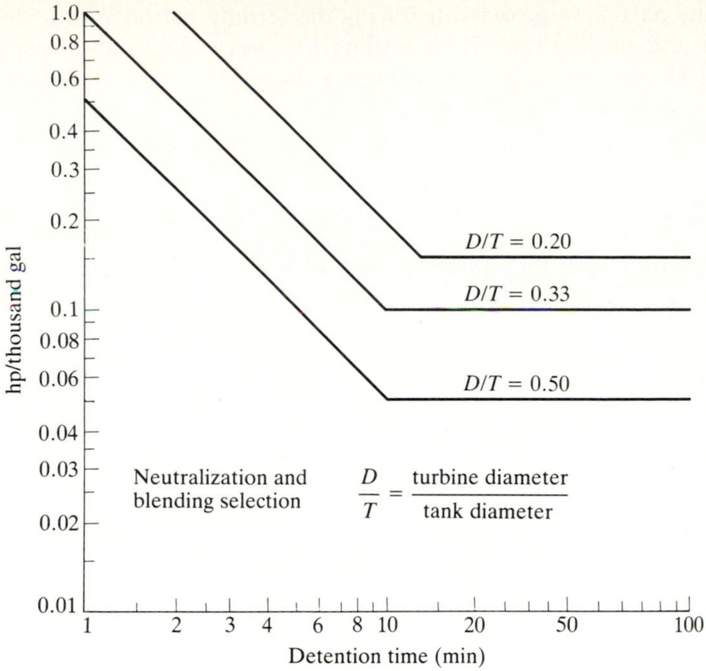

Note: hp/thousand gal = 198 W/m^3

FIGURE 3-9

The average lime dosage is 2700 lb/d (1.35 ton/d) (1230 kg/d). With this dosage and type of lime each basin should be designed with a detention time of 5 min.

$$\text{Volume} = 100 \text{ gal/min} \times 5 \text{ min} = 500 \text{ gal } (1.9 \text{ m}^3)$$

Use two tanks, 4.6 ft (1.40 m) diameter × 4.1 ft (1.25 m) deep, so that diameter and depth are approximately equal.

To maintain proper mixing in the reactor tanks, the power level required for 5 min detention time, $D/T = 0.33$, is 0.2 hp/thousand gal (40 W/m^3) (Fig. 3-9). Use 0.1 hp (75 W) mixers in each reaction tank.

Either one or two standard wall baffles, 180° apart, and one-twelfth to one-twentieth of the width of the tank diameter, located 24 in (61 cm) from the periphery of the impeller, are recommended for this operation.

3.3 SEDIMENTATION

Sedimentation is employed for the removal of suspended solids from wastewaters. The process can be considered in three basic classifications, depending on the nature of the solids present in the suspension: discrete, flocculent, and zone settling. In discrete settling, the particle maintains its individuality and does not change in size, shape, or density during the settling process. Flocculent settling occurs when the particles agglomerate during the settling period with a resulting change in size and settling rate. Zone settling involves a flocculated suspension which forms a lattice structure and settles as a mass, exhibiting a distinct interface during the settling process. Compaction of the settled sludge occurs in all sedimentation but will be considered separately under thickening.

Discrete Settling

A particle will settle when the impelling force of gravity exceeds the inertia and viscous forces. The terminal settling velocity of a particle is defined by the relationship

$$v = \sqrt{\frac{4g(\rho_s - \rho_l)D}{3C_d \rho_l}} \tag{3.3}$$

where ρ_l = density of the fluid
ρ_s = density of the particle
v = terminal settling velocity of the particle
D = diameter of the particle
C_d = drag coefficient, which is related to the Reynolds number and particle shape
g = acceleration due to gravity

When the Reynolds number is small, i.e., less than 1.0 (small particles at low velocity), the viscous forces are predominant and

$$C_d = \frac{24}{N_{\mathrm{Re}}} \tag{3.4}$$

where

$$N_{\mathrm{Re}} = \frac{vD\rho_l}{\mu} \tag{3.5}$$

ρ_l and μ are the density and viscosity of the liquid respectively. Substitution of Eq. (3.4) in Eq. (3.3) yields Stokes' law:

$$v = \frac{\rho_s - \rho_l}{18\mu} gD^2 \tag{3.6}$$

As the Reynolds number increases, a transition zone occurs in which both inertia and viscous forces are effective. This occurs over a Reynolds number range of 1 to 1000, where

$$C_d = \frac{18.5}{N_{\mathrm{Re}}^{0.6}} \tag{3.7}$$

Above a Reynolds number of 1000, viscous forces are not significant and the coefficient of drag is constant at 0.4.

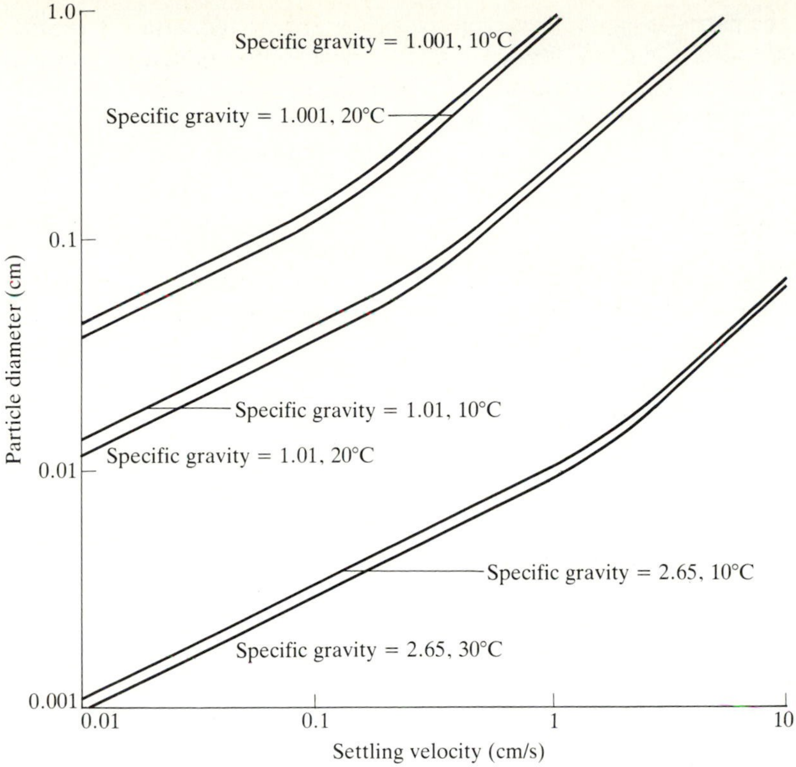

FIGURE 3-10
Settling properties of discrete particles.

The settling velocities of discrete particles, as related to diameter and specific gravity, are shown in Fig. 3-10.

Hazen[4] and Camp[5] developed relationships applicable to the removal of discrete particles in an ideal settling tank, based on the premises that the particles entering the tanks are uniformly distributed over the influent cross section and that a particle is considered removed when it hits the bottom of the tank. The settling velocity of a particle that settles through a distance equal to the effective depth of the tank in the theoretical detention period can be considered as an overflow rate:

$$v_o = \frac{Q}{A} \tag{3.8}$$

where Q = rate of flow through the tank
A = tank surface area

All particles with settling velocities greater than v_o will be completely removed, and particles with settling velocities less than v_o will be removed in the ratio v/v_o,

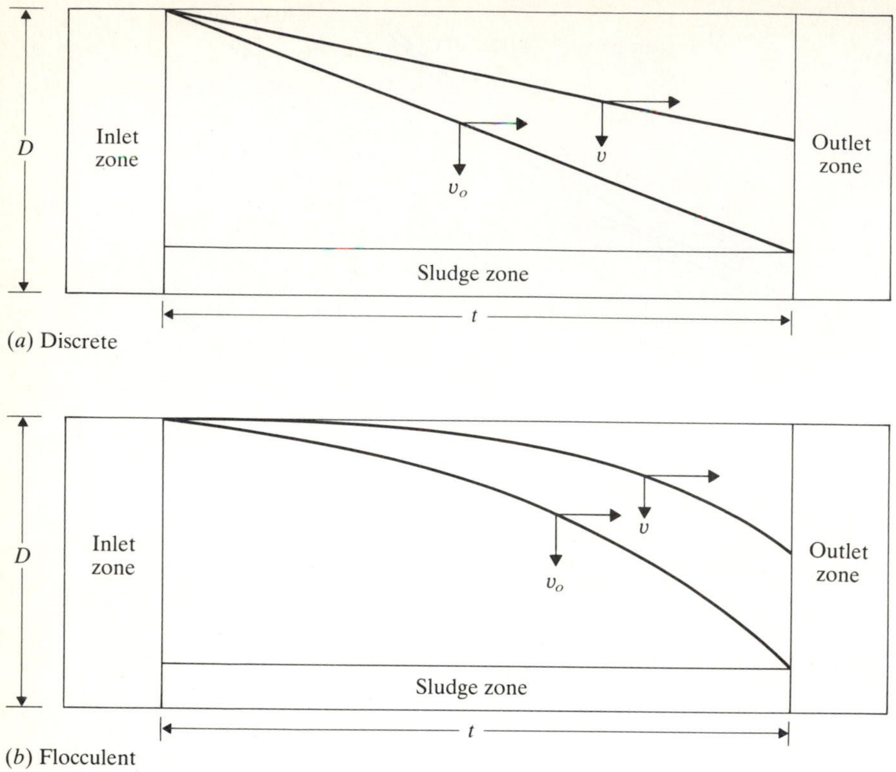

FIGURE 3-11
Ideal settling tank.

as shown in Fig. 3-11*a*. The removal of discrete particles is independent of tank depth and is a function only of the overflow rate.

When the suspension to be removed has a wide range of particle sizes, the total removal is defined by the relationship

$$\text{Total removal} = (1 - f_o) + \frac{1}{v_o}\int_o^{f_o} v\,df \tag{3.9}$$

where f_o is the fraction of particles having a settling velocity equal to or less than v_o. Equation (3.9) must usually be solved by graphical integration to obtain the total removal.

The foregoing analysis is based on the performance of an ideal settling tank under quiescent conditions. In practice, however, short-circuiting, turbulence, and bottom scour will affect the degree of solids removal. Dobbins[6] and Camp[5] have developed a relationship to compensate for decreased removals due to turbulence (Fig. 3-12). Figure 3-12 shows the effect of turbulence in reducing the removal ratio of particles having a settling velocity v from the theoretical value of v/v_o, where v_o refers to the overflow rate. The ratio $vH/2E$ is a parameter of turbulent

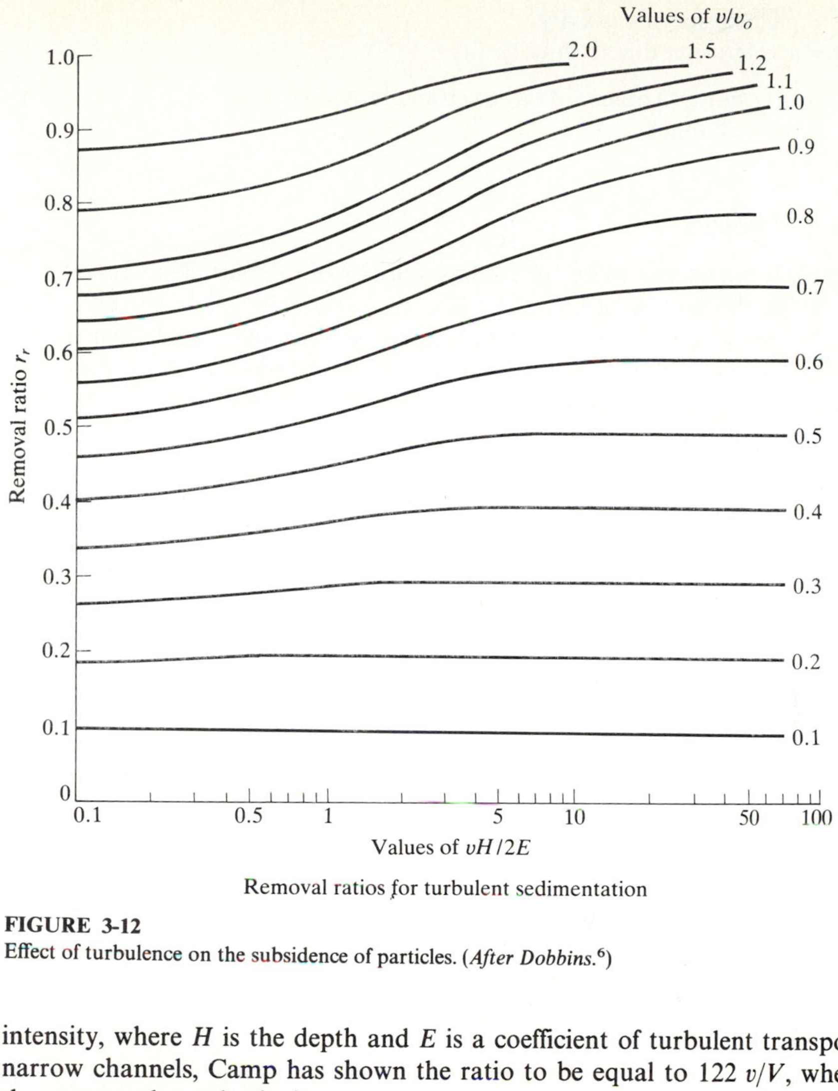

Removal ratios for turbulent sedimentation

FIGURE 3-12
Effect of turbulence on the subsidence of particles. (*After Dobbins.*[6])

intensity, where H is the depth and E is a coefficient of turbulent transport. For narrow channels, Camp has shown the ratio to be equal to 122 v/V, where V is the average channel velocity.

Scour occurs when the flow-through velocity is sufficient to resuspend previously settled particles; it is defined by the relationship:

$$v_c = \sqrt{\frac{8\beta}{f} gD(S-1)} \tag{3.10}$$

where v_c = velocity of scour
β = constant (0.04 for unigranular sand, 0.06 for nonuniform sticky material)
f = Weisbach-D'Arcy friction factor, 0.03 for concrete

S = particle specific gravity
g = acceleration due to gravity

Scour is usually not a problem in large settling tanks but can occur in grit chambers and narrow channels.

Flocculent Settling

Flocculent settling occurs when the settling velocity of the particle increases as it settles through the tank depth, because of coalescence with other particles. This increases the settling rate, yielding a curvilinear settling path, as shown in Fig. 3-11*b*. Most of the suspended solids in industrial wastes are of a flocculent nature. For discrete particles, the efficiency of removal is related only to the overflow rate, but when flocculation occurs, both overflow rate and detention time become significant.

Since a mathematical analysis is not possible in the case of flocculent suspensions, a laboratory settling analysis is required to establish the necessary parameters. The laboratory settling study can be conducted in a column of the type shown in Fig. 3-13. A minimum diameter of 5 in (12.7 cm) is recommended to minimize wall effects. Taps are located at 2-ft (0.61-m) depth intervals.

The concentration of suspended solids must be uniform at the start of the test; sparging air into the bottom of the column for a few minutes will accomplish this. It is also essential that the temperature be maintained constant throughout the test period to eliminate settling interference by thermal currents. Suspended solids are determined on samples drawn off at selected time intervals up to 120 min. The data collected from the 2- (0.61-), 4- (1.22-), and 6-ft (1.83-m) depth taps are used to develop the settling rate–time relationships.

The results obtained are expressed in terms of percent removal of suspended solids at each tap and time interval. These removals are then plotted against their respective depths and times, as shown in Fig. 3-14. Smooth curves are drawn connecting points of equal removal. The curves thus drawn represent the limiting or maximum settling path for the indicated percent; i.e., the specified percent suspended solids will have a settling path equal to or greater than that shown, and would therefore be removed in an ideal settling tank of the same depth and detention time. The calculation of removal can be illustrated from the data of Fig. 3-14.

The overflow rate v_o is the effective depth, 6 ft (1.83 m), divided by the time required for a given percent to settle this distance. All particles having a settling velocity equal to or greater than v_o will be removed in an ideal settling basin having an overflow rate of v_o. Particles with a lesser settling velocity v will be removed in the proportion v/v_o. For example, referring to Fig. 3-14*a*, at a detention period of 60 min and a 6-ft (1.83-m) settling depth [v_o = 6 ft/h (1.83 m/h)], 50 percent of the suspended solids are completely removed; that is, 50 percent of the particles have a settling velocity equal to or greater than 6 ft/h (1.83 m/h). Particles in each additional 10 percent range will be removed in the proportion v/v_o

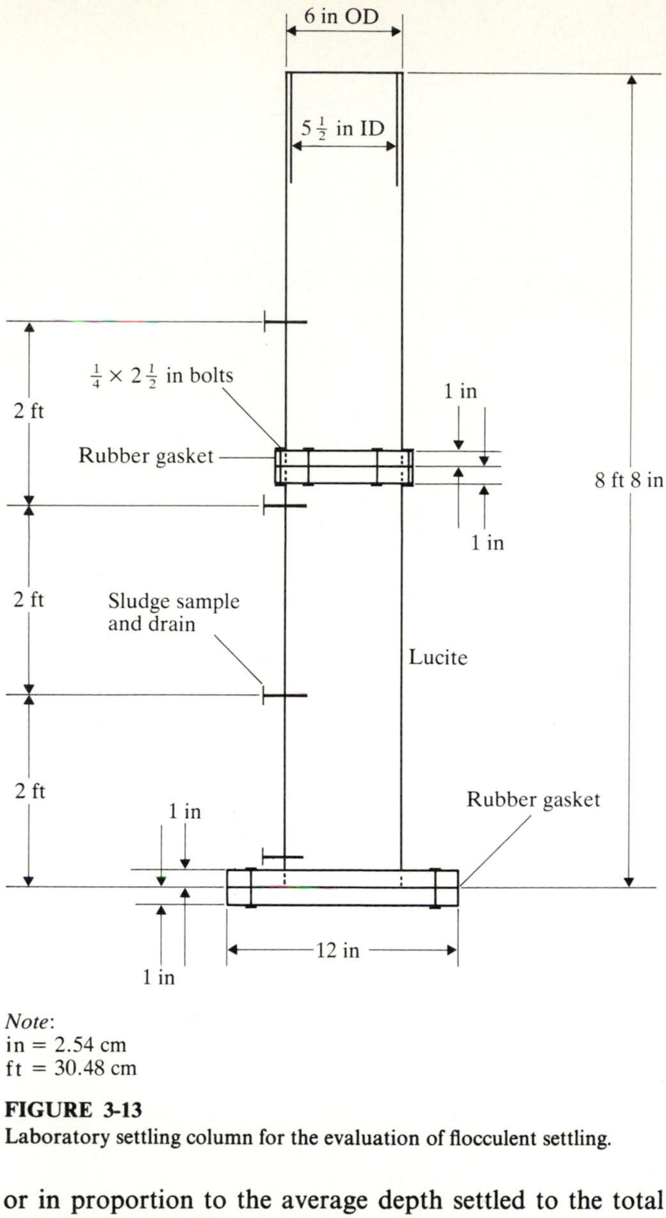

Note:
in = 2.54 cm
ft = 30.48 cm

FIGURE 3-13
Laboratory settling column for the evaluation of flocculent settling.

or in proportion to the average depth settled to the total depth of 6 ft (1.83 m). The average depth to which the 50 to 60 percent range has settled in Fig. 3-14 is 3.8 ft (1.16 m). The percent removal of this fraction is therefore 3.8 ft/6.0 ft (1.16 m/1.83 m), or 63 percent of the 10 percent. Each subsequent percent range is computed in a similar manner, and the total removal developed as shown in Table 3.3.

The total removal of suspended solids of 62.4 percent is effected at an overflow rate of 6 ft/h = 1080 gal/(d · ft^2) [44 m^3/(d · m^2)] at a retention period of 60

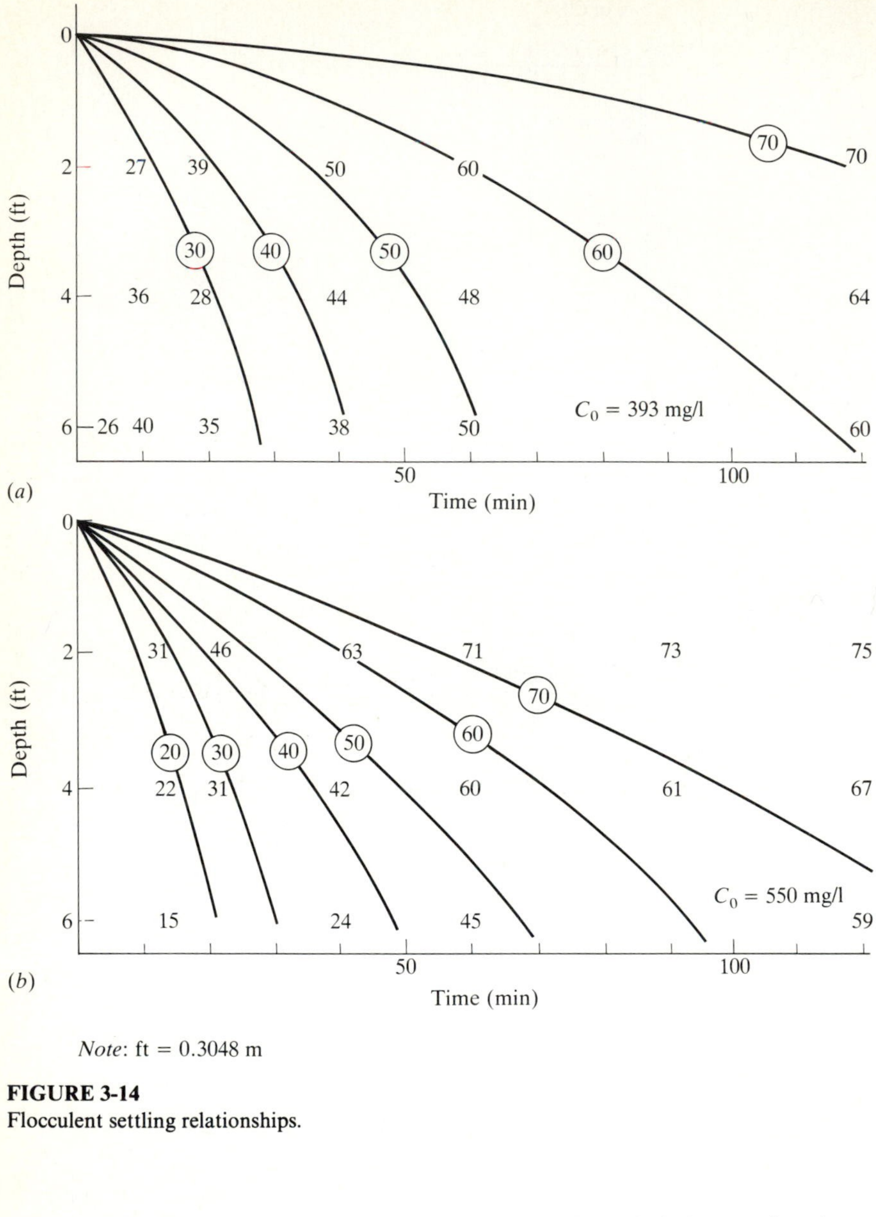

Note: ft = 0.3048 m

FIGURE 3-14
Flocculent settling relationships.

min. In a similar manner various percent removals and their associated overflow rates and detention periods can be computed.

Since the degree of flocculation is influenced by initial concentration of suspended solids, settling tests should be run over the anticipated range of suspended solids in the influent waste. In many wastes, a fraction of the measured suspended solids is not removed by settling, so that the curves developed from the laboratory analysis will approach the removal fraction as a limit.

TABLE 3.3
Percent ranges for total removal of suspended solids

SS range, %	d/d_o	SS removal, %
0–50	1.0	50
50–60	0.64	6.4
60–70	0.25	2.5
70–100	0.05	1.5
	Total removal	62.4

Since the data obtained from the laboratory analysis represent ideal settling conditions, criteria for prototype design must account for the effects of turbulence, short-circuiting, and inlet and outlet losses. The net effect of these factors is a decrease in the overflow rate and an increase in the detention time over that derived from the laboratory analysis. As a general rule, the overflow rate will be decreased by a factor of 1.25 to 1.75, and the detention period increased by a factor of 1.50 to 2.00. The development of these relationships is shown in Example 3.5.

Example 3.5. Laboratory data were obtained on the settling of a paper-mill waste (Table 3.4). Design a settling tank to produce a maximum effluent suspended solids of 150 mg/l.

TABLE 3.4
Settling data

	Removal, %		
Time, min	**2 ft**	**4 ft**	**6 ft**
Initial solids, 393 mg/l			
5			26
10	27	36	40
20	39	28	35
40	50	44	38
60	60	48	50
120	70	64	60
Initial solids, 550 mg/l			
15	31	22	15
20	46	31	
40	63	42	24
60	71	60	45
90	73	61	
120	75	67	59

Note: ft = 30.48 cm.

TABLE 3.5

Time, min	Velocity, ft/h	Removal of SS, %	Overflow rate, gal/(d·ft²)
		C_0 = 393 mg/l	
27.5	13.1	41.6	2360
42.0	8.6	49.5	1550
62.0	5.8	57.3	1050
115.0	3.1	65.8	560
		C_0 = 550 mg/l	
20	18.0	36.7	3250
27	13.3	46.8	2400
47	7.7	56.5	1400
66	5.4	62.5	980
83	3.8	70.8	690

Note:
ft/h = 0.305 m/h
gal/(d · ft²) = 4.07×10^{-2} m³/(d · m²)

With the values given in Table 3.4, Fig. 3-14 can be drawn and Table 3.5 can be constructed calculating the percent removal, the velocity, and the overflow rate of different times as indicated above.

Plotting the SS removal versus overflow rate and time, Figs. 3-15 and 3-16 can be drawn.

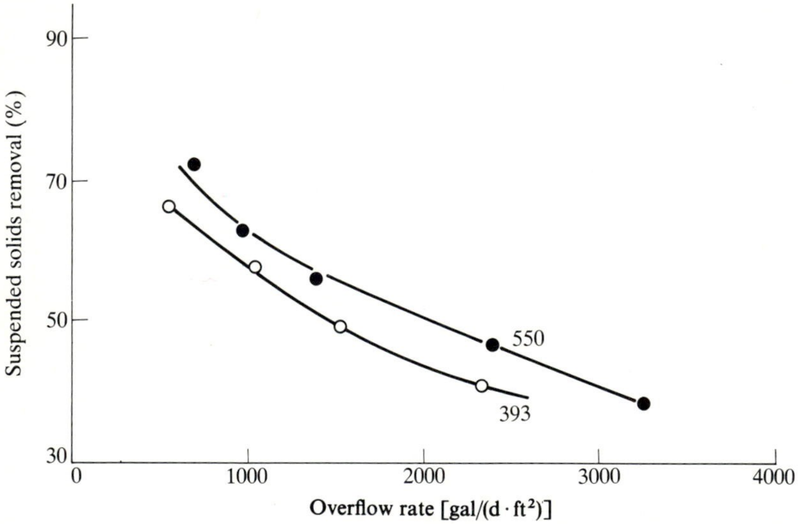

Note: gal/(d · ft²) = 4.07×10^{-2} m³/(d · m²)

FIGURE 3-15

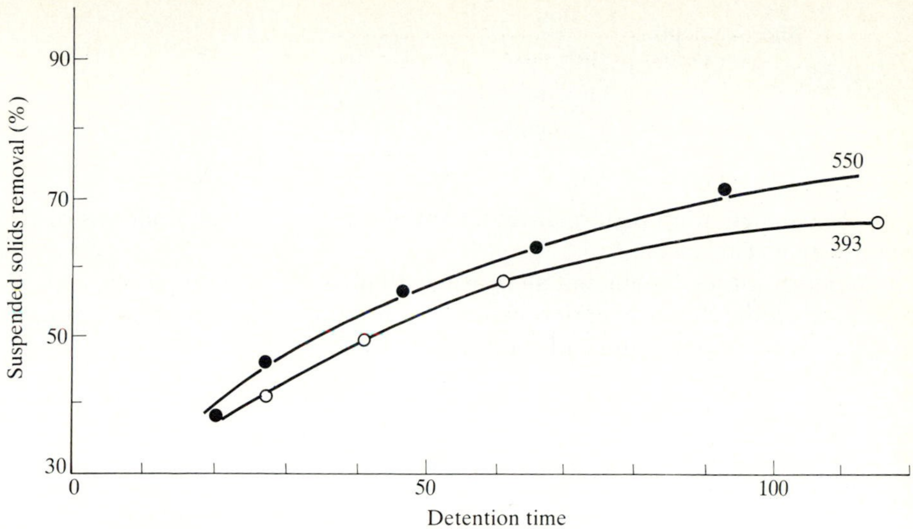

FIGURE 3-16

From Figs. 3-15 and 3-16:

For 393 mg/l:

$$\text{Percent removal} = \frac{393 - 150}{393} \times 100 = 62$$

$$\text{Overflow rate} = 770 \text{ gal/(d} \cdot \text{ft}^2\text{)[31.4 m}^3\text{/(d} \cdot \text{m}^2\text{)]}$$

$$\text{Detention time} = 74 \text{ min}$$

For 550 mg/l:

$$\text{Percent removal} = \frac{550 - 150}{550} \times 100 = 73$$

$$\text{Overflow rate} = 540 \text{ gal/(d} \cdot \text{ft}^2\text{)[22.0 m}^3\text{/(d} \cdot \text{m}^2\text{)]}$$

$$\text{Detention time} = 104 \text{ min}$$

Design:

$$\text{Overflow rate} = \frac{540}{1.5} = 360 \text{ gal/(d} \cdot \text{ft}^2\text{)[14.7 m}^3\text{/(d} \cdot \text{m}^2\text{)]}$$

$$\text{Detention time } t = \frac{104 \text{ min}}{60 \text{ min/h}} \times 1.75 = 3\text{h}$$

For 1 million gal/d:

$$\text{Area } A = \frac{\text{flow}}{\text{overflow rate}} = \frac{10^6 \text{ gal/d}}{360 \text{ gal/(d} \cdot \text{ft}^2\text{)}} = 2780 \text{ ft}^2 \text{ (258 m}^2\text{)}$$

$$\text{Effective depth} = \frac{t \times \text{flow}}{A}$$

$$= \frac{3\ \text{h} \times 10^6\ \text{gal/d}}{2780\ \text{ft}^2} \times \frac{0.134\ \text{ft}^3/\text{gal}}{24\ \text{h/d}}$$

$$= 6\ \text{ft}\ (1.8\ \text{m})$$

The sedimentation performances for various pulp and paper-mill wastes are summarized in Table 3.6.

Tanning wastes[8] with initial suspended solids of 1200 mg/l showed 69 percent reduction with a retention period of 2 h. BOD removals of 86.9 percent were obtained from the settling of cornstarch wastes.[9]

TABLE 3.6
Settling characteristics of pulp and paper-mill wastes[7]

Type of waste	Flow, million gal/day	Raw SS, ppm	Raw BOD, ppm	Temperature, °F	Removal, % SS	Removal, % BOD	Detention time, h	OR, gal/(d·ft²)
Paperboard	4.5	2,500	450	85	90	67	5.35	504
	0.75	136		85	90	50	1.15	940
	1.36	10,000	360	62	85	24	5.40	430
	2.5	1,185	395		96.1	19	5.3	525
	31	524	195	110	42	25	9.4	438
	30	850	250	95	80	25	0.5	1910
	3.3	2,000		90	85		2.6	1028
	0.25	50	100	100	80	25	4.5	39
	0.301	1,150	250	110	98	50		90
	35	4,000	200	100	90	10–15	1.5	374
Specialty	9.4	203	97	81	94	86	2.56	832
	2.2	6,215	120	120		90	1.5	157
	1.8	665	620	95	91	58	0.5	406
	50	120	85	100	80	16	18.2	477
Fine paper	6	200		65	95	90	3.9	695
	6.0	254	235	90	50	35	2.2	890
	9.9	500	364	70–100	90	35	2.4	1120
	3.5	300	250	65	95	48	6.0	372
	7.5–9.0	560	126	65	80	42	4.0	670
Miscellaneous	7	430	250	70	70	20	1.8	505
	14	1,000	330	73	65	60		911
	25	75	100		90	0.0	6.9	17
	17	100	425		95	50	5.9	846
	0.5	200	200	85	90		1.9	1590
	1.0	1,000	900	100		95	2.9	509

Note:
million gal/d = 3.78×10^3 m³/d
°C = $\frac{5}{9}$(°F − 32)
gal/(d·ft²) = 4.07×10^{-2} m³/(d·m²)

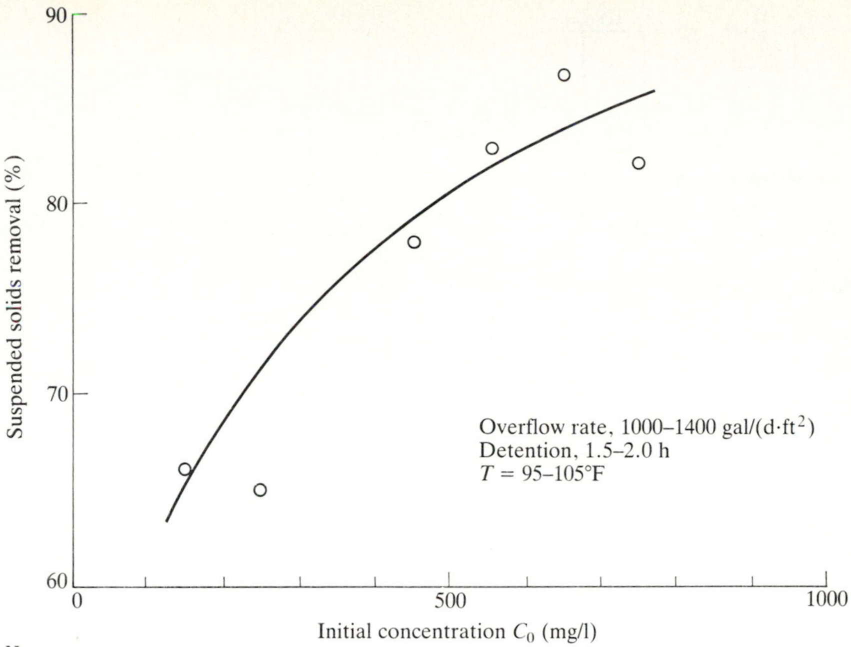

Note:
gal/(d·ft^2) = 4.07×10^{-2} m^3/(d·m^2)
°C = $\frac{5}{9}$ (°F − 32)

FIGURE 3-17
Effect of initial suspended solids concentration on percent removal from a pulp and paper-mill wastewater.

The effect of initial concentration of suspended solids on the sedimentation efficiency of a pulp-wood waste is shown in Fig. 3-17.

Zone Settling

Zone settling is characterized by activated sludge and flocculated chemical suspensions when the concentration of solids exceeds approximately 500 mg/l. The floc particles adhere together and the mass settles as a blanket, forming a distinct interface between the floc and the supernatant. The settling process is distinguished by three zones, as shown in Fig. 3-18. Initially, all the sludge is at a uniform concentration, *A* in Fig. 3-18.

During the initial settling period the sludge settles at a uniform velocity. The settling rate is a function of the initial solids concentration *A*. As settling proceeds, the collapsed solids on the bottom of the settling unit *D* build up at a constant rate. *C* is a zone of transition through which the settling velocity decreases as the result of an increasing concentration of solids. The concentration of solids in the zone settling layer remains constant until the settling interface approaches the rising layers of collapsed solids, III, and a transition zone occurs. Through the transition zone *C*, the settling velocity will decrease because of the increasing density and viscosity of the suspension surrounding the particles.

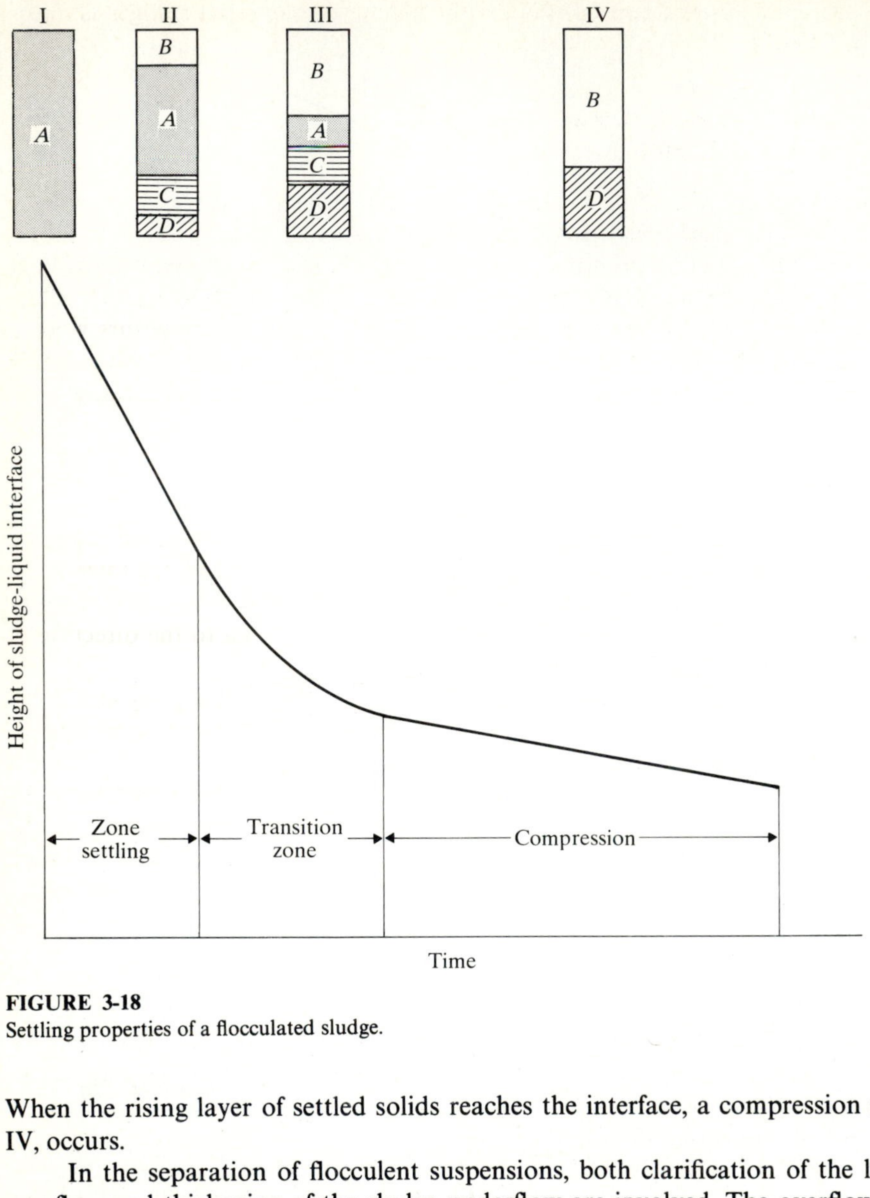

FIGURE 3-18
Settling properties of a flocculated sludge.

When the rising layer of settled solids reaches the interface, a compression zone, IV, occurs.

In the separation of flocculent suspensions, both clarification of the liquid overflow and thickening of the sludge underflow are involved. The overflow rate for clarification requires that the average rise velocity of the liquid overflowing the tank be less than the zone settling velocity of the suspension. The tank surface area requirements for thickening the underflow to a desired concentration level are related to the solids loading to the unit and are usually expressed in terms of a mass loading (pounds of solids per square foot per day or kilograms per square meter per day) or a unit area (square feet per pound of solids per day or square meters per kilogram per day).

The mass loading concept for the thickening of industrial sludges is developed in Chap. 11.

Laboratory Evaluation of Zone Settling and Calculation of Solids Flux

The settling properties of flocculated sludges can be evaluated in a liter cylinder equipped with a slow-speed stirrer rotating at a speed of 4 to 5 revolutions per hour (r/h). The effect of the stirrer is to simulate the hydraulic motion and rake action in a clarifier and to break the stratification and arching action of the settling sludge. In some cases, an initial flocculation of the sludge occurs when it is added to the graduated cylinder. The settling and compaction curve is developed by plotting the height of the sludge interface versus the time of settling.

Clarifiers

Clarifiers may either be rectangular or circular. In most rectangular clarifiers, scraper flights extending the width of the tank move the settled sludge toward the inlet end of the tank at a speed of about 1 ft/min (0.3 m/min). Some designs move the sludge toward the effluent end of the tank, corresponding to the direction of flow of the density current. A typical unit is shown in Fig. 3-19.

Circular clarifiers may employ either a center-feed well or a peripheral inlet. The tank can be designed for center sludge withdrawal or vacuum withdrawal over the entire tank bottom.

Circular clarifiers are of three general types. With the center-feed type, the waste is fed into a center well and the effluent is pulled off at the weir along the outside. With a peripheral-feed tank, the effluent is pulled off at the tank center. With a rim-flow clarifier, the peripheral feed and effluent drainoff are also along the clarifier rim, but this type is usually used for larger clarifiers.

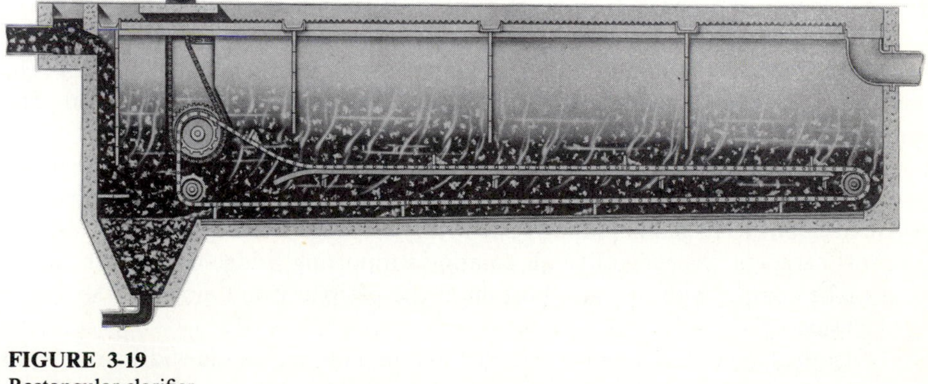

FIGURE 3-19
Rectangular clarifier.

FIGURE 3-20
Circular clarifier.

The circular clarifier usually gives the optimal performance. Rectangular tanks may be desired where construction space is limited. In addition, a series of rectangular tanks would be cheaper to construct due to the "shared wall" concept.

The reactor clarifier is another variation where the functions of chemical mixing, flocculation, and clarification are combined in the highly efficient solids-contact unit. This combination achieves the highest overflow rate and the highest effluent quality of all clarifier designs.

The circular clarifier can be designed for center sludge withdrawal or vacuum withdrawal over the entire tank bottom. Center sludge withdrawal requires a minimum bottom slope of 1 in/ft (8.3 cm/m). The flow of sludge to the center well is largely hydraulically motivated by the collection mechanism, which serves to overcome inertia and avoid sludge adherence to the tank bottom. The vacuum drawoff is particularly adaptable to secondary clarification and thickening of activated sludge. A circular clarifier is shown in Fig. 3-20.

The mechanisms can be of the plow type or the rotary-hoe type. The plow-type mechanism employs staggered plows attached to two opposing arms that move at about 10 ft/min (3 m/min). The rotary-hoe mechanism consists of a series of short scrapers suspended from a rotating supporting bridge on endless chains that make contact with the tank bottom at the periphery and move to the center of the tank.

An inlet device is designed to distribute the flow across the width and depth of the settling tank. The outlet device is likewise designed to collect the effluent uniformly at the outlet end of the tank. Well-designed inlets and outlets will reduce the short-circuiting characteristics of the tank. Increased weir length can be provided by extending the effluent channels back into the basin or by providing multiple effluent channels. In circular basins, inboard or radial weirs will ensure low takeoff velocities. Relocation of weirs is sometimes necessary to minimize solids carryover induced by density currents resulting in upwelling swells of sludge at the end of the settling tank. The installation of a plate below the effluent weir extending 18 in (45.7 cm) into the clarifier will deflect rising solids and permit them to resettle. Improved performance of secondary clarifiers have been achieved using this modification.

Tube settlers offer increased removal efficiency at higher loading rates and lower detention times. An immediate advantage is that modules of inclined tubes constructed of plastic can be installed in existing clarifiers to upgrade performance.

There are two types of tube clarifiers, the slightly inclined and the steeply inclined units. The slightly inclined unit usually has the tubes inclined at a 5° angle. For the removal of discrete particles, an inclination of 5° has proven most efficient.

The steeply inclined unit is less efficient in removal of discrete particles, but can be operated continuously. When the tubes are inclined greater than 45°, the sludge is deposited and slides back out of the tube, forming a countercurrent flow. In practice most wastes are flocculent in nature, and removal efficiency is improved when the tubes are inclined to 60° to take advantage of the increased flocculation that occurs as the solids slide back out of the tube. Steeply inclined units are usually used where sedimentation units are being upgraded.

The hydraulic characteristics of a settling tank can be defined by a dispersion test in which dye or tracer is injected into the influent as a slug, and the concentration measured in the effluent as a function of time.

It is frequently possible to improve the performance of an existing settling basin by making modifications based on the results of a dispersion test. A comparison of a hydraulically overloaded center-feed tank converted to a peripheral-feed tank is shown in Fig. 3-21.

3.4 OIL SEPARATION

In an oil separator, free oil is floated to the surface of a tank and then skimmed off. The same conditions holding for the subsidence of particles apply, except that the lighter-than-water oil globules rise through the liquid. The design of gravity separators as specified by the American Petroleum Institute[10] is based on the removal of all free oil globules larger than 0.015 cm.

The Reynolds number is less than 0.5 so Stokes' law applies. A design procedure considering short-circuiting and turbulence has been developed.[10] Typically, effluent oil concentrations in the order of 50 mg/l are achieved.

Plate separators include parallel plates and corrugated plates (CPS). Plate separators are designed to separate oil droplets larger than 0.006 cm. It has been

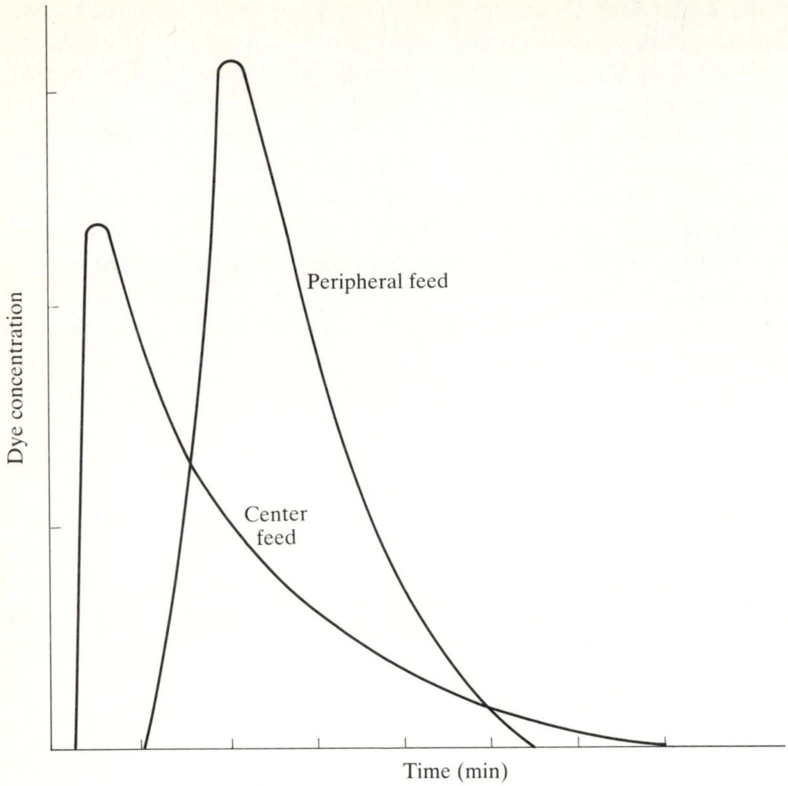

FIGURE 3-21
Dispersion characteristics of two settling tanks.

found by experience that a 0.006-cm separation will generally produce a 10-mg/l free oil nonemulsified effluent. This quality can usually be met when the influent oil content is less than 1 percent. One problem with the CPS is that reduced efficiency results from high oil loadings resulting from oil-droplet shear and reentrainment of the oil droplet. This is largely overcome using a cross-flow corrugated plate separator in which the separated oil rises across the direction of flow rather than against the direction of flow (which have plates angled at 45° and spaced at 10 mm). The hydraulic loading varies with temperature and the specific gravity of the oil. Nominal flow rates are specified for a temperature of 20°C and a specific gravity of 0.9 for the oil. A hydraulic loading of 0.5 $m^3/(h \cdot m^2)$ of actual plate area will usually result in separation of 0.006-cm droplets. A 50 percent safety factor is usually employed for design purposes. A plate separator is shown in Fig. 3-22.

Emulsified oily materials require special treatment to break the emulsions so that the oily materials will be free and can be separated by gravity, coagulation, or air flotation. The breaking of emulsions is a complex art and may

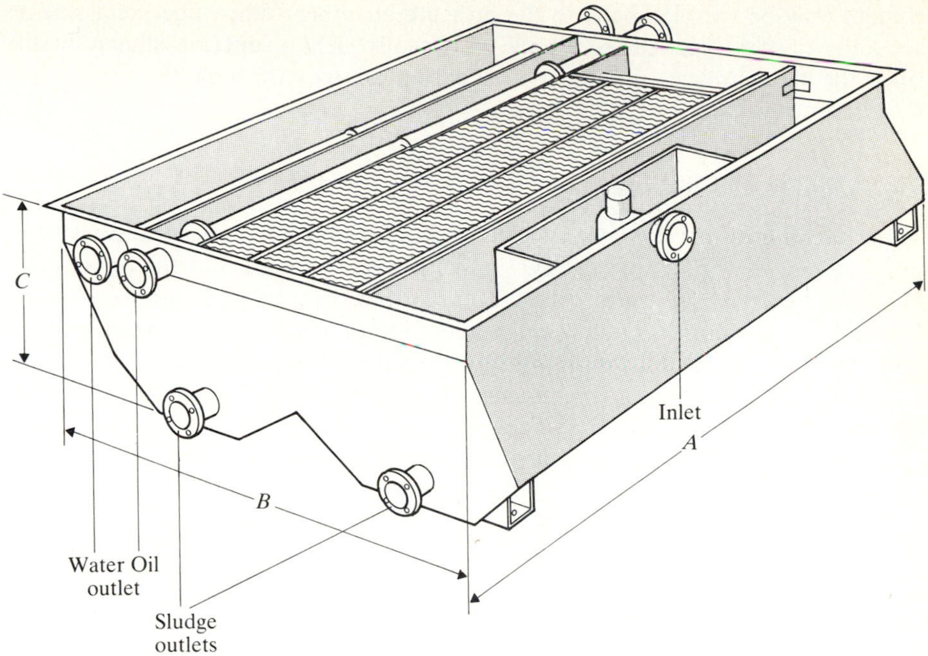

FIGURE 3-22
Cross-flow plate separator.

require laboratory or pilot-scale investigations prior to developing a final process design.

Emulsions can be broken by a variety of techniques. Quick-breaking detergents form unstable emulsions which break in 5 to 60 min to 95 to 98 percent completion. Emulsions can be broken by acidification, the addition of alum or iron salts, or the use of emulsion-breaking polymers. The disadvantage of alum or iron is the large quantities of sludge generated.

3.5 FLOTATION

Flotation is used for the removal of suspended solids and oil and grease from wastewaters and for the separation and concentration of sludges. The waste flow or a portion of clarified effluent is pressurized to 50 to 70 lb/in^2 (345 to 483 kPa or 3.4 to 4.8 atm) in the presence of sufficient air to approach saturation. When this pressurized air-liquid mixture is released to atmospheric pressure in the flotation unit, minute air bubbles are released from solution. The sludge flocs, suspended solids, or oil globules are floated by these minute air bubbles, which attach themselves to and become enmeshed in the floc particles. The air-solids mixture rises to the surface, where it is skimmed off. The clarified liquid is removed from the bottom of the flotation unit; at this time a portion of the effluent may be recycled back to the pressure chamber. When flocculent sludges are to be clarified, pressurized recycle will usually yield a superior effluent quality since the flocs are not subjected to shearing stresses through the pumps and pressurizing system.

Air Solubility and Release

The saturation of air in water is directly proportional to pressure and inversely proportional to temperature. Pray[11] and Frohlich[12] found that the oxygen and nitrogen solubility in water follow Henry's law over a wide pressure range. Vrablick[13] has shown that although a linear relationship exists between pressure and solubility for most industrial wastes, the slope of the curve varies, depending on the nature of the waste constituents present. The solubility of air in water at atmospheric pressure is shown in Table 3.7.

The quantity of air that will theoretically be released from solution when the pressure is reduced to 1 atm can be computed from

$$s = s_a \frac{P}{P_a} - s_a \tag{3.11}$$

where s = air released at atmospheric pressure per unit volume at 100 percent saturation, cm^3/l

s_a = air saturation at atmospheric pressure, cm^3/l

P = absolute pressure

P_a = atmospheric pressure

The actual quantity of air released will depend upon the turbulent mixing conditions at the point of pressure reduction and on the degree of saturation obtained in the pressurizing system. Since the solubility in industrial wastes may

TABLE 3.7
Air characteristics and solubilities

Temperature		Volume solubility		Weight solubility		Density	
°C	°F	ml/l	CF/thousand gal	mg/l	lb/thousand gal	g/l	lb/CF
0	32	28.8	3.86	37.2	0.311	1.293	0.0808
10	50	23.5	3.15	29.3	0.245	1.249	0.0779
20	68	20.1	2.70	24.3	0.203	1.206	0.0752
30	86	17.9	2.40	20.9	0.175	1.166	0.0727
40	104	16.4	2.20	18.5	0.155	1.130	0.0704
50	122	15.6	2.09	17.0	0.142	1.093	0.0682
60	140	15.0	2.01	15.9	0.133	1.061	0.0662
70	158	14.9	2.00	15.3	0.128	1.030	0.0643
80	176	15.0	2.01	15.0	0.125	1.000	0.0625
90	194	15.3	2.05	14.9	0.124	0.974	0.0607
100	212	15.9	2.13	15.0	0.125	0.949	0.0591

Values presented in absence of water vapor and at 14.7 lb/in^2 abs pressure (1 atm).

be less than that in water, a correction may have to be applied to Eq. (3.11). Retention tanks will generally yield 85 to 90 percent saturation. Equation (3.11) can be modified to account for air saturation:

$$s = s_a \left(\frac{fP}{P_a} - 1 \right) \tag{3.12}$$

where f is the fraction of saturation in the retention tank.

The performance of a flotation system depends upon having sufficient air bubbles present to float substantially all of the suspended solids. An insufficient quantity of air will result in only partial flotation of the solids, and excessive air will yield no improvement. The performance of a flotation unit in terms of effluent quality and solids concentration in the float can be related to an air/solids ratio, which is usually defined as mass of air released per mass of solids in the influent waste:

$$\frac{A}{S} = \frac{s_a R}{S_a Q} \left(\frac{fP}{P_a} - 1 \right) \tag{3.13}$$

where Q = wastewater flow
R = pressurized recycle
S_a = influent oil and/or suspended solids

The relationship between the air/solids ratio and effluent quality is shown in Fig. 3-23. It should be noted that the shape of the curve obtained will vary with the nature of the solids in the feed.

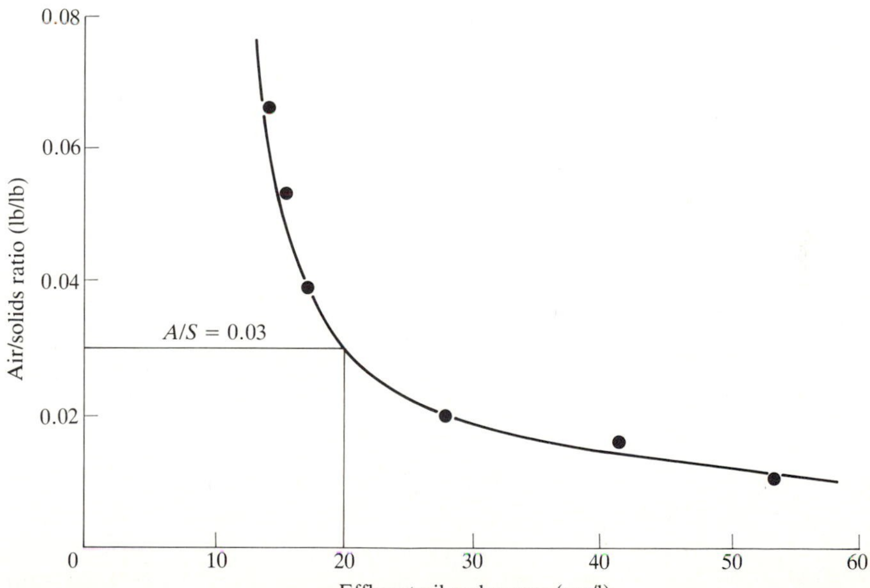

FIGURE 3-23
Effects of A/S on effluent quality.

Vrablick[13] has shown that the bubbles released after pressurization (20 to 50 lb/in^2 or 1.36 to 3.40 atm) range in size from 30 μm to 120 μm. The rise velocity closely follows Stokes' law. The rise velocity of a solids-air mixture has been observed to vary from 1 to 5 in/min (2.56 to 12.7 cm/min) and will increase with an increasing air/solids ratio. In the flotation of activated sludge of 0.91 percent solids and 40 lb/in^2 (276 kPa or 2.72 atm), Hurwitz and Katz[14] observed free rises of 0.3 (9), 1.2 (37), and 1.8 ft/min (55 cm/min) for recycle ratios of 100, 200, and 300 percent, respectively. The initial rise rate will vary with the character of the solids.

The primary variables for flotation design are pressure, recycle ratio, feed solids concentration, and retention period. The effluent suspended solids decrease and the concentration of solids in the float increase with increasing retention period. When the flotation process is used primarily for clarification, a detention period of 20 to 30 min is adequate for separation and concentration. Rise rates of 1.5 to 4.0 gal/(min · ft^2) [0.061 to 0.163 m^3/(min · m^2)] are commonly employed. When the process is employed for thickening, longer retention periods are necessary to permit the sludge to compact.

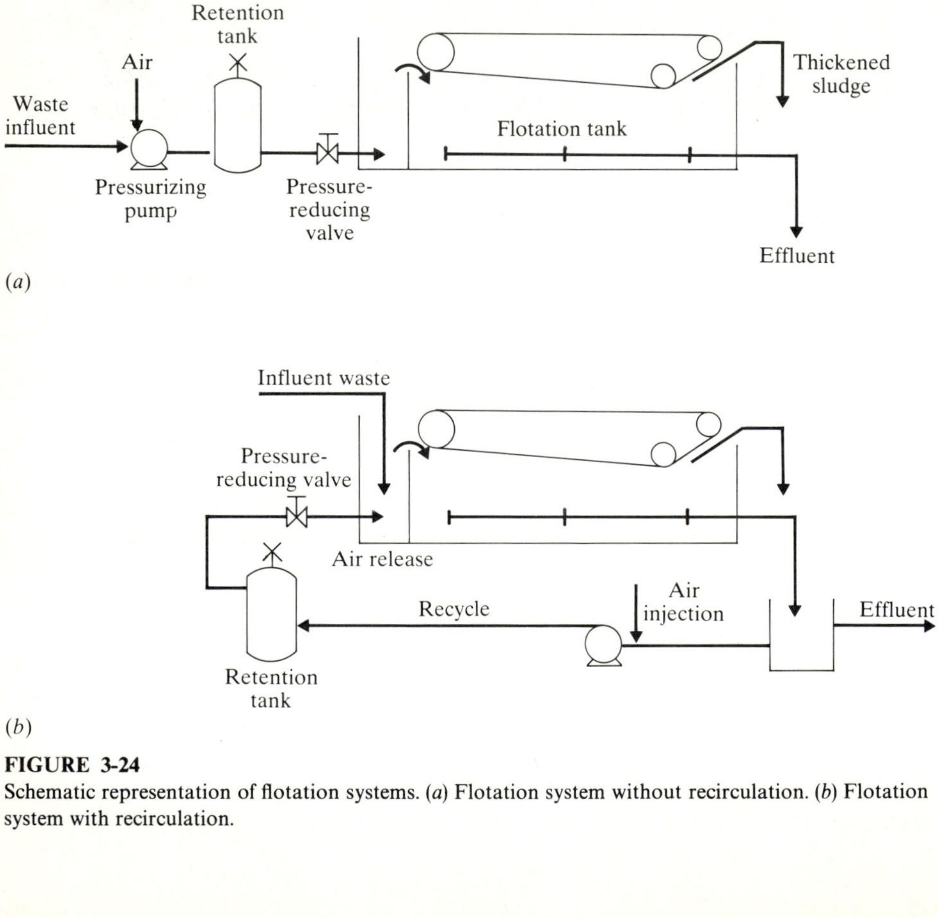

FIGURE 3-24
Schematic representation of flotation systems. (*a*) Flotation system without recirculation. (*b*) Flotation system with recirculation.

FIGURE 3-25
Clarifier flotation unit.

The principal components of a flotation system are a pressurizing pump, air-injection facilities, a retention tank, a backpressure regulating device, and a flotation unit, as shown in Fig. 3-24. The pressurizing pump creates an elevated pressure to increase the solubility of air. Air is usually added through an injector on the suction side of the pump or directly to the retention tank.

The air and liquid are mixed under pressure in a retention tank with a detention time of 1 to 3 min. A backpressure regulating device maintains a constant head on the pressurizing pump. Various types of valves are used for this purpose. The flotation unit may be either circular or rectangular with a skimming device to remove the thickened, floated sludge, as shown in Fig. 3-25.

The induced air flotation system, shown in Fig. 3-26, operates on the same principles as a pressurized air DAF unit. The gas, however, is self-induced by a rotor-disperser mechanism. The rotor, the only moving part of the mechanism that is submerged in the liquid, forces the liquid through the disperser openings, thereby creating a negative pressure. It pulls the gas downward into the liquid, causing the desired gas-liquid contact. The liquid moves through a series of four cells before leaving the tank, and the float skimmings pass over the overflow weirs on each side of the unit. This type of system offers the advantages of signifi-

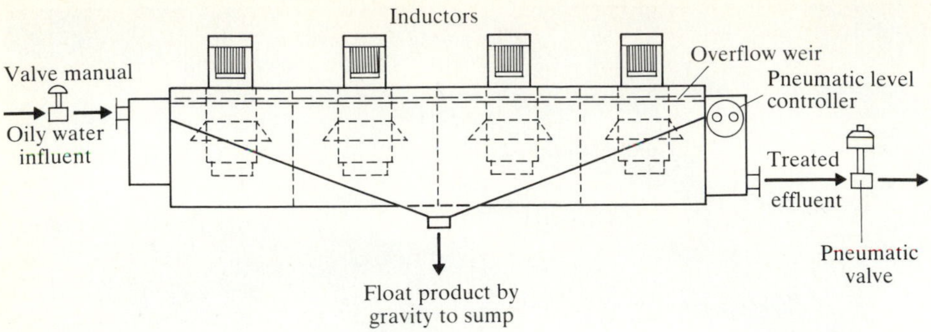

FIGURE 3-26
Induced air flotation system (*Wemco Envirotech Company*).

cantly lower capital cost and smaller space requirements than the pressurized system, and current performance data indicate that these systems have the capacity to effectively remove free oil and suspended materials.

The disadvantages include higher connected power requirements than the pressurized system, performance dependent on strict hydraulic control, less chemical addition and flocculation flexibility, and relatively high volumes of float skimmings as a function of liquid throughput (3 to 7 percent of the incoming flow for induced air systems in common compared to less than one percent for pressurized air systems).

It is possible to estimate the flotation characteristics of a waste by the use of a laboratory flotation cell, as shown in Fig. 3-27. The procedure is as follows:

1. Partially fill the calibrated cylinder with waste or flocculated sludge mixture and the pressure chamber with clarified effluent or water.
2. Apply compressed air to the pressure chamber to attain the desired pressure.
3. Shake the air-liquid mixture in the pressure chamber for 1 min and allow to stand for 3 min to attain saturation. Maintain the pressure on the chamber for this period.
4. Release a volume of pressurized effluent to the cylinder and mix with the waste or sludge. The volume to be released is computed from the desired recycle ratio. The velocity of release through the inlet nozzle should be of such a magnitude as not to shear the suspended solids in the feed mixture but to maintain adequate mixing.
5. Measure the rise of the sludge interface with time. Correction must be applied to scale-up the height of rise in the test cylinder to the depth of the prototype unit.
6. After a detention time of 20 min, the clarified effluent and the floated sludge are drawn off through a valve in the bottom of the cylinder.

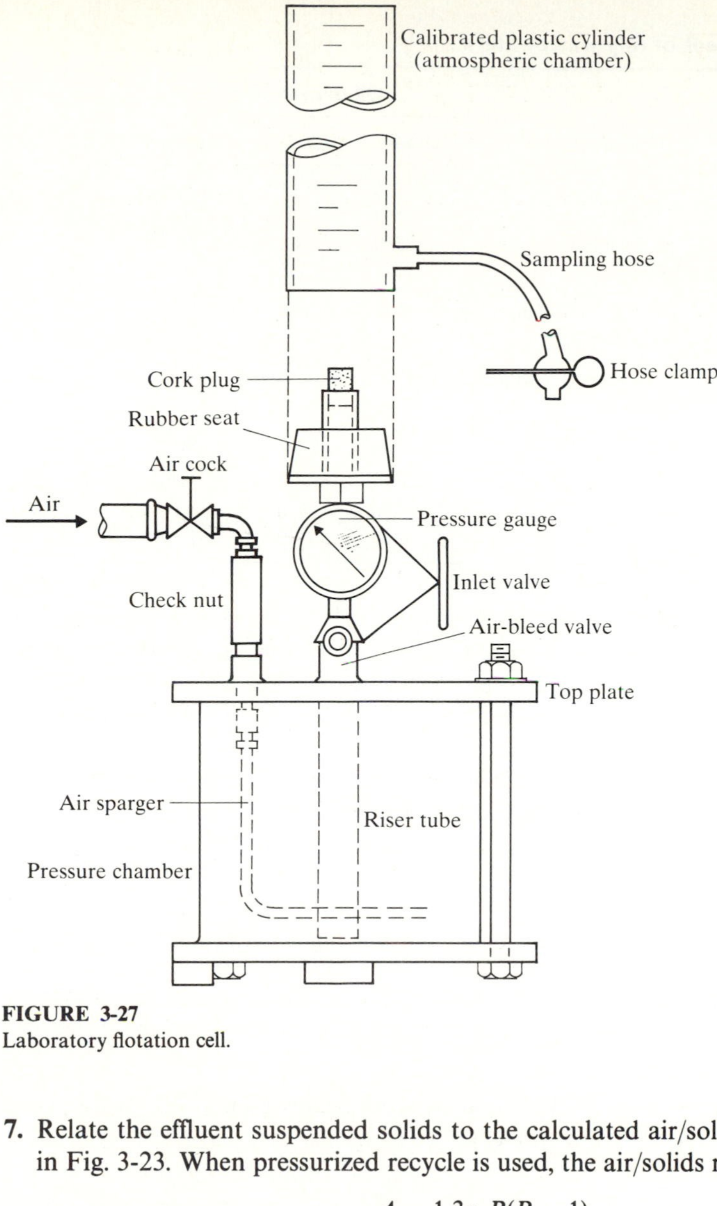

FIGURE 3-27
Laboratory flotation cell.

7. Relate the effluent suspended solids to the calculated air/solids ratio as shown in Fig. 3-23. When pressurized recycle is used, the air/solids ratio is computed:

$$\frac{A}{S} = \frac{1.3 s_a R(P - 1)}{Q S_a}$$

where s_a = air saturation, cm^3/l
R = pressurized volume, l
P = absolute pressure, atm
Q = waste flow, l
S_a = influent suspended solids, mg/l

TABLE 3.8
Air flotation treatment of oily wastewaters

Wastewater	Coagulant, mg/l	Oil concentration, mg/l Influent	Effluent	Removal, %
Refinery	0	125	35	72
	100 alum	100	10	90
	130 alum	580	68	88
	0	170	52	70
Oil tanker ballast water	100 alum + 1 mg/l polymer	133	15	89
Paint manufacture	150 alum + 1 mg/l polymer	1900	0	100
Aircraft maintenance	30 alum + 10 mg/l activated silica	250–700	20–50	>90
Meat packing		3830	270	93
		4360	170	96

High-degree clarification frequently requires that flocculating chemicals be added to the influent before it is mixed with the pressurized recycle. Alum or polyelectrolytes are used as flocculating agents. Katz[15] has shown that the rise rate of chemical floc will vary from 0.65 to 2.0 ft/min (20 to 61 cm/min), depending on the floc size and characteristics.

Flotation design for an oily wastewater is shown in Example 3.6. Flotation units have been used for the clarification of wastewaters, for the removal of oil, and for the concentration of waste sludges. Some reported data are summarized in Table 3.8.

Example 3.6. A wastestream of 150 gal/min (0.57 m^3/min) and a temperature of 103°F (39.4°C) contains significant quantities of nonemulsified oil and nonsettleable suspended solids. The concentration of oil is 120 mg/l. Reduce the oil to less than 20 mg/l. Laboratory studies showed:

Alum dose = 50 mg/l
Pressure = 60 lb/in^2 gage (515 kPa absolute or 4.1 relative atm)
Sludge production = 0.64 mg/mg alum
Sludge = 3 percent by weight

Calculate:

(*a*) The recycle rate
(*b*) Surface area of the flotation unit
(*c*) Sludge quantities generated

The air/solids ratio for effluent oil and grease of 20 mg/l is found from Fig. 3-23:

$$\frac{A}{S} = 0.03 \text{ lb air released/lb solids applied}$$

At 103°F (39.4°C) the weight solubility of air is 18.6 mg/l from Table 3.7. The value of f is assumed to be 0.85.

Solution

(*a*) The recycle rate is

$$R = \frac{(A/S)QS_a}{s_a(fP/P_a - 1)}$$

$$= \frac{0.03 \times 150 \times 120}{18.6(0.85 \times 515/101.3) - 1}$$

$$= 8.75 \text{ gal/min } (33.1 \text{ l/min})$$

(*b*) The hydraulic loading for oil removal is determined using Fig. 3-28 and is 2.6 gal/(min · ft^2) [0.11 m^3/(min · m^2)] for an effluent of 20 mg/l. The required surface area is

$$A = \frac{Q + R}{\text{loading}}$$

$$= \frac{150 + 8.75}{2.6}$$

$$= 61 \text{ ft}^2 \ (5.7 \text{ m}^2)$$

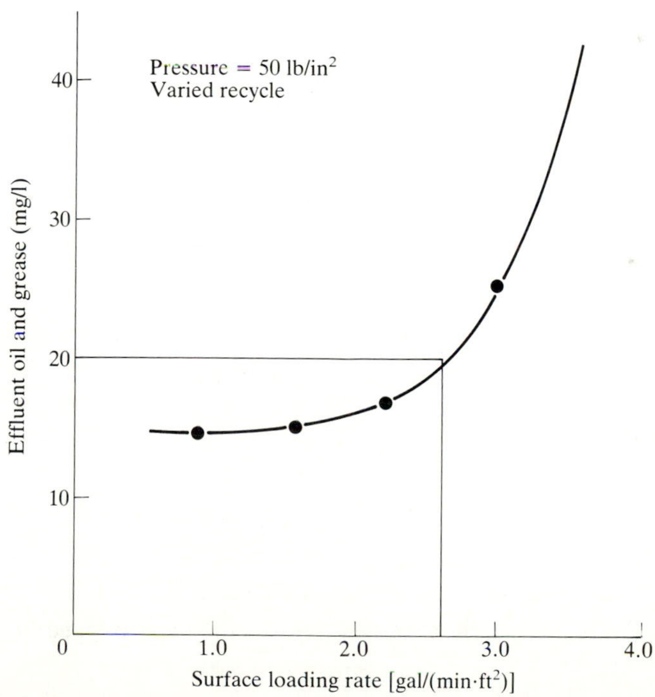

FIGURE 3-28

(*c*) Sludge quantities generated:

$$\text{Oil sludge} = (120 - 20)\ \text{mg/l} \times 150\ \text{gal/min} \times 1440\ \text{min/d}$$
$$\times\ (\text{million gal}/10^6\ \text{gal})\left(8.34\ \frac{\text{lb/million gal}}{\text{mg/l}}\right)$$
$$= 180\ \text{lb/d}\ (82\ \text{kg/d})$$

$$\text{Alum sludge} = 0.64\ \text{mg sludge/mg alum} \times 50\ \text{mg/l alum}$$
$$\times\ 150\ \text{gal/min}$$
$$\times\ (1440\ \text{min/d})(\text{million gal}/10^6\ \text{gal})(8.34)$$
$$= 58\ \text{lb/d}\ (26\ \text{kg/d})$$

$$\text{Total sludge} = 238\ \text{lb/d}\ (108\ \text{kg/d})$$

$$\text{Total sludge volume} = 238/0.03\ \text{lb/d}\ (\text{gal}/8.34\ \text{lb})(\text{day}/1440\ \text{min})$$
$$= 0.66\ \text{gal/min}\ (2.5\ \text{l/min})$$

PROBLEMS

3.1. An industry discharging to a POTW is required to equalize its wastewater and discharge it such that the load in BOD/day to the treatment plant is constant over the 24 h. The sewage flow is 6.47 million gal/d, with a BOD of 200 mg/l. The diurnal variation in flow is shown in Fig. P3-1. The industrial waste flow is 3.17 million gal/d, with a BOD of 1200 mg/l, and is constant over 10 h (8 A.M. to 6 P.M.).

(*a*) Compute the volume of equalization (or holding) tank required.
(*b*) Plot the discharge curve of the industrial waste over 24 h.

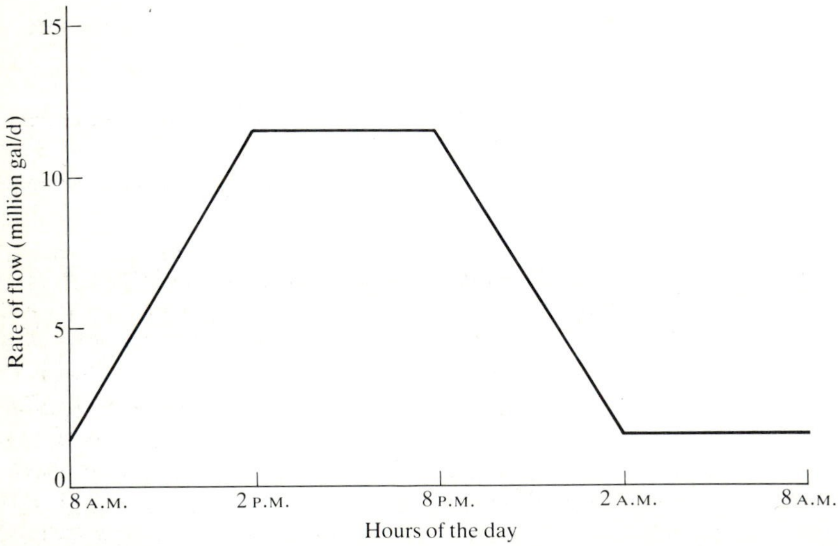

FIGURE P3-1

3.2. A survey of the discharge from a chemical plant showed the following results:

	Flow, gal/min	TOC, mg/l
8 A.M.	450	920
10 A.M.	620	1130
12 noon	840	1475
2 P.M.	800	1525
4 P.M.	340	910
6 P.M.	270	512
8 P.M.	570	1210
10 P.M.	1100	1520
12 midnight	1200	1745
2 A.M.	800	820
4 A.M.	510	410
6 A.M.	570	490

Compute the volume of an equalization basin to yield a peaking factor of 1.2 based on concentration for a flow-through system and on mass discharge for a constant discharge rate. For the variable volume case, the low level in the basin will be 10 percent of the daily flow.

3.3. An acidic industrial wastewater of 150 gal/min (0.57 m^3/min) with a peak factor of 1.2 is to be neutralized to pH 6.0 using lime. The titration curve is as follows:

pH	mg lime/l waste
1.8	0
1.9	500
2.05	1000
2.25	1500
3.5	2000
4.1	2100
5.0	2150
7.0	2200

Consider providing the necessary capacity for a maximum 2-week lime requirement for pH = 7.0 with an hourly requirement 20 percent greater.

Determine:

(*a*) The lime feed rate for each stage in a two-stage system.

(*b*) The lime storage capacity (ft^3) based on the controlling requirement of either the mean monthly requirement or the maximum two-week requirement. Assume the use of pebbled quicklime (CaO), which has a bulk density of 65 lb/ft^3 (1060 kg/m^3).

(*c*) The capacity of the lime slaker and mechanism for transporting the bulk lime based on the maximum estimated requirements.

(*d*) The mean and maximum slaking water requirements assuming a 10 percent by weight slurry.

(*e*) The size of the slurry control tank based on a minimum resident time of 5 minutes.

(*f*) The size of the caustic storage tank based on 24 hours caustic feed at the maximum extended usage equivalent to a hydrated lime requirement of 1000 mg/l. Assume NaOH is available having a purity of 98.9 percent and a solubility of 2.5 lb/gal (300 kg/m^3).

(*g*) The maximum caustic (NaOH) feed rate for backup of the lime system.

3.4. A wastewater flow of 100 gal/min (0.38 m^3/min) with 0.1 normal H_2SO_4 requires neutralization prior to secondary treatment. This flow is to be neutralized to a pH of 7.0 using a limestone bed. Figure P3-4 presents the results of a series of laboratory pilot tests using a one-foot diameter limestone bed. These data were for upflow units with the effluent being aerated to remove residual CO_2. Assume limestone is 60 percent reactive.

Design a neutralization system specifying:

(*a*) The most economical bed depth of limestone

(*b*) The amount in pounds of acid per day to be neutralized

(*c*) The limestone requirements on an annual basis

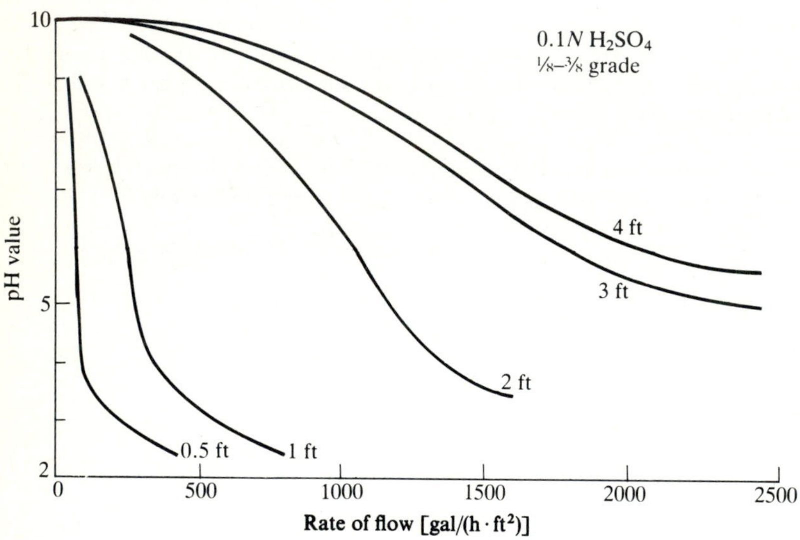

FIGURE P3-4

3.5. A laboratory settling analysis for a pulp and paper-mill effluent gave the following results for $C_0 = 430$ mg/l, $T = 29°C$:

	Suspended solids removed at indicated depth, %		
Time, min	**2 ft**	**4 ft**	**6 ft**
10	47	27	16
20	50	34	43
30	62	48	47
45	71	52	46
60	76	65	48

Note: ft = 30.5 cm.

(*a*) Design a settling tank to remove 70 per cent of the suspended solids for 1 million gal/day flow (3785 m^3/d) (Apply appropriate factors and neglect initial solids effects.)

(*b*) What removal will be attained if the flow is increased to 2 million gal/day (7570 m^3/d)?

3.6. A wastewater has a flow of 250 gal/min (0.95 m^3/min) and a temperature of 105°F (40.5°C). The concentration of oil and grease is 150 mg/l and the suspended solids 100 mg/l. An effluent concentration of 20 mg/l is required. An alum dosage of 30 mg/l is required. The required A/S is 0.04 and the surface loading rate is 2 gal/(min · ft^2) [0.081 m^3/(min · m^2)]. The operating pressure is 65 lb/in^2 (gage 4.4 rel. atm).

Compute:

(*a*) The required recycle flow

(*b*) The surface area of the unit

(*c*) The volume of sludge produced if the skimmings are 3 percent by weight

REFERENCES

1. Novotny, V., and A. J. England: *Water Res.*, vol. 8, p. 325, 1974.
2. Patterson, J. W., and J. P. Menez: *Am. Inst. Chem. Engrs Env. Prog.*, vol. 3, p. 2, 1984.
3. Tully, T. J.: *Sewage Ind. Wastes*, vol. 30, p. 1385, 1958.
4. Hazen, A.: *Trans. ASCE*, vol. 53, p. 45, 1904.
5. Camp, T. R.: *Trans. ASCE*, vol. 111, p. 909, 1946.
6. Dobbins, W. E.: "Advances in Sewage Treatment Design," Sanitary Engineering Division, Met. Section, Manhattan College, May 1961.
7. Committee on Industrial Waste Practice of SED: *J. Sanit. Engng Div. ASCE*, December 1964.
8. Sutherland, R.: *Ind. Eng. Chem.*, p. 630, May 1947.
9. Greenfield, R. E., and G. N. Cornell: *Ind. Eng. Chem.*, p. 583, May 1947.
10. American Petroleum Institute: "Manual on Disposal of Refinery Wastes," vol. 1, New York, 1959.
11. Pray, H. A.: *Ind. Eng. Chem.*, vol. 44, pt. 1, p. 146, 1952.
12. Frolich, R.: *Ind. Eng. Chem.*, vol. 23, p. 548, 1931.
13. Vrablick, E. R.: *Proc. 14th Ind. Waste Conf.*, 1959, Purdue University.
14. Hurwitz, E., and W. J. Katz: "Laboratory Experiments on Dewatering Sewage Sludges by Dissolved Air Flotation," unpublished report, Chicago, 1959.
15. Katz, W. J.: *Proc. 10th Ind. Waste. Conf.*, 1955, Purdue University.

CHAPTER 4

COAGULATION AND PRECIPITATION

4.1 COAGULATION

Coagulation is employed for the removal of waste materials in suspended or colloidal form. Colloids are presented by particles over a size range of 1 nm (10^{-7} cm) to 0.1 nm (10^{-8} cm). These particles do not settle out on standing and cannot be removed by conventional physical treatment processes.

Colloids present in wastewater can be either hydrophobic or hydrophilic. The hydrophobic colloids (clays, etc.) possess no affinity for the liquid medium and lack stability in the presence of electrolytes. They are readily susceptible to coagulation. Hydrophilic colloids, such as proteins, exhibit a marked affinity for water. The absorbed water retards flocculation and frequently requires special treatment to achieve effective coagulation.[1]

Colloids possess electrical properties which create a repelling force and prevent agglomeration and settling. Stabilizing ions are strongly adsorbed to an inner fixed layer which provides a particle charge that varies with the valence and number of adsorbed ions. Ions of an opposite charge form a diffuse outer layer which is held near the surface by electrostatic forces. The psi (ψ) potential is defined as the potential drop between the interface of the colloid and the body of solution. The zeta potential (ζ) is the potential drop between the slipping plane and the body of solution and is related to the particle charge and the thickness of the double layer. The thickness of the double layer (χ) is inversely proportional to the concentration and valence of nonspecific electrolytes, as shown in Fig. 4-1. A van der Waals attractive force is effective in close proximity to the colloidal particle.

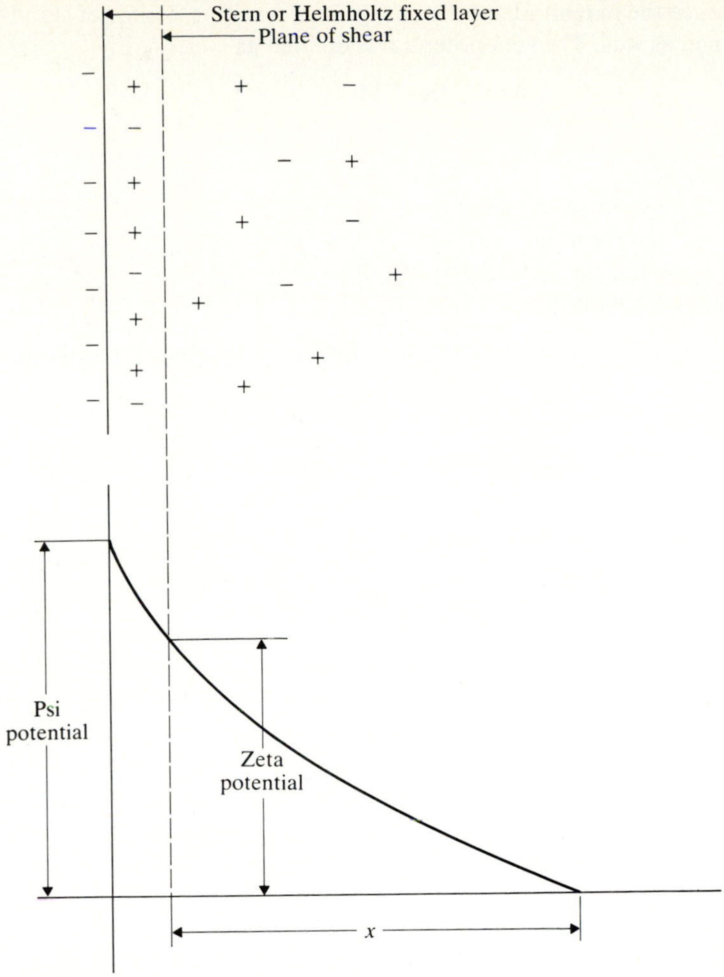

FIGURE 4-1
Electrochemical properties of a colloidal particle.

The stability of a colloid is due to the repulsive electrostatic forces and in the case of hydrophilic colloids to solvation in which an envelope of water retards coagulation.

Zeta Potential

Since the stability of a colloid is primarily due to electrostatic forces, neutralization of this charge is necessary to induce flocculation and precipitation. Although it is not possible to measure the psi potential, the zeta potential can be determined, and hence the magnitude of the charge and resulting degree of stability can be determined as well. The zeta potential is defined as

$$\zeta = \frac{4\pi v}{\varepsilon X} = \frac{4\pi\eta \text{ EM}}{\varepsilon} \tag{4.1}$$

where v = particle velocity
ε = dielectric constant of the medium
η = viscosity of the medium
X = applied potential per unit length of cell
EM = electrophoretic mobility

For practical usage in the determination of the zeta potential, Eq. (4.1) can be reexpressed:

$$\zeta \text{ (mV)} = \frac{113{,}000}{\varepsilon} \eta\text{(poise) EM} \left(\frac{\mu\text{m/s}}{\text{V/cm}}\right) \tag{4.2}$$

where EM = electrophoretic mobility, (μm/s)/(V/cm).

At 25°C, Eq. (4.2) reduces to

$$\zeta = 12.8 \text{ EM} \tag{4.3}$$

The zeta potential is determined by measurement of the mobility of colloidal particles across a cell, as viewed through a microscope.[2,3] Several types of apparatus are commercially available for this purpose. Recently a Lazer Zee meter has been developed which does not track individual particles but rather adjusts the image to produce a stationary cloud of particles using a rotating prism technique. This apparatus is shown in Fig. 4-2. The computations involved in determining the zeta potential are illustrated in Example 4.1.

Example 4.1. In an electrophoreses cell 10 cm in length, grid divisions are 160 μm at $\times$ 6 magnification. Compute the zeta potential at an impressed voltage of 35 V. The time of travel between grid divisions is 42 s and the temperature is 20°C.
At 20°C:

$$\eta = 0.01 \text{ poise}$$

$$\varepsilon = 80.36$$

$$\text{EM} = \frac{v}{X} = \frac{160\ \mu\text{m}/42\ \text{s}}{35\ \text{V}/10\ \text{cm}} = 1.09 \left(\frac{\mu\text{m/s}}{\text{V/cm}}\right)$$

$$\zeta\text{(mV)} = 113{,}000 \times \frac{\eta\text{(poise) EM } (\mu\text{m/s})/(\text{V/cm})}{\varepsilon}$$

$$= 113{,}000 \times \frac{0.01 \times 1.09}{80.36}$$

$$= 15.3 \text{ mV}$$

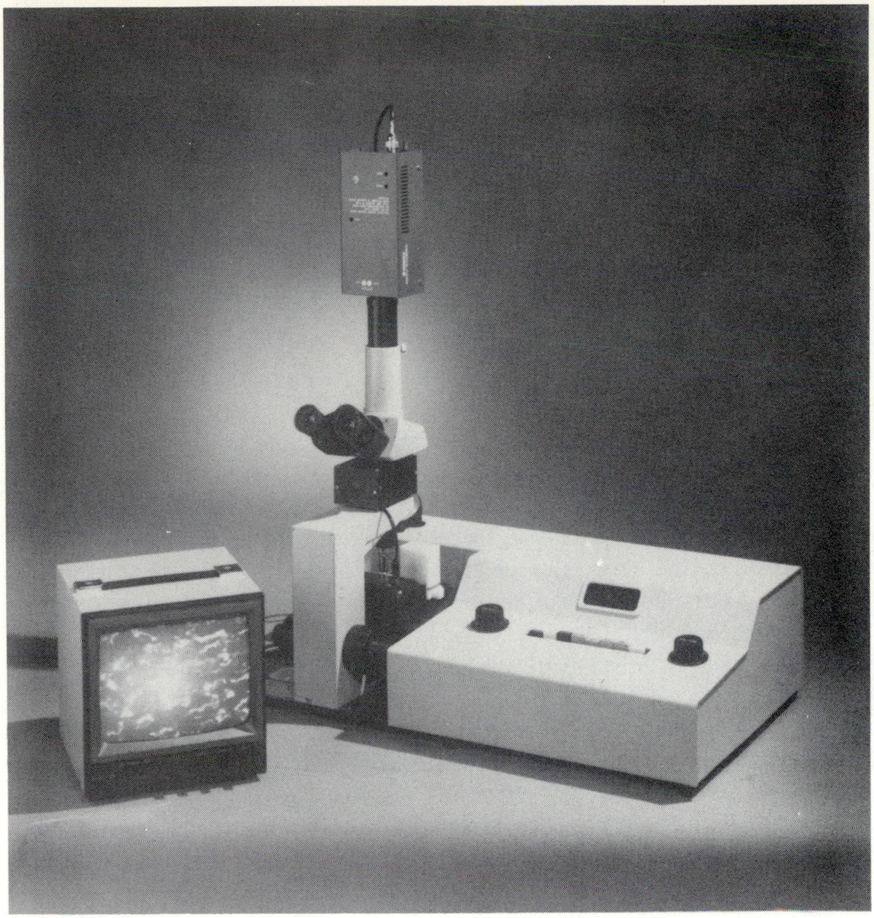

FIGURE 4-2
Lazer Zee meter for zeta potential measurement (*Courtesy of Penkem Inc.*).

Since there will usually be a statistical variation in the mobility of individual particles, around 20 to 30 values should be averaged for any one determination. The magnitude of the zeta potential for water and waste colloids has been found to average from -16 to -22 mV with a range of -12 to -40 mV.[3] The zeta potential is unaffected by pH over a range of pH 5.5 to 9.5.

The zeta potential is lowered by:

1. Change in the concentration of the potential determining ions
2. Addition of ions of opposite charge
3. Contraction of the diffuse part of the double layer by increase in the ion concentration in solution

Since the vast majority of colloids in industrial wastes possess a negative charge, the zeta potential is lowered and coagulation is induced by the addition of high-valence cations. The precipitating power of effectiveness of cation valence in the precipitation of arsenious oxide is

$$Na^+ : Mg^{2+} : Al^{3+} = 1:63:570$$

Optimum coagulation will occur when the zeta potential is zero; this is defined as the isoelectric point. Effective coagulation will usually occur over a zeta potential range of ± 0.5 mV.

Mechanism of Coagulation

Coagulation results from two basic mechanisms: perikinetic or electrokinetic coagulation, in which the zeta potential is reduced by ions or colloids of opposite charge to a level below the van der Waals attractive forces, and orthokinetic coagulation, in which the micelles aggregate and form clumps which agglomerate the colloidal particles.

The addition of high-valence cations depresses the particle charge and the effective distance of the double layer, thereby reducing the zeta potential. As the coagulant dissolves, the cations serve to neutralize the negative charge on the colloids. This occurs before visible floc formation, and rapid mixing which "coats" the colloid is effective in this phase. Microflocs are then formed which retain a positive charge in the acid range because of the adsorption of H^+. These microflocs also serve to neutralize and coat the colloidal particle. Flocculation agglomerates the colloids with a hydrous oxide floc. In this phase, surface adsorption is also active. Colloids not initially adsorbed are removed by enmeshment in the floc.

Riddick[3] has outlined a desired sequence of operation for effective coagulation. If necessary, alkalinity should first be added. (Bicarbonate has the advantage of providing alkalinity without raising the pH.) Alum or ferric salts are added next; they coat the colloid with Al^{3+} or Fe^{3+} and positively charged microflocs. Coagulant aids, such as activated silica and/or polyelectrolyte for floc buildup and zeta potential control, are added last. After addition of alkali and coagulant, a rapid mixing of 1 to 3 min is recommended, followed by flocculation, with addition of coagulant aid, for 20 to 30 min. Destabilization can also be accomplished by the addition of cationic polymers, which can bring the system to the isoelectric point without a change in pH. Although polymers are 10 to 15 times as effective as alum as a coagulant they are considerably more expensive. The mechanism of the coagulation process is shown in Fig. 4-3.

Properties of Coagulants

The most popular coagulant in waste-treatment application is aluminum sulfate, or alum $[Al_2(SO_4)_3 \cdot 18H_2O]$, which can be obtained in either solid or liquid form. When alum is added to water in the presence of alkalinity, the reaction is

$$Al_2(SO_4)_3 \cdot 18H_2O + 3Ca(OH)_2 \rightarrow 3CaSO_4 + 2Al(OH)_3 + 18H_2O$$

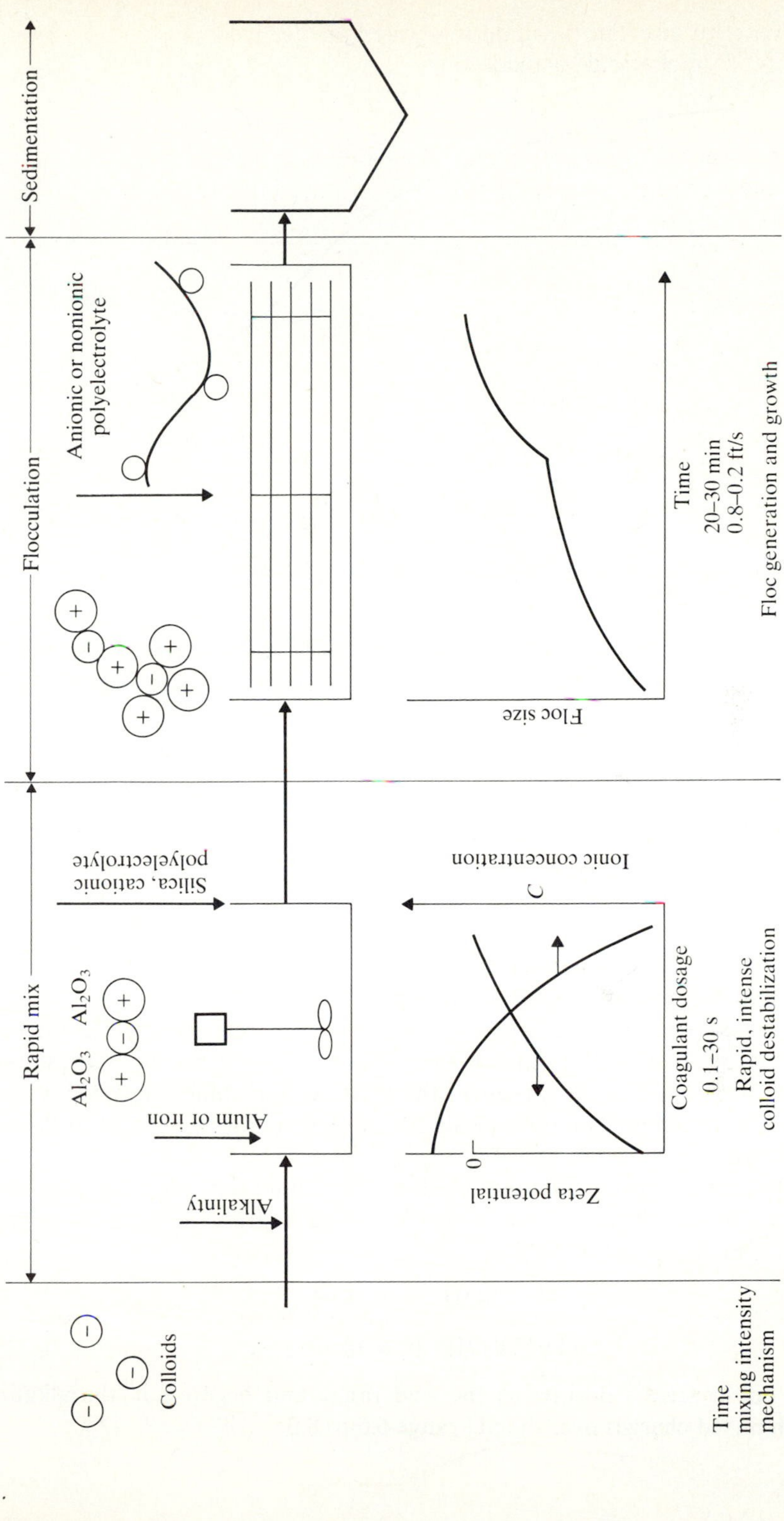

Note: ft/s = 30.48 cm/s

FIGURE 4-3
Mechanisms of coagulation.

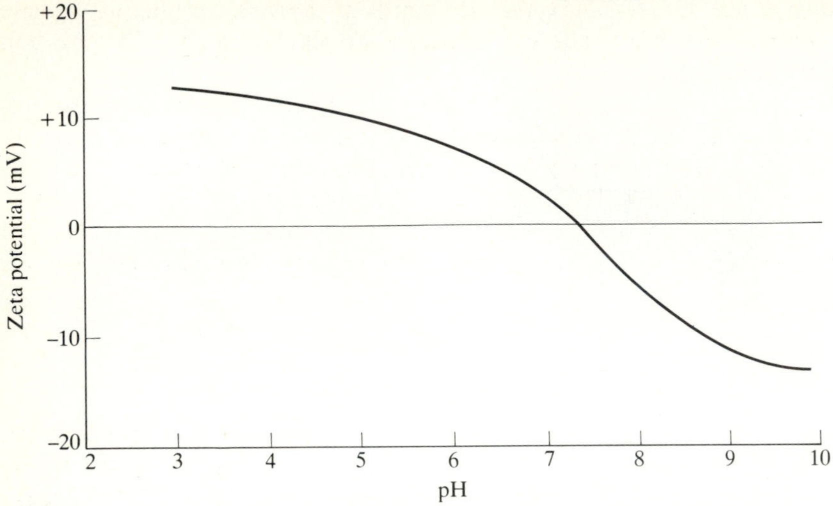

FIGURE 4-4
Zeta potential–pH plot for electrolytic aluminum hydroxide. (*After Riddick.*[3])

The aluminum hydroxide is actually of the chemical form $Al_2O_3 \cdot xH_2O$ and is amphoteric in that it can act as either an acid or a base. Under acidic conditions

$$[Al^{3+}][OH^-]^3 = 1.9 \times 10^{-33}$$

At pH 4.0, 51.3 mg/l of Al^{3+} is in solution. Under alkaline conditions, the hydrous aluminum oxide dissociates:

$$Al_2O_3 + 2OH^- \rightarrow 2AlO_2^- + H_2O$$

$$[AlO_2^-][H^+] = 4 \times 10^{-13}$$

At pH 9.0, 10.8 mg/l of aluminum is in solution.

The alum floc is least soluble at a pH of approximately 7.0. The floc charge is positive below pH 7.6 and negative above pH 8.2. Between these limits the floc charge is mixed. These relationships with respect to the zeta potential are shown in Fig. 4-4. High alum dosages used in the treatment of some industrial wastes may bring about postprecipitation of alum floc, depending on the pH of flocculation.

Ferric salts are also commonly used as coagulants but have the disadvantage of being more difficult to handle. An insoluble hydrous ferric oxide is produced over a pH range of 3.0 to 13.0:

$$Fe^{3+} + 3OH^- \rightarrow Fe(OH)_3$$

$$[Fe^{3+}][OH^-]^3 = 10^{-36}$$

The floc charge is positive in the acid range and negative in the alkaline range, with mixed charges over the pH range 6.5 to 8.0.

The presence of anions will alter the range of effective flocculation. Sulfate ion will increase the acid range but decrease the alkaline range. Chloride ion increases the range slightly on both sides.

Lime is not a true coagulant but reacts with bicarbonate alkalinity to precipitate calcium carbonate and with *ortho*-phosphate to precipitate calcium hydroxyapatite. Magnesium hydroxide precipitates at high pH levels. Good clarification usually requires the presence of some gelatinous $Mg(OH)_2$ but this does make the sludge more difficult to dewater. Lime sludge can frequently be thickened, dewatered, and calcined to convert calcium carbonate to lime for reuse.

Coagulant Aids

The addition of some chemicals will enhance coagulation by promoting the growth of large, rapid-settling flocs. Activated silica is a short-chain polymer that serves to bind together particles of microfine aluminum hydrate. At high dosages, silica will inhibit floc formation because of its electronegative properties. The usual dosage is 5 to 10 mg/l.

Polyelectrolytes are high-molecular-weight polymers which contain adsorbable groups and form bridges between particles or charged flocs. Large flocs (0.3 to 1 mm) are thus created when small dosages of polyelectrolyte (1 to 5 mg/l) are added in conjunction with alum or ferric chloride. The polyelectrolyte is substantially unaffected by pH and can serve as a coagulant itself by reducing the effective charge on a colloid. There are three types of polyelectrolytes: a cationic, which adsorbs on a negative colloid or floc particle; an anionic, which replaces the anionic groups on a colloidal particle and permits hydrogen bonding between the colloid and the polymer; and a nonionic, which adsorbs and flocculates by hydrogen bonding between the solid surfaces and the polar groups in the polymer. The general application of coagulants is shown in Table 4.1.

Laboratory Control of Coagulation

Because of the complex reactions involved, laboratory experimentation is essential to establish the optimum pH and coagulant dosage for coagulation of a wastewater. Two procedures can be followed for this purpose: the jar test, in which pH and coagulant dosage are varied to attain the optimum operating conditions; and zeta potential control as proposed by Riddick,[3] in which coagulant is added to zero zeta potential. The procedures to determine the optimum coagulant dosage using these two tests are outlined below:

1. By zeta potential measurement:[3]
   (*a*) Place 1000 ml of sample in a beaker.
   (*b*) Add the coagulant in known increments. (The optimum pH should be established either by zeta potential or by a jar-test procedure.)

TABLE 4.1
Chemical coagulant applications

Chemical process	Dosage range, mg/l	pH	Comments
Lime	150–500	9.0–11.0	For colloid coagulation and P removal Wastewater with low alkalinity, and high and variable P Basic reactions: $Ca(OH)_2 + Ca(HCO_3)_2 \rightarrow 2CaCO_3 + 2H_2O$ $MgCO_3 + Ca(OH)_2 \rightarrow Mg(OH)_2 + CaCO_3$
Alum	75–250	4.5–7.0	For colloid coagulation and P removal Wastewater with high alkalinity and low and stable P Basic reactions: $Al_2(SO_4)_3 + 6H_2O \rightarrow 2Al(OH)_3 + 3H_2SO_4$
$FeCl_3$, $FeCl_2$	35–150	4.0–7.0	For colloid coagulation and P removal
$FeSO_4 \cdot 7H_2O$	70–200	4.0–7.0	Wastewater with high alkalinity and low and stable P Where leaching of iron in the effluent is allowable or can be controlled Where economical source of waste iron is available (steel mills, etc.) Basic reactions: $FeCl_3 + 3H_2O \rightarrow Fe(OH)_3 + 3HCl$
Cationic polymers	2–5	No change	For colloid coagulation or to aid coagulation with a metal Where the buildup of an inert chemical is to be avoided
Anionic and some nonionic polymers	0.25–1.0	No change	Use as a flocculation aid to speed flocculation and settling and to toughen floc for filtration
Weighting aids and clays	3–20	No change	Used for very dilute colloidal suspensions for weighting

(*c*) Rapid-mix the sample for 3 min after each addition of coagulant; follow with a slow mix.

(*d*) Determine the zeta potential after each reagent addition and plot the results as shown below. To maintain constant volume, return the sample after each determination (Fig. 4-5).

(*e*) If a polyelectrolyte is to be used as a coagulant aid, it should be added last.

2. By jar-test procedure:

(*a*) Using 200 ml of sample on a magnetic stirrer, add coagulant in small increments at a pH of 6.0. After each addition, provide a 1-min rapid mix

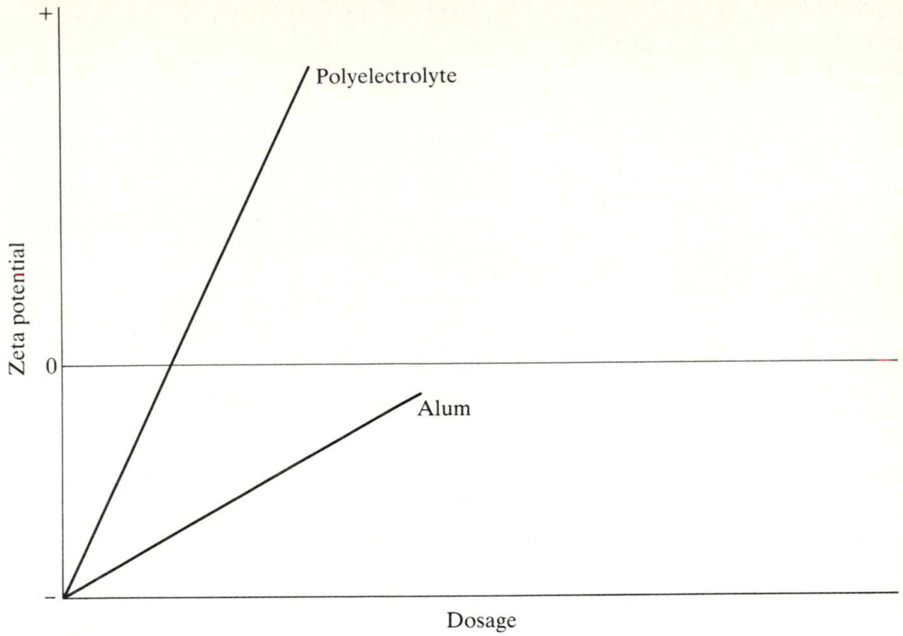

FIGURE 4-5

followed by a 3-min slow mix. Continue addition until a visible floc is formed.

(*b*) Using this dosage, place 1000 ml of sample in each of six beakers.

(*c*) Adjust the pH to pH 4.0, 5.0, 6.0, 7.0, 8.0, and 9.0 with standard alkali.

(*d*) Rapid-mix each sample for 3 min; follow this with 12 min flocculation at slow speed.

(*e*) Measure the effluent concentration of each settled sample.

(*f*) Plot the percent removal of characteristic versus pH and select the optimum pH (Fig. 4-6).

(*g*) Using this pH, repeat steps (*b*), (*d*), and (*e*), varying the coagulant dosage.

(*h*) Plot the percent removal versus the coagulant dosage and select the optimum dosage (Fig. 4-6).

(*i*) If a polyelectrolyte is used, repeat the procedure, adding polyelectrolyte toward the end of a rapid mix.

Coagulation Equipment

There are two basic types of equipment adaptable to the flocculation and coagulation of industrial wastes. The conventional system uses a rapid-mix tank, followed by a flocculation tank containing longitudinal paddles which provide slow mixing. The flocculated mixture is then settled in conventional settling tanks.

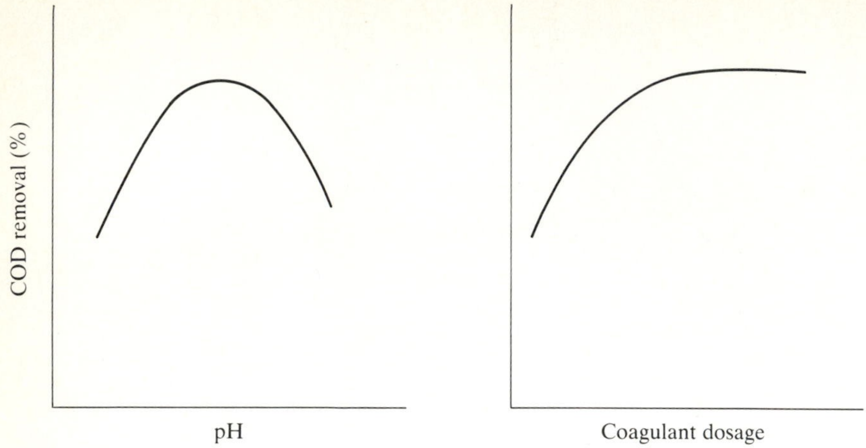

FIGURE 4-6
Characteristic plots of jar-test analysis.

A sludge-blanket unit combines mixing, flocculation, and settling in a single unit. Although colloidal destabilization might be less effective than in the conventional system, there are distinct advantages in recycling preformed floc. With lime and a few other coagulants the time required to form a settleable floc is a function of the time necessary for calcium carbonate or other calcium precipitates to form a nuclei on which other calcium materials can deposit and grow large enough to settle. It is possible to reduce both coagulant dosage and the time of floc formation by seeding the influent wastewater with previously formed nuclei or by recycling a portion of the precipitated sludge. Recycling preformed floc can frequently reduce chemical dosages, the blanket serves as a filter for improved effluent clarity, and denser sludges are frequently attainable. A sludge-blanket unit is shown in Fig. 4-7.

Coagulation of Industrial Wastes

Coagulation may be used for the clarification of industrial wastes containing colloidal and suspended solids. Paperboard wastes can be effectively coagulated with low dosages of alum. Silica or polyelectrolyte will aid in the formation of a rapid-settling floc. Typical data reported are summarized in Table 4.2.

Wastes containing emulsified oils can be clarified by coagulation.[7] An emulsion can consist of droplets of oil in water. The oil droplets are of approximately 10^{-5} cm and are stabilized by adsorbed ions. Emulsifying agents include soaps and anion-active agents. The emulsion can be broken by "salting out" with the addition of salts, such as $CaCl_2$. Flocculation will then effect charge neutralization and entrainment, resulting in clarification. An emulsion can also frequently be broken by lowering of the pH of the waste solution. An example of such a

TABLE 4.2
Chemical treatment of paper and paperboard wastes

	Influent		Effluent			Coagulants						
Waste	BOD, ppm	SS, ppm	BOD, ppm	SS, ppm	pH	Alum, ppm	Silica, ppm	Other, ppm	Detention, h	Sludge, % solids	Remarks	Reference
Board		350–450		15–60		3	5		1.7	2–4		4
Board		140–420		10–40		1		10†	0.3	2	Flotation	4
Board		240–600		35–85					2.0	2–5	950 gal/(d · ft²)	5
Board‡	127	593	68	44	6.7	10–12	10		1.3	1.76		5
Tissue	140	720	36	10–15		2	4					6
Tissue	208		33		6.6		4					6

† Glue.
‡ 15,000 gal/ton waste paper.
Note:
gal/(d · ft²) = 4.075×10^{-2} m³/(d · m²)
gal/ton = 4.17×10^{-3} m³/t

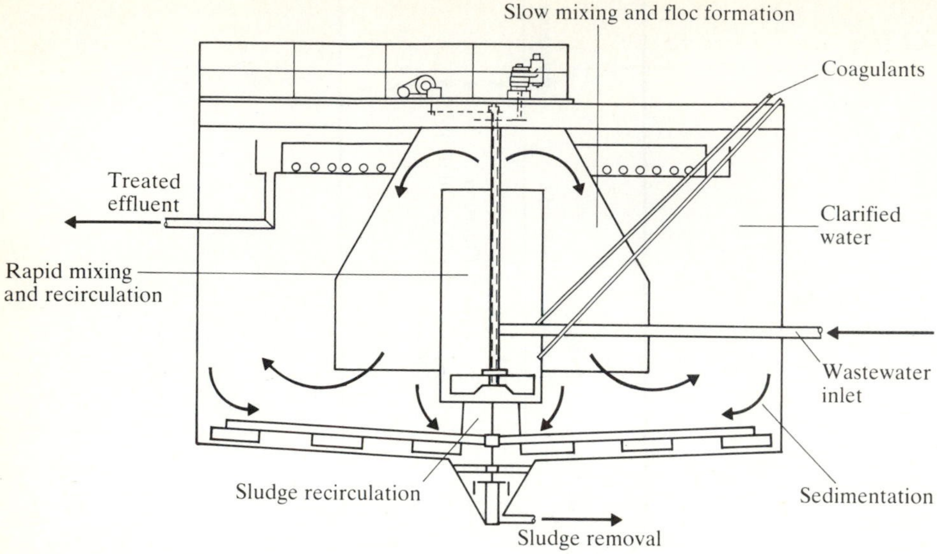

FIGURE 4-7
A reactor clarifier designed for both coagulation and settling.

waste is that produced by ball-bearing manufacture, which contains cleaning soaps and detergents, water-soluble grinding oils, cutting oils, and phosphoric acid cleaners and solvents. Treatment of this waste has been effected by the use of 800 mg/l alum, 450 mg/l H_2SO_4, and 45 mg/l polyelectrolyte. The results obtained are summarized in Table 4.3*a*.

The presence of anionic surface agents in a waste will increase the coagulant dosage. The polar head of the surfactant molecule enters the double layer and stabilizes the negative colloids. Industrial laundry wastes have been treated with H_2SO_4 followed by lime and alum; this has resulted in a reduction of COD of 12,000 mg/l to 1800 mg/l and a reduction of suspended solids of 1620 mg/l to 105 mg/l. Chemical dosages of 1400 mg/l H_2SO_4, 1500 mg/l lime, and 300 mg/l alum were required, yielding 25 percent by volume of settled sludge.

Laundromat wastes containing synthetic detergent have been coagulated with a cationic surfactant to neutralize the anionic detergent and the addition of a calcium salt to provide a calcium phosphate precipitate for flocculation. Typical results obtained are summarized in Table 4.3*b*. Operating the system above pH 8.5 will obtain nearly complete phosphate removal.

Polymer waste from latex manufacture has been coagulated with 500 mg/l ferric chloride and 200 mg/l lime at pH 9.6. COD and BOD reductions of 75 and 94 percent, respectively, were achieved from initial values of 1000 mg/l and 120 mg/l. The resulting sludge was 1.2 percent solids by weight, containing 101 lb solids/thousand gal waste (12 kg/m^3) treated. Waste from the manufacture of latex base paints[8] has been coagulated with 345 mg/l alum at pH 3.0 to 4.0,

TABLE 4.3
Coagulation of industrial wastewaters

(*a*)

	Analysis	
	Influent	Effluent
pH	10.3	7.1
Suspended solids, mg/l	544	40
Oil and grease, mg/l	302	28
Fe, mg/l	17.9	1.6
PO_4, mg/l	222	8.5

(*b*)

	Influent, mg/l	Effluent, mg/l
ABS	63	0.1
BOD	243	90
COD	512	171
PO_4	267	150
$CaCl_2$	480	
Cationic surfactant	88	
pH	7.1	7.7

(*c*)

	Influent, mg/l	Effluent, mg/l
COD	4340	178
BOD	1070	90
Total solids	2550	446

yielding 20.5 lb sludge/thousand gal waste (2.5 kg/m^3) of 2.95 percent solids by weight. The treatment results are summarized in Table 4.3*c*. Wastes from paint-spray booths in automobile assembly plants have been clarified with 400 mg/l $FeSO_4$ at pH 7.0, yielding 8 percent sludge by weight.

Synthetic rubber wastes have been treated at pH 6.7 with 100 mg/l alum, with 2 percent sludge of the original waste volume. COD was reduced from 570 to 100 mg/l, and BOD from 85 to 15 mg/l.

Vegetable-processing wastes have been coagulated[9] with lime, yielding BOD removals of 35 to 70 percent with lime dosages of approximately 0.5 lb lime/lb influent BOD (0.5 kg/kg). Results from the coagulation of textile wastewaters are shown in Table 4.4 and color removal from pulp and paper-mill effluents in Table 4.5.

BOD removals of 90 percent have been achieved on laundry wastes at pH 6.4 to 6.6 with coagulant dosages of 2 lb $Fe_2(SO_4)_3$/thousand gal waste[12] (0.24 kg/m^3).

TABLE 4.4
Coagulation of textile wastewaters[10]

				Color†		COD	
Plant	Coagulant	Dosage, mg/l	pH	Influent	Removal, %	Influent, mg/l	Removal, %
1	$Fe_2(SO_4)_3$	250	7.5–11.0	0.25	90	584	33
	Alum	300	5–9		86		39
	Lime	1200			68		30
2	$Fe_2(SO_4)_3$	500	3–4, 9–11	0.74	89	840	49
	Alum	500	8.5–10		89		40
	Lime	2000			65		40
3	$Fe_2(SO_4)_3$	250	9.5–11	1.84	95	825	38
	Alum	250	6–9		95		31
	Lime	600			78		50
4	$Fe_2(SO_4)_3$	1000	9–11	4.60	87	1570	31
	Alum	750	5–6		89		44
	Lime	2500			87		44

† Color sum of absorbances at wavelengths of 450, 550, and 650 nm.

TABLE 4.5
Color removal from pulp and paper-mill effluents[11]

				Color		COD	
Plant	Coagulant	Dosage, mg/l	pH	Influent	Removal, %	Influent	Removal, %
1	$Fe_2(SO_4)_3$	500	3.5–4.5	2250	92	776	60
	Alum	400	4.0–5.0		92		53
	Lime	1500	—		92		38
2	$Fe_2(SO_4)_3$	275	3.5–4.5	1470	91	480	53
	Alum	250	4.0–5.5		93		48
	Lime	1000	—		85		45
3	$Fe_2(SO_4)_3$	250	4.5–5.5	940	85	468	53
	Alum	250	5.0–6.5		91		44
	Lime	1000	—		85		40

BOD removals of 75 to 80 percent have been obtained on wool-scouring wastes with 1 to 3 lb $CaCl_2$/lb BOD (1 to 3 kg/kg). Carbonation was used for pH control.[13]

4.2 HEAVY METALS REMOVAL

Precipitation is employed for the removal of heavy metals from wastewaters. Heavy metals are generally precipitated as the hydroxide through the addition of lime or caustic to a pH of minimum solubility. However, several of these com-

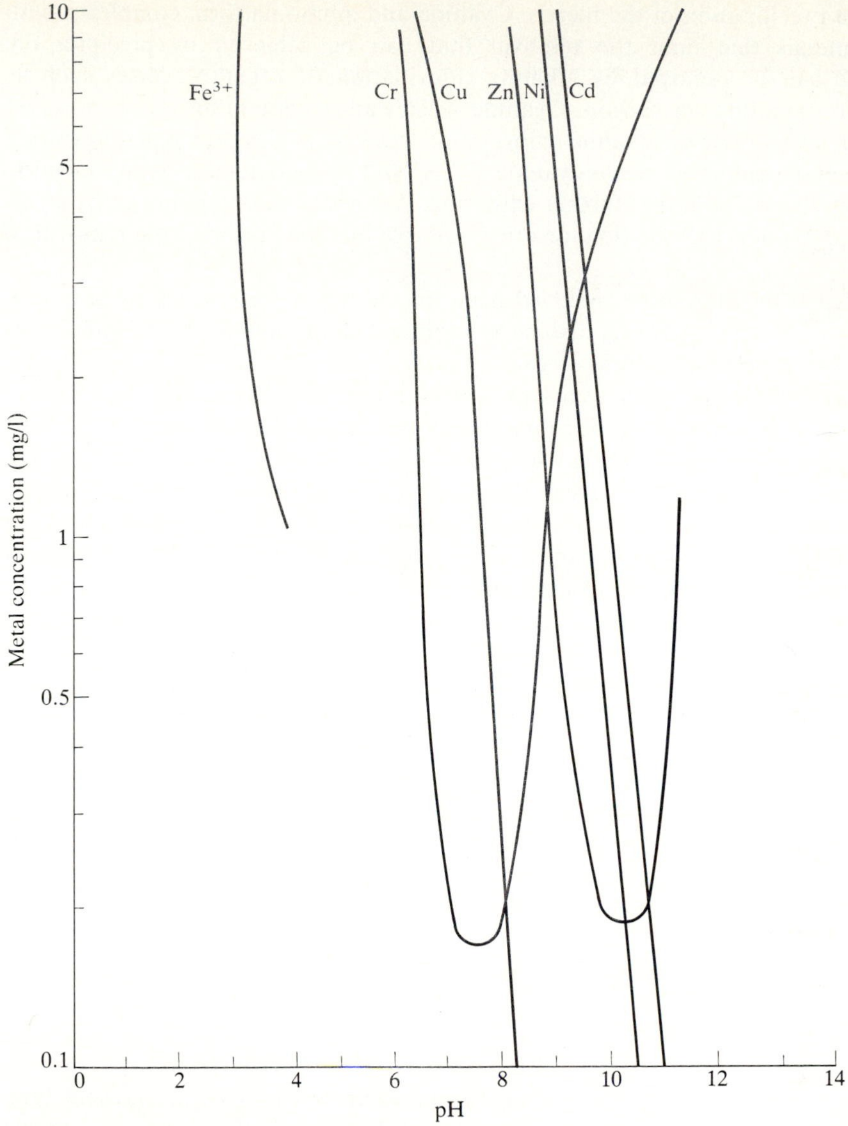

FIGURE 4.8
Solubility of metals versus pH.

pounds are amphoteric and exhibit a point of minimum solubility. The pH of minimum solubility varies with the metal in question, as shown in Fig. 4-8.

The solubilities for chromium and zinc are minimum at pH 7.5 and 10.2, respectively, and show a significant increase in concentration above these pH values.

When treating industrial wastewaters containing metals, it is frequently necessary to pretreat the wastewaters to remove substances that will interfere with the precipitation of the metals. Cyanide and ammonia form complexes with many metals that limit the removal that can be achieved by precipitation. Cyanide can be removed by alkaline chlorination or other processes such as catalytic oxidation on carbon. Cyanide wastewaters containing nickel or silver are difficult to treat by alkaline chlorination because of the slow reaction rate of these metal complexes. Ferrocyanide [$Fe(CN)_6^{4-}$] is oxidized to ferricyanide [$Fe(CN)_6^{3-}$] which resists further oxidation. Ammonia can be removed by stripping, break-point chlorination, or other suitable methods prior to the removal of metals.

Heavy metals can be removed from an industrial wastewater by lime precipitation. Heavy metals may also be precipitated as the sulfide and in some cases as the carbonate, as in the case of lead.

For many metals such as arsenic and cadmium, coprecipitation with iron or aluminum is highly effective for removal to low residual levels. In order to meet low effluent requirements, it may be necessary in some cases to provide filtration to remove floc carried over from the precipitation process. Using precipitation and clarification alone, effluent metals concentrations may be as high as 1 to 2 mg/l. Filtration should reduce these concentrations to 0.5 mg/l or less. For chromium wastes treatment hexavalent chromium must first be reduced to the trivalent state Cr^{3+} and then precipitated with lime. This is referred to as the process of reduction and precipitation.

The reducing agents commonly used for chromium wastes are ferrous sulfate, sodium *meta*-bisulfite, or sulfur dioxide. Ferrous sulfate and sodium *meta*-bisulfite may be dry- or solution-fed; SO_2 is diffused into the system directly from gas cylinders. Since the reduction of chromium is most effective at acidic pH values, a reducing agent with acidic properties is desirable. When ferrous sulfate is used as the reducing agent, the Fe^{2+} is oxidized to Fe^{3+}; if *meta*-bisulfite or sulfur dioxide is used, the negative radical SO_3^{2-} is converted to SO_4^{2-}.

The general reactions are

$$Cr^{6+} + Fe^{2+} \text{ or } SO_2 \text{ or } Na_2S_2O_5 + H^+ \rightarrow Cr^{3+} + Fe^{3+} \text{ or } SO_4^{2-}$$

$$Cr^{3+} + 3OH^- \rightarrow Cr(OH)_3\downarrow$$

Ferrous ion reacts with hexavalent chromium in an oxidation-reduction reaction, reducing the chromium to a trivalent state and oxidizing the ferrous ion to the ferric state. This reaction occurs rapidly at pH levels below 3.0. The acidic properties of ferrous sulfate are low at high dilution; acid must therefore be added for pH adjustment. The use of ferrous sulfate as a reducing agent has the disadvantage that a contaminating sludge of $Fe(OH)_3$ is formed when an alkali is added. In order to obtain a complete reaction, an excess dosage of 2.5 times the theoretical addition of ferrous sulfate must be used.

Reduction of chromium can also be accomplished by using either sodium *meta*-bisulfite or SO_2. In either case reduction occurs by reaction with the H_2SO_3 produced in the reaction. The H_2SO_3 ionizes according to the mass action:

$$\frac{(H^+)(HSO_3^-)}{(H_2SO_3)} = 1.72 \times 10^{-2}$$

Above pH 4.0, only 1 percent of sulfite is present as H_2SO_3 and the reaction is very slow. During this reaction, acid is required to neutralize the NaOH formed. The reaction is highly dependent on both pH and temperature. At pH levels below 2.0 the reaction is practically instantaneous at close to the theoretical requirements.

The theoretical quantities of chemicals required for 1 ppm Cr are:

2.81 ppm $Na_2S_2O_5$ (97.5 percent)
1.52 ppm H_2SO_4
2.38 ppm lime (90 percent)
1.85 ppm SO_2

At pH levels above 3, when a basic chrome sulfate is produced, the quantities of lime required for subsequent neutralization are reduced. At pH 8.0 to 9.9, $Cr(OH)_3$ is virtually insoluble. Experimental investigations have shown that the sludge produced will compact to 1 to 2 percent by weight.

Since dissolved oxygen is usually present in wastewaters, an excess of SO_2 must be added to account for the oxidation of the SO_3 to SO_4:

$$H_2SO_3 + \tfrac{1}{2}O_2 \rightarrow H_2SO_4$$

An excess dosage of 35 ppm SO_2 will usually be sufficient for reaction with the dissolved oxygen present.

The acid requirements for the reduction of Cr^{6+} depend upon the acidity of the original waste, the pH of the reduction reaction, and the type of reducing agent used (for example, SO_2 produces an acid but *meta*-bisulfite does not). Since it is difficult, if not impossible, to predict these requirements, it is usually necessary to titrate a sample to the desired pH endpoint with standardized acid.

Many small plating plants have a total daily volume of waste of less than 30,000 gal/d (114 m^3/d). The most economical system for such plants is a batch treatment in which two tanks are provided, each with a capacity of one day's flow. One tank is undergoing treatment while the other is filling. Accumulated sludge is either drawn off and hauled to disposal or dewatered on sand drying beds. A spadable dry cake can be obtained after 48 hours on a sand bed. A typical batch treatment system is schematically shown in Fig. 4-9.

When the daily volume of waste exceeds 30,000 to 40,000 gal (114 to 151 m^3), batch treatment is usually not feasible because of the large tankage required. Continuous treatment requires a tank for acidification and reduction, then a mixing tank, for lime addition, and a settling tank. The retention time in the reduction tank is dependent on the pH employed but should be at least 4 times the theoretical time for complete reduction. Twenty minutes will usually be adequate for flocculation. Final settling should not be designed for an overflow rate in excess of 500 gal/(d · ft^2) [20 $m^3/(d \cdot m^2)$].

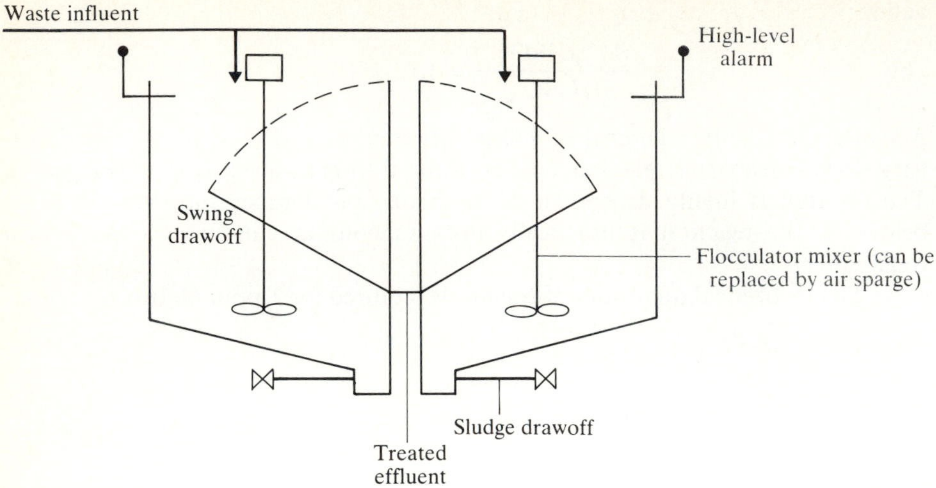

FIGURE 4-9
Batch treatment of chromium wastewaters.

In cases where the chrome content of the rinsewater varies markedly, equalization should be provided before the reduction tank to minimize fluctuations in the chemical feed system. The fluctuation in chrome content can be minimized by provision of a drain station before the rinse tanks.

Successful operation of a continuous chrome reduction process requires instrumentation and automatic control. pH and redox control are provided for the reduction tank. The addition of lime should be modulated by a second pH control system. An example of metals removal is shown in Example 4.2.

Example 4.2. 30,000 gal/d (114 m^3/d) of a waste containing 49 mg/l Cr^{6+}, 11 mg/l Cu, and 12 mg/l Zn is to be treated daily by using SO_2. Compute the chemical requirements and the daily sludge production. (Assume the waste contains 5 mg/l O_2.)

Solution

(*a*) SO_2 requirements are as follows:
For Cr^{6+}:

$$1.85\left(\frac{\text{mg } SO_2}{\text{mg } Cr^{6+}}\right) \times 49\ (\text{mg } Cr^{6+}/\text{l}) \times 8.34\left(\frac{\text{lb/million gal}}{\text{mg/l}}\right)$$

$$\times\ 0.03 \text{ million gal/d} = 22.7 \text{ lb/d } (10.3 \text{ kg/d})$$

and for O_2, where 1 part of O_2 requires 4 parts of SO_2:

$$4\left(\frac{\text{mg } SO_2}{\text{mg } O_2}\right) \times 5\ (\text{mg } O_2/\text{l}) \times 8.34 \times 0.03 = 5.0 \text{ lb/d } (2.3 \text{ kg/d})$$

Total = 27.7 lb/d (12.6 kg/d)

(*b*) Lime requirements are as follows:
For Cr^{3+}:

$$2.38\left(\frac{\text{mg lime}}{\text{mg Cr}^{3+}}\right) \times 49 \times 8.34 \times 0.03 = 29.2 \text{ lb/d } (13.3 \text{ kg/d})$$

and for Cu and Zn (each part of Cu and Zn requiring 1.3 parts of 90% lime for precipitation):

$$1.3\left(\frac{\text{mg lime}}{\text{mg Cu or Zn}}\right) \times 23 \text{ (mg Cu and Zn/l)} \times 8.34 \times 0.03 = 7.5 \text{ lb/d } (3.4 \text{ kg/d})$$

Total = 36.7 lb/d (16.7 kg/d)

(*c*) Sludge production is:

$$1.98\left(\frac{\text{mg Cr(OH)}_3}{\text{mg Cr}^{6+}}\right) \times 49 \times 8.34 \times 0.03 = 24.3 \text{ lb/d Cr(OH)}_3 \text{ (11 kg/d)}$$

$$1.53\left(\frac{\text{mg sludge}}{\text{mg Cu or Zn}}\right) \times 23 \times 8.34 \times 0.03$$

$$= 8.8 \text{ lb/d Cu(OH}_2\text{) and Zn(OH}_2\text{) (4 kg/d)}$$

Total 33.1 lb/d (15 kg/d)

If the sludge concentrates to 1.5 percent by weight, the volume that will require disposal each day can be calculated as follows:

$$\frac{33.1 \text{ lb/d}}{0.15 \text{ lb solids/lb sludge} \times 8.34 \text{ lb/gal}} = 26.5 \text{ gal/d } (0.1 \text{ m}^3\text{/d})$$

It should be noted that some of the copper and zinc will be soluble unless the final pH after lime addition exceeds pH 9.0.

Arsenic

Arsenic and arsenical compounds are present in wastewaters from the metallurgical industry, glassware and ceramic production, tannery operation, dye stuff, pesticide manufacture, some organic and inorganic chemicals manufacture, petroleum refining, and the rare-earth industry. Arsenic is removed from wastewater by chemical precipitation. Effluent arsenic levels of 0.05 mg/l are obtainable by precipitation of the arsenic as the sulfide by the addition of sodium or hydrogen sulfide at pH of 6 to 7. In order to meet reported effluent levels, polishing of the effluent by filtration would usually be required.

Arsenic present in low concentrations can also be reduced by filtration through activated carbon. Effluent concentrations of 0.06 mg/l arsenic have been reported from an initial concentration of 0.2 mg/l. Arsenic is removed by coprecipitation with a ferric hydroxide floc that ties up the arsenic and removes it from solution. Effluent concentrations of less than 0.005 mg/l have been reported from this process.

Barium

Barium is present in wastewaters from the paint and pigment industry, the metallurgical industry, glass, ceramics, and dye manufacturers, and in the vulcanizing of rubber. Barium has also been reported in explosives manufacturing wastewater. Barium is removed from solution by precipitation as barium sulfate.

Barium sulfate is extremely insoluble, having a maximum theoretical solubility at 25°C of approximately 1.4 mg/l as barium at stoichiometric concentrations of barium and sulfate. The solubility level of barium can be reduced in the presence of excess sulfate. Coagulation of barium salts as the sulfate would be capable of reducing barium to effluent levels of 0.5 mg/l. Barium can also be removed from solution by ion exchange and electrodialysis, although these processes would be more expensive than chemical precipitation.

Cadmium

Cadmium is present in wastewaters from metallurgical alloying, ceramics, electroplating, photography, pigment works, textile printing, chemical industries, and lead mine drainage. Cadmium is removed from wastewaters by precipitation or ion exchange. In some cases, electrolytic and evaporative recovery processes can be employed providing the wastewater is in a concentrated form. Cadmium forms an insoluble and highly stable hydroxide at an alkaline pH. Cadmium in solution is approximately 1 mg/l at pH 8 and 0.05 mg/l at pH 10 to 11. Coprecipitation with iron hydroxide precipitation at pH 6.5 will reduce cadmium to 0.008 mg/l; iron hydroxide at pH 8.5 reduces cadmium to 0.05 mg/l. Cadmium is not precipitated in the presence of complexing ions, such as cyanide. In these cases, it is necessary to pretreat the wastewater to destroy the complexing agent. In the case of cyanide, cyanide destruction is necessary prior to cadmium precipitation. A hydrogen peroxide oxidation precipitation system has been developed that simultaneously oxidizes cyanides and forms the oxide of cadmium, thereby yielding cadmium where recovery of the cadmium is feasible.

Copper

The primary sources of copper in industrial wastewaters are metal-process pickling baths and plating baths. Copper may also be present in wastewaters from a variety of chemical manufacturing processes employing copper salts or a copper catalyst. Copper is removed from wastewaters by precipitation or recovery processes, which include ion exchange, evaporation, and electrodialysis. The value of recovered copper metal will frequently make recovery processes attractive. Ion exchange or activated carbon are feasible treatment methods for wastewaters containing copper at concentrations of less than 200 mg/l. Copper is precipitated as a relatively insoluble metal hydroxide at alkaline pH. In the presence of high sulfates, calcium sulfate will also be precipitated that will interfere with the recovery value of the copper sludge. This may dictate the use of a more expensive alkali such as NaOH to obtain a pure sludge. Cupric oxide has a minimum solubility between pH 9.0 and 10.3 with a reported solubility of 0.01 mg/l. Field practice has indicated that the maximum technically feasible treatment level for copper by chemical precipitation is 0.02 to 0.07 mg/l as soluble copper. Precipitation with sulfide at pH 8.5 will result in effluent copper concentrations of 0.01 to 0.02 mg/l. Low residual concentrations of copper are difficult to achieve in the presence of complexing agents such as cyanide and ammonia. Removal of the complexing agent by pretreatment is essential for high copper removal. Copper cyanide is effectively removed on activated carbon.

Fluorides

Fluorides are present in wastewaters from glass manufacturing, electroplating, steel and aluminum, and pesticide and fertilizer manufacture. Fluoride is removed by precipitation with lime as calcium fluoride. Effluent concentrations in the order of 10 to 20 mg/l are readily obtainable. Enhanced removal of fluoride has been reported in the presence of magnesium. The increased removal is attributed to adsorption of the fluoride ion into the magnesium hydroxide floc, resulting in effluent fluoride concentrations of less than 1.0 mg/l. Low concentrations of fluoride can be removed by ion exchange. Fluoride removal through ion exchange pretreated and regenerated with aluminum salts is attributable to aluminum hydroxide precipitated in a column bed. Fluoride is removed through contact beds of activated alumina, which may be employed as a polishing unit to follow lime precipitation. Fluoride concentrations of 30 mg/l from the lime precipitation process have been reduced to approximately 2 mg/l upon passage through an activated alumina contact bed.

Iron

Iron is present in a wide variety of industrial wastewaters including mining operations, ore milling, chemical industrial wastewater, dye manufacture, metal processing, textile mills, petroleum refining, and others. Iron exists in the ferric or ferrous form, depending upon pH and dissolved oxygen concentration. At neutral pH and the presence of oxygen, soluble ferrous iron is oxidized to ferric iron, which readily hydrolyzes to form the insoluble ferric hydroxide precipitate. At high pH values ferric hydroxide will solubilize through the formation of the $Fe(OH)_4^-$ complex. Ferric and ferrous iron may also be solubilized in the presence of cyanide due to the formation of ferro- and ferricyanide complexes. The primary removal process for iron is conversion of the ferrous to the ferric state and precipitation of ferric hydroxide at a pH of near 7, corresponding to minimum solubility. Conversion of ferrous to ferric iron occurs rapidly upon aeration at pH 7.5. In the presence of dissolved organic matter, the iron oxidation rate is reduced.

Lead

Lead is present in wastewaters from storage-battery manufacture. Lead is generally removed from wastewaters by precipitation. Lead is precipitated as the carbonate, $PbCO_3$, or the hydroxide, $Pb(OH)_2$. Lead is effectively precipitated as the carbonate by the addition of soda ash, resulting in effluent-dissolved lead concentrations of 0.01 to 0.03 mg/l at a pH of 9.0 to 9.5. Precipitation with lime at pH 11.5 resulted in effluent concentrations of 0.019 to 0.2 mg/l. Precipitation as the sulfide can be accomplished with sodium sulfide at a pH of 7.5 to 8.5.

Manganese

Manganese and its salts are found in wastewaters from steel alloy, dry-cell battery manufacture, glass and ceramics, paint and varnish, ink and dye works. Among the many forms and compounds of manganese only the manganous salts and the highly oxidized permanganate anion are appreciably soluble. The latter is a strong oxidant that is reduced under normal circumstances to insoluble manganese dioxide. Treatment technology for the removal of manganese involves conversion of the soluble manganous ion to an insoluble precipitate. Removal is effected by oxidation of the manganous ion and separation of the resulting insoluble oxides and hydroxides. Manganous ion has a low reactivity with oxygen and simple aeration is not an effective technique below pH 9. It has been reported that even at high pH levels, organic matter in solution can combine with manganese and prevent its oxidation by simple aeration. A reaction pH above 9.4 is required to achieve significant manganese reduction by precipitation. The use of chemical oxidants to convert manganous ion to insoluble manganese dioxide in conjunction with coagulation and filtration has been employed. The presence of copper ion enhances air oxidation of manganese, and chlorine dioxide rapidly oxidizes manganese to the insoluble form. Permanganate has successfully been employed in the oxidation of manganese. Ozone has been employed in conjunction with lime for the oxidation and removal of manganese. The drawback in the application of ion exchange is the nonselective removal of other ions which increases operating costs.

Mercury

The major consumptive user of mercury in the United States is the chlor-alkali industry. Mercury is also used in the electrical and electronics industry, explosives manufacturing, photographic industry, and the pesticide and preservative industry. Mercury is used as a catalyst in the chemical and petrochemical industry. Mercury is also found in most laboratory wastewaters. Power generation is a large source of mercury release into the environment through the combustion of fossil fuel. When scrubber devices are installed on thermal power plant stacks for sulfur dioxide removal, accumulation of mercury is possible if extensive

TABLE 4.6
Effluent levels

Technology	Effluent, µg/l
Sulfide precipitation	10–20
Alum coprecipitation	1–10
Iron coprecipitation	0.5–5
Ion exchange	1–5
Carbon adsorption	
Influent	
High	20
Moderate	2
Low	0.25

recycle is practiced. Mercury can be removed from wastewaters by precipitation, ion exchange, and adsorption. Mercury ions can be reduced upon contact with other metals such as copper, zinc, or aluminum. In most cases mercury recovery can be achieved by distillation. For precipitation, mercury compounds must be oxidized to the mercuric ion. Table 4.6 shows effluent levels achievable by candidate technology.

Nickel

Wastewaters containing nickel originate from the metal-processing industries, steel foundries, motor vehicle, and aircraft industries, printing, and, in some cases, the chemicals industry. In the presence of complexing agents such as cyanide, nickel may exist in a soluble complex form. The presence of nickel cyanide complexes interferes with both cyanide and nickel treatment. Nickel forms insoluble nickel hydroxide upon the addition of lime, resulting in a minimum solubility of 0.12 mg/l at pH 10 to 11. Nickel can also be precipitated as the carbonate or the sulfate associated with recovery systems. In practice, lime addition (pH 11.5) may be expected to yield residual nickel concentrations in the order of 0.15 mg/l after sedimentation and filtration. Recovery of nickel can be accomplished by ion exchange or evaporative recovery, providing the nickel concentrations in the wastewaters are at a sufficiently high level.

Selenium

Selenium may be present in various types of paper, fly-ash, and in metallic sulfide ores. The selenious ion appears to be the most common form of selenium in wastewater except for pigment and dye waste, which contain the selenide (yellow cadmium selenide). Selenium can be removed from wastewaters by precipitation as the sulfide at a pH of 6.6. Effluent levels of 0.05 mg/l are reported.

Silver

Soluble silver, usually in the form of silver nitrate, is found in wastewaters from the porcelain, photographic, electroplating, and ink manufacturing industries. Treatment technology for the removal of silver usually considers recovery because of the high value of the metal. Basic treatment methods include precipitation, ion exchange, reductive exchange, and electrolytic recovery. Silver is removed from wastewater by precipitation as silver chloride, which is an extremely insoluble precipitate resulting in the maximum silver concentration at 25°C of approximately 1.4 mg/l. An excess of chloride ion will reduce this value, but greater excess concentrations will increase the solubility of silver through the formation of soluble silver chloride complexes. Silver can be selectively precipitated as silver chloride from a mixed-metal wastestream without initial wastewater segregation or concurrent precipitation of other metals. If the treatment conditions are alkaline, resulting in precipitation of hydroxides of other metals along with the silver chloride, acid washing of the precipitated sludge will remove contaminated metal ions, leaving the insoluble silver chloride. Plating wastes contain silver in the form of silver cyanide, which interferes with the precipitation of silver chloride. Therefore, cyanide removal is necessary prior to precipitation of silver as the chloride salt. Oxidation of the cyanide with chlorine releases chloride ions into solution, which in turn react to form silver chloride directly. Sulfide

TABLE 4.7
Effluent levels achievable in heavy metal removals[14]

Metal	Achievable effluent concentration, mg/l	Technology
Arsenic	0.05	Sulfide ppt with filtration
	0.06	Carbon adsorption
	0.005	Ferric hydroxide co-ppt
Barium	0.5	Sulfate ppt
Cadmium	0.05	Hydroxide ppt at pH 10–11
	0.05	Co-ppt with ferric hydroxide
	0.008	Sulfide precipitation
Copper	0.02–0.07	Hydroxide ppt
	0.01–0.02	Sulfide ppt
Mercury	0.01–0.02	Sulfide ppt
	0.001–0.01	Alum co-ppt
	0.0005–0.005	Ferric hydroxide co-ppt
	0.001–0.005	Ion exchange
Nickel	0.12	Hydroxide ppt at pH 10
Selenium	0.05	Sulfide ppt
Zinc	0.1	Hydroxide ppt at pH 11

will precipitate silver from photographic solutions as the extremely insoluble silver sulfide. Ion exchange has been employed for the removal of soluble silver from wastewaters. Activated carbon will remove low concentrations of silver. The mechanism reported is one of reductive recovery by formation of elemental silver at the carbon surface. Reported results indicate that the carbon is capable of retaining silver to 9 percent of its weight at a pH of 2.1 and 12 percent of its weight at a pH of 5.4.

Zinc

Zinc is present in wastewater streams from steelworks, rayon yarn and fiber manufacture, ground wood-pulp production, and recirculating cooling water systems employing cathodic treatment. Zinc is also present in wastewaters from the plating and metal-processing industry. Zinc can be removed by precipitation as zinc hydroxide with either lime or caustic. The disadvantage of lime addition is the concurrent precipitation of calcium sulfate in the presence of high sulfate levels in the wastewater. An effluent soluble zinc of less than 0.1 mg/l has been achieved at pH 11.0.

Metals removal is summarized in Table 4.7. A detailed discussion of metals removal has been presented by Patterson.[14]

PROBLEM

4.1. A metal-finishing plant has a wastewater flow of 72,000 gal/d (273 m^3/d) with the following characteristics:

Cr^{6+} 75 mg/l
Cu 10 mg/l
Ni 8 mg/l

Design a reduction and precipitation plant for

1. Continuous flow
2. Batch flow

(*a*) Develop the ORP control points using SO_2 as the reducing agent.
(*b*) Compute the quantity of sludge and drying bed area assuming the sludge concentrates to 2 percent is applied to the bed to a depth of 18 in (0.46 m) and is removed every 5 days at 12 percent.
(*c*) Compute the residual soluble metal if a terminal pH of 8.5 is used for the precipitation.

REFERENCES

1. Mysels, K. J.: *Introduction to Colloid Chemistry*, Interscience Publishers, New York, 1959.
2. Black, A. P., and H. L. Smith: *J. Am. Water Works Ass.*, vol. 54, p. 371, 1962.
3. Riddick, T. M.: *Tappi*, vol. 47, pt. 1, p. 171A, 1964.
4. Palladino, A. J.: *Proc. 10th Ind. Waste Conf.*, May 1955, Purdue University.
5. Knack, M. F.: *Proc. 4th Ind. Waste Conf.*, 1949, Purdue University.

6. Leonard, A. G., and R. G. Keating: *Proc. 13th Ind. Waste Conf.*, 1946, Purdue University.
7. Bloodgood, D., and W. J. Kellenher: *Proc. 7th Ind. Waste Conf.*, 1952, Purdue University.
8. Eckenfelder, W. W., and D. J. O'Connor: *Proc. 10th Ind. Waste Conf.*, May 1955, p. 17, Purdue University.
9. Webster, R. A.: *Sewage Ind. Wastes*, vol. 25, pt. 12, p. 1432, December 1953.
10. Olthof, M. G., and W. W. Eckenfelder: *Textile Chemist and Colorist*, vol. 8, pt. 7, p. 18, 1976.
11. Olthof, M. G., and W. W. Eckenfelder: *Water Res.*, vol. 9, p. 853, 1975.
12. Southgate, B. A.: *Treatment and Disposal of Industrial Waste Waters*, His Majesty's Stationery Office, London, 1948, p. 186.
13. McCarthy, Joseph A.: *Sewage Works J.*, vol. 21, pt. 1, p. 75, January 1949.
14. Patterson, J. W.: *Industrial Wastewater Treatment Technology*, Butterworth Publishers, Boston, 1985.

CHAPTER 5

AERATION AND MASS TRANSFER

5.1 MECHANISM OF OXYGEN TRANSFER

Aeration is used for transferring oxygen to biological-treatment processes, for stripping solvents from wastewaters, and for removing volatile gases such as H_2S and NH_3.

Aeration is a gas-liquid mass-transfer process in which interphase diffusion occurs when a driving force is created by a departure from equilibrium. In the gas phase, the driving force is a partial pressure gradient; in the liquid phase, a concentration gradient.

The rate of molecular diffusion of a dissolved gas in a liquid is dependent on the characteristics of the gas and the liquid, the temperature, the concentration gradient, and the cross-sectional area across which diffusion occurs. The diffusional process is defined by Fick's law:

$$N = -D_L A \frac{dc}{dy} \tag{5.1}$$

where N = mass transfer per unit time

A = cross-sectional area through which diffusion occurs

$\frac{dc}{dy}$ = concentration gradient perpendicular to cross-sectional area

D_L = diffusion coefficient

If it is assumed that equilibrium conditions exist at the interface, the mass-transfer process can be reexpressed as

$$N = \left(-D_g A \frac{dp}{dy}\right)_1 = \left(-D_L A \frac{dc}{dy}\right)_2 = \left(-D_e A \frac{dc}{dy}\right)_3 \qquad (5.2)$$

where D_g = coefficient of diffusivity through the gas film
D_e = eddy diffusion coefficient of the gas in the body of the liquid

Since the systems dealt with in waste treatment involve high degrees of turbulence, the eddy diffusivity will be several orders of magnitude greater than the coefficients of molecular diffusivity, and this need not be considered as a rate-controlling step. An exception may be large aerated lagoons or aeration in flowing rivers.

Lewis and Whitman[1] developed the two-film concept which considers stagnant films at the gas and liquid interfaces through which mass transfer must occur. Equation (5.2) can be reexpressed in terms of a liquid and gas film:

$$N = K_L A(C_s - C_L) = K_g A(P_g - P) \qquad (5.3)$$

where C_s = oxygen saturation concentration
C_L = concentration of oxygen in the liquid
K_L = liquid film coefficient defined as D_L/Y_L
K_g = gas film coefficient defined as D_g/Y_g
where Y_L = liquid film thickness
Y_g = gas film thickness

For sparingly soluble gases, such as oxygen and CO_2, the liquid-film resistance controls the rate of mass transfer; for highly soluble gases, such as ammonia, the gas-film resistance controls the transfer rate. Most of the mass-transfer applications in waste treatment are liquid-film controlled. Increasing the fluid turbulence will decrease the film thickness and hence increase K_L. Danckwertz[2] has defined the liquid-film coefficient as the square root of the product of the diffusivity and the rate of surface renewal:

$$K_L = \sqrt{D_L r} \qquad (5.4)$$

The rate of surface renewal r can be considered as the frequency with which fluid with a solute concentration C_L is replacing fluid from the interface with a concentration C_s. High degrees of fluid turbulence will increase r.

Dobbins[3] has proposed a relationship that describes the aforementioned transfer mechanisms:

$$K_L = (D_L r)^{1/2} \coth \left(\frac{rY_L^2}{D_L}\right)^{1/2} \qquad (5.5)$$

When the surface-renewal rate is zero, K_L is equal to D_L/Y_L and the transfer is controlled by molecular diffusion through the surface film. As r increases, K_L becomes equal to $\sqrt{D_L r}$ and the transfer is a function of the rate of surface renewal.

For liquid-film-controlled processes, Eq. (5.3) can be reexpressed in concentration units:

$$\frac{1}{V} N = \frac{dc}{dt} = K_L \frac{A}{V} (C_s - C_L) \tag{5.6}$$

where $K_L \frac{A}{V} = K_L a$

$K_L a$ is an overall film coefficient and is usually used to compute the transfer rate.

The most important application of aeration in waste treatment is the transfer of oxygen to biological-treatment processes and the natural reaeration of streams and other watercourses.

The equilibrium concentration of oxygen in contact with water, C_s, is defined by Henry's law:

$$C_s = Hp \tag{5.7}$$

where p = partial pressure of oxygen in the gas phase
H = Henry's constant, which is proportional to temperature and is influenced by the presence of dissolved solids

The solubility of oxygen in water at various temperatures is summarized in Table 5.1. As temperature and dissolved solids concentrations increase, Henry's constant also increases, thereby reducing C_s. It is therefore usually necessary in industrial wastes to measure solubility experimentally.

In aeration tanks, where air is released at a 12- to 15-ft (3.7- to 4.6-m) liquid depth, the solubility of oxygen is influenced both by the increased partial pressure of the air entering the aeration tank and by the decreasing partial pressure in the air bubble as oxygen is absorbed. For these cases, a mean saturation value corresponding to the aeration tank middepth is used:

$$C_{s,m} = C_s \times \tfrac{1}{2}\left(\frac{P_b}{P_a} + \frac{O_t}{20.9}\right) \tag{5.8}$$

where P_a = atmospheric pressure
P_b = absolute pressure at the depth of air release
O_t = percent concentration of oxygen in the air leaving the aeration tank

Mueller *et al.*[4] has shown that oxygen saturation is a function not only of submergence but also of diffuser type. Coarse bubble units provided lower clean water saturation values than the fine bubble or jet diffusers. Saturation appears to be related to both bubble size and mixing pattern.

To account for wastewater constituents, a factor is used:

$$\beta = \frac{C_s \text{ wastewater}}{C_s \text{ tap water}}$$

The ASCE committee on oxygen transfer has recommended using a TDS correction shown in Table 5.1 to determine β.[5]

TABLE 5.1
Solubility of oxygen (mg/l) at various temperatures, elevations, and total dissolved solid levels

Temperature		Elevation, ft							TDS (at sea level), ppm			
°F	°C	0	1000	2000	3000	4000	5000	6000	400	800	1500	2500
32.0	0	14.6	14.1	13.6	13.1	12.6	12.1	11.7	—	—	—	—
35.6	2	13.8	13.3	12.8	12.4	11.9	11.5	11.1	13.74	13.68	13.58	13.42
39.2	4	13.1	12.6	12.2	11.8	11.4	10.9	10.5	13.04	12.98	12.89	12.75
42.8	6	12.5	12.0	11.6	11.2	10.8	10.4	10.0	12.44	12.38	12.29	12.15
46.4	8	11.9	11.4	11.0	10.6	10.2	9.9	9.5	11.85	11.80	11.70	11.58
50.0	10	11.3	10.9	10.5	10.1	9.8	9.4	9.1	11.25	11.20	11.12	11.00
53.6	12	10.8	10.4	10.1	9.7	9.4	9.0	8.6	10.76	10.71	10.64	10.52
57.2	14	10.4	10.0	9.6	9.3	8.9	8.6	8.3	10.36	10.32	10.25	10.15
60.8	16	10.0	9.6	9.2	8.9	8.6	8.3	8.0	9.96	9.92	9.85	9.75
64.4	18	9.5	9.2	8.9	8.5	8.2	7.9	7.6	9.46	9.43	9.36	9.27
68.0	20	9.2	8.8	8.5	8.2	7.9	7.6	7.3	9.16	9.13	9.06	8.97
7.16	22	8.8	8.5	8.2	7.9	7.6	7.3	7.1	8.77	8.73	8.68	8.60
75.2	24	8.5	8.2	7.9	7.6	7.3	7.1	6.8	8.47	8.43	8.38	8.30
78.8	26	8.2	7.9	7.6	7.3	7.1	6.8	6.6	8.17	8.13	8.08	8.00
82.4	28	7.9	7.6	7.4	7.1	6.8	6.6	6.3	7.87	7.83	7.78	7.70
86.0	30	7.6	7.4	7.1	6.9	6.6	6.4	6.1	7.57	7.53	7.48	7.40
89.6	32	7.4	7.1	6.9	6.6	6.4	6.2	5.9	7.4	—	—	—
93.2	34	7.2	6.9	6.7	6.4	6.2	6.0	5.8	7.2	—	—	—
96.8	36	7.0	6.7	6.5	6.3	6.0	5.8	5.6	7.0	—	—	—
100.4	38	6.8	6.6	6.3	6.1	5.9	5.6	5.4	6.8	—	—	—
104.0	40	6.6	6.4	6.1	5.9	5.7	5.5	5.3	6.6	—	—	—

Note: ft = 0.3048 m.

The oxygen-transfer coefficient $K_L a$ is affected by the physical and chemical variables characteristic of the aeration system:

1. Temperature. The liquid-film coefficient will increase with increasing temperature. When air bubbles are involved, changes in liquid temperature will also affect the size of bubbles generated in the system. The effect of temperature on the coefficient is

$$K_L(T) = K_L(20°\text{C})\Theta^{(T-20)} \tag{5.9}$$

   The effect of temperature on $K_L a$ is shown in Fig. 5-1.

2. Turbulent mixing. Increasing the degree of turbulent mixing will increase the overall transfer coefficient.
3. Liquid depth. The effect of liquid depth (H) on $K_L a$ will depend in large measure on the method of aeration. For most types of bubble-diffusion systems, $K_L a$ will vary with depth according to the relationship

$$\frac{K_L a(H_1)}{K_L a(H_2)} = \left(\frac{H_1}{H_2}\right)^n \tag{5.10}$$

   The exponent n has a value near 0.7 for most systems.

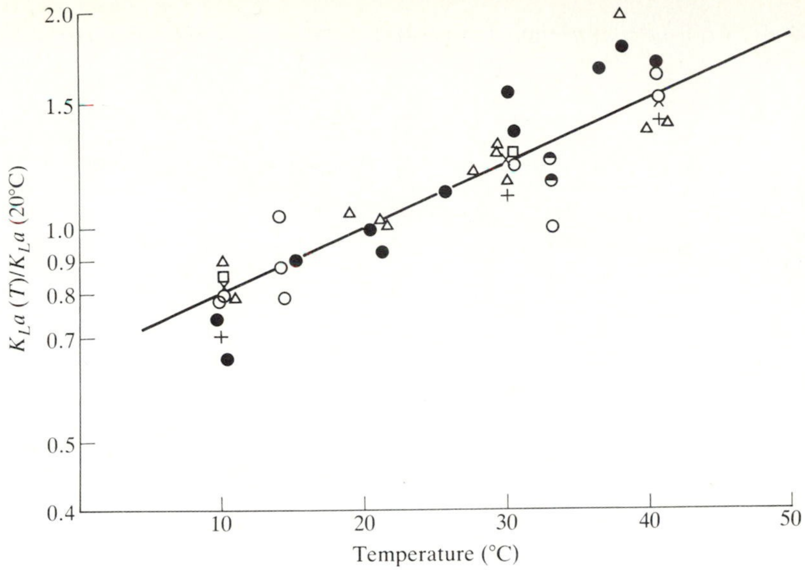

Symbol	Substance	Unit
○	Water	Spinnerette 20 holes, 0.035 mm diameter
+	Water	Spinnerette 10 holes, 0.05 mm diameter
●	Water	Aloxite stone
△	1% KCl	Aloxite stone
□	1% KCl	Spinnerette 10 holes, 0.05 mm diameter
X	50 ppm heptanoic acid	Spinnerette 20 holes, 0.035 mm diameter
◓	Water	Data of Carpani and Roxburgh[6]
△	Water	Data of Gameson and Robertson[7]

FIGURE 5-1
Relationship between the overall transfer coefficient $K_L a$ and temperature.

4. Waste characteristics. The presence of surface-active agents and other organics will have a profound effect on both K_L and A/V. Molecules of surface-active materials will orient themselves on the interfacial surface and create a barrier to diffusion. The excess surface concentration is related to the change in surface tension, as defined by the Gibbs equation, such that small concentrations of surface-active material will depress K_L while large concentrations will exert no further effect. The absolute effect of surfactants on K_L will also depend on the nature of the aeration surface. Less effect would be exerted at a highly turbulent liquid surface, since the short life of any interface would restrict the formation of an adsorbed film. Conversely, a greater effect would be exerted at a bubble surface because of the relatively long life of the bubble as it rises through an aeration tank. A decrease in surface tension will decrease the size of bubbles generated from an air-diffusion system. This in turn will increase A/V. In some cases, the increase in A/V will exceed the decrease in

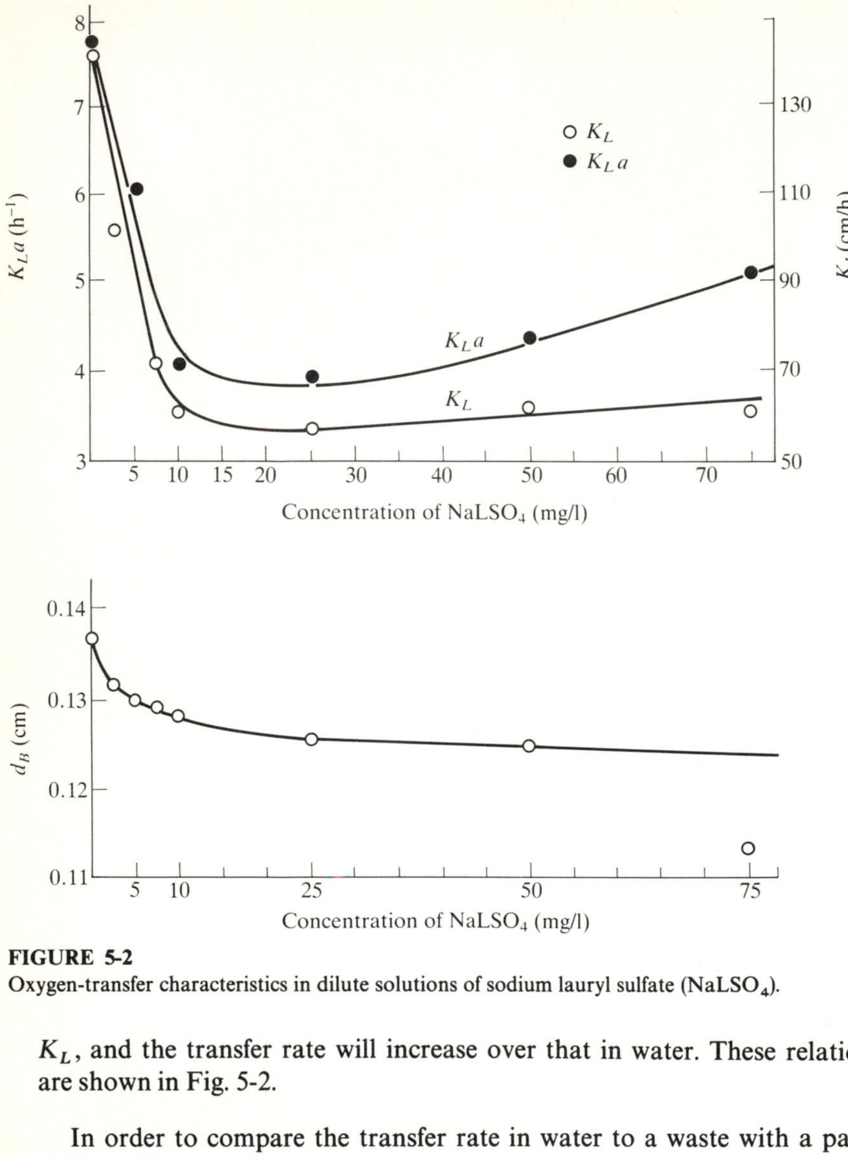

FIGURE 5-2
Oxygen-transfer characteristics in dilute solutions of sodium lauryl sulfate ($NaLSO_4$).

K_L, and the transfer rate will increase over that in water. These relationships are shown in Fig. 5-2.

In order to compare the transfer rate in water to a waste with a particular aeration device, a coefficient α has been defined as $K_L a$ waste/$K_L a$ water.

The coefficient α can be expected to increase or decrease and approach unity during the course of biooxidation, since the substances affecting the transfer rate are being removed in the biological process; Fig. 5-3 shows an increase case.

In view of the effects previously mentioned, the type of aeration will have a profound effect on α. For the fine-bubble diffuser systems, values are generally lower than for coarse-bubble or surface aeration systems. Variation in mixed liquor suspended solids (MLSS) from 2000 to 7000 mg/l with dome diffusers had no significant effect on $K_L a$. Reported values of α for different aeration devices are summarized in Table 5.2.

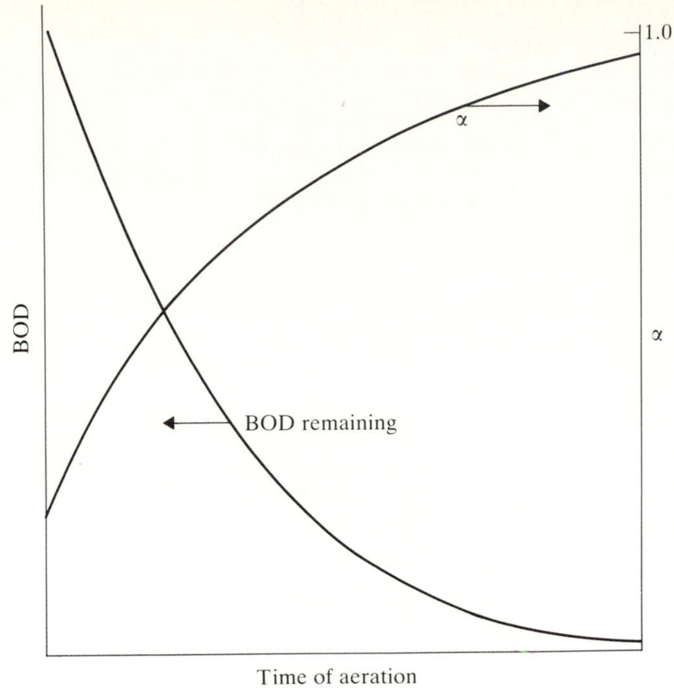

FIGURE 5-3
Change in transfer coefficient α in a biological oxidation process.

In diffused aeration systems, air bubbles are formed at an orifice from which they break off and rise through the liquid, finally bursting at the liquid surface. The velocity and shape of the air bubbles is related to a modified Reynolds number. At N_{Re} of 300 to 4000, the bubbles assume an ellipsoidal shape and rise with a rectilinear rocking motion. At N_{Re} greater than 4000, the bubbles form spherical caps. The rising velocity of the bubble is increased at high air flows because of the proximity of other bubbles and the resulting disturbances of the bubble wakes.

A general correlation for oxygen transfer from air bubbles rising through a still water column has been developed by Eckenfelder,[8] as shown in Fig. 5-4. The general relationship obtained is

$$\frac{K_L d_B}{D_L} H^{1/3} = C\left(\frac{d_B v_B \rho}{\mu}\right) \tag{5.11}$$

where C = constant × ρ lb air/(ft$^3 \cdot$ min)
d_B = bubble diameter
v_B = bubble velocity
ρ = liquid density
μ = liquid viscosity

TABLE 5.2
Values of α for different aeration devices

Aeration device	Alpha factor	Wastewater
Fine bubble diffuser	0.4–0.6	Tap water containing detergent
Brush	0.8	Domestic wastewater
Coarse bubble diffuser, sparger	0.7–0.8	Domestic wastewater
Coarse bubble diffuser, wide band	0.65–0.75	Tap water with detergent
Coarse bubble diffuser, sparger	0.55	Activated sludge contact tank
Static aerator	0.60–0.95	Activated sludge treating high-strength industrial waste
Static aerator	1.0–1.1	Tap water with detergent
Surface aerators	0.6–1.2	Alpha factor tends to increase with increasing power (tap water containing detergent and small amounts of activated sludge)
Turbine aerators	0.6–1.2	Alpha factor tends to increase with increasing power, 25, 50, 190 gal tanks (tap water containing detergent)

Equation (5.11) can be expressed in terms of the overall coefficient $K_L a$ if A/V for air bubbles in an aeration tank is considered as

$$\frac{A}{V} = \frac{6G_s H}{d_B v_B V} \tag{5.12}$$

where G_s is the air flow. Equation (5.11) neglects the aeration tank liquid surface as being small compared with the interfacial bubble surface.

Over the range of air flows normally encountered in aeration practice,

$$d_B \sim G_s^n \tag{5.13}$$

Equations (5.11) to (5.13) can be combined to yield a general relationship for oxygen transfer from air-diffusion systems:

$$K_L a = \frac{C' H^{2/3} G_s^{(1-n)}}{V} \tag{5.14}$$

Equation (5.14) can be reexpressed in terms of the mass of oxygen transferred per diffusion unit:

$$N = C' H^{2/3} G_s^{(1-n)} (C_s - C_L) \tag{5.15}$$

where N is the mass of O_2 per hour per unit transferred.

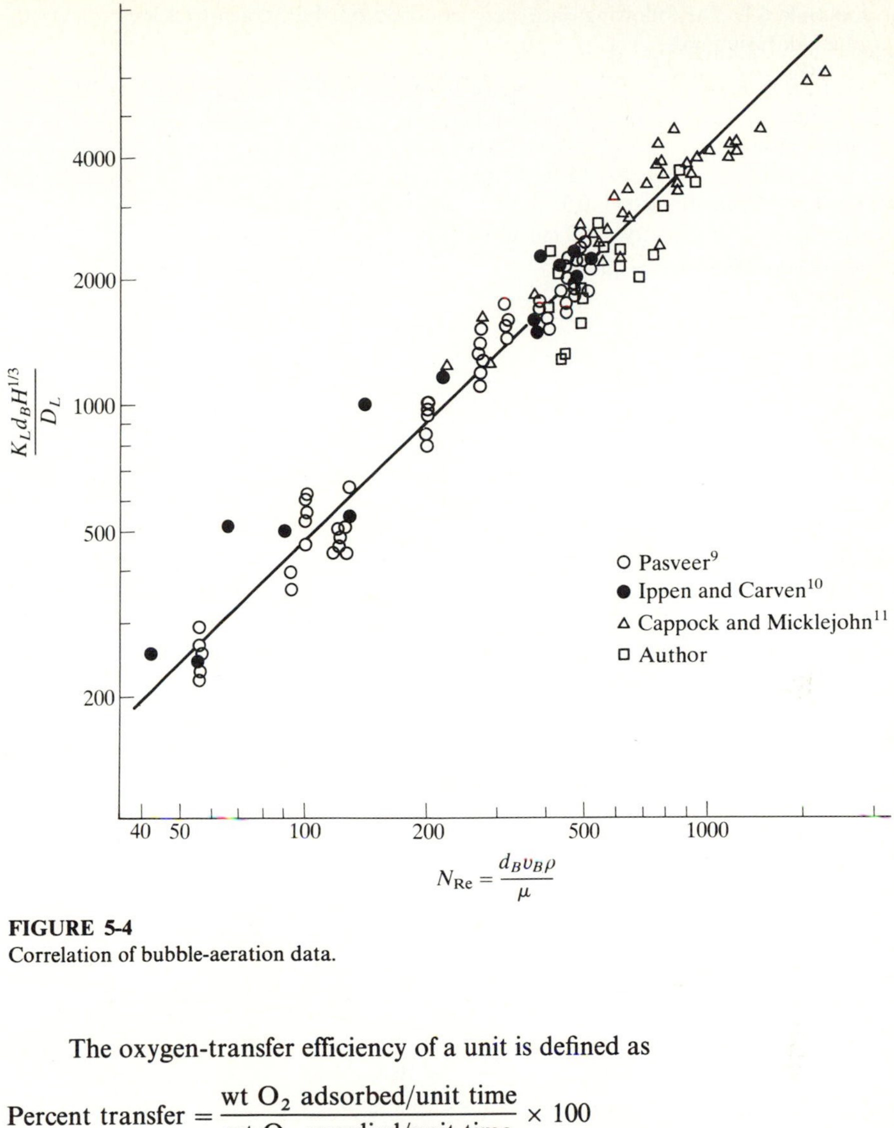

FIGURE 5-4
Correlation of bubble-aeration data.

The oxygen-transfer efficiency of a unit is defined as

$$\text{Percent transfer} = \frac{\text{wt } O_2 \text{ adsorbed/unit time}}{\text{wt } O_2 \text{ supplied/unit time}} \times 100$$

$$= \frac{K_L a(C_s - C_L) \times V}{G_s \text{ (standard ft}^3\text{/min)} \times 0.232 \text{ lb } O_2\text{/lb air} \times \rho \text{ lb air/(ft}^3 \cdot \text{min)}} \times 100 \qquad (5.16)$$

where $\rho = 0.0808 \dfrac{P_b'}{14.7} \times \dfrac{492}{T_f}$

P_b' is the absolute pressure of the air at the tank middepth and T_f, the temperature of the air expressed in kelvin.

Example 5.1. The following data were obtained on the oxygen-transfer capacity of an air-diffusion unit:

Air flow rate	25 standard ft^3/min · thousand ft^3)[25 l/(min · m^3)]
Volume	1000 ft^3 (28.3 m^3)
Temperature	54°F (12°C)
Liquid depth	15 ft (4.6 m)
Average bubble diameter	0.3 cm
Average bubble velocity	32 cm/s

Time, min	C_L, mg/l
3	0.6
6	1.6
9	3.1
12	4.3
15	5.4
18	6.0
21	7.0

(*a*) Compute $K_L a$ and K_L.
(*b*) Compute the mass of O_2 per hour transferred per unit volume at 20°C and zero dissolved oxygen, and the oxygen-transfer efficiency.
(*c*) How much oxygen will be transferred to a waste with an α of 0.82, a temperature of 32°C, and an operating dissolved oxygen of 1.5 mg/l?

Solution

(*a*) At a temperature of 54°F (12°C) saturation is 10.8 mg/l. The mean saturation in the aeration tank, assuming 10 percent oxygen absorption, is

$$C_{s,m} = C_s\left(\frac{P_b}{29.4} + \frac{O_t}{42}\right)$$

where

$$P_b = \frac{15 \text{ ft}}{2.3 \text{ ft/(lb} \cdot \text{in}^2)} + 14.7 = 21.2 \text{ lb/in}^2 \text{ (1.44 atm)}$$

$$O_t = \frac{21\,(1 - 0.1)}{21(1 - 0.1) + 79} \times 100 = 19.3\%\ O_2$$

$$C_{s,m} = 10.8\left(\frac{21.2}{29.4} + \frac{19.3}{42}\right) = 10.8 \times 1.18$$

$$= 12.7 \text{ mg/l}$$

Time, min	$C_{s,m} - C_L$
3	12.1
6	11.1
9	9.6
12	8.4
15	7.3
18	6.7
21	5.7

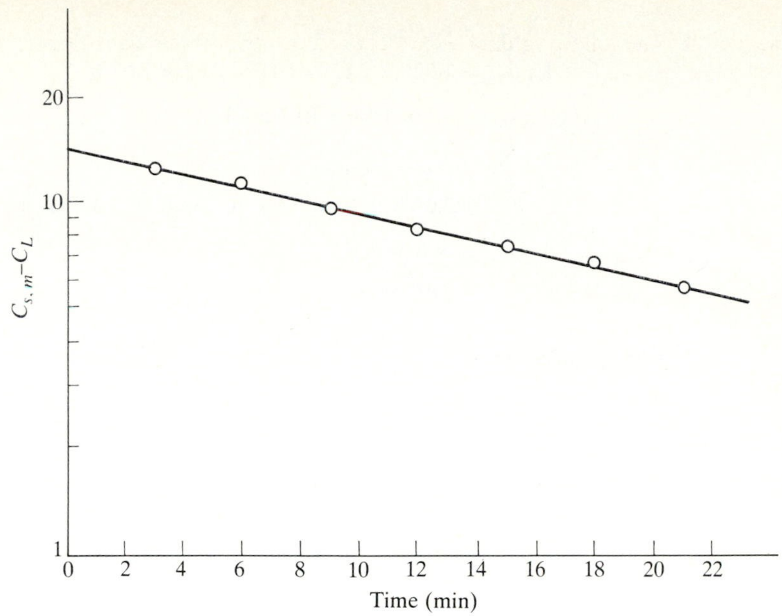

FIGURE 5-5

From Eq. (5.22),

$$\log (C_s - C_L) = \log (C_s - C_o) - \frac{K_L a}{2.3} t$$

which represents a straight line on a semilog plot of log $(C_s - C_L)$ versus L. From Fig. 5-15 $K_L a$ can be computed:

$$K_L a = 2.3 \frac{\log (14/9)}{10} \times 60 = 2.63/\text{h}$$

The interfacial area/volume ratio A/V is

$$\frac{A}{V} = \frac{6G_s H}{d_B v_B V} = \frac{6 \times 25 \times 15 \times 60}{(0.3/30.5) \times (32/30.5) \times 3600 \times 1000} = 3.65 \text{ ft}^2/\text{ft}^3 \text{ (12 m}^2/\text{m}^3)$$

where d_B = average bubble diameter
H = aeration liquid depth
v_B = average bubble velocity
V = volume

$$K_L = \frac{K_L a}{A/V}$$

$$= \frac{2.63}{3.65} = 0.72 \text{ ft/h (21.96 cm/h)}$$

(*b*)
$$K_L a_T = K_L a_{20} \times 1.02^{(T-20)}$$
$$K_L a_{20} = 1.02^8 \times 2.63 = 1.17 \times 2.63 = 3.07/\text{h}$$
$$C_{s,\,m(20^\circ)} = 9.1 \times 1.18 = 10.7 \text{ mg/l}$$
$$N_o = K_L a V C_{s,\,m}$$
$$= 3.07/\text{h} \times 1000 \text{ ft}^3 \times 10.7 \text{ mg/l} \times 7.48 \text{ gal/ft}^3 \times 8.34 \times 10^{-6}$$
$$= 2.05 \text{ lb } O_2/\text{h } (0.93 \text{ kg } O_2/\text{h})$$
$$\% \ O_2\text{-transfer efficiency} = \frac{2.05 \text{ lb } O_2/\text{h transferred}}{25 \times 60 \times 0.0746 \times 0.232 \text{ lb } O_2/\text{h supplied}} \times 100$$
$$= 8.0$$

(*c*) [Eq. (5.17)]
$$N = N_o\left(\frac{\beta C_{s,\,m} - C_L}{C_{s,\,m(20^\circ)}}\right)\alpha \times 1.02^{(T-20)}$$
$$= 2.05\left(\frac{0.99 \times 7.4 \times 1.18 - 1.5}{10.7}\right) \times 0.82 \times 1.02^{12}$$
$$= 1.42 \text{ lb/h } (0.64 \text{ kg/h})$$

5.2. AERATION EQUIPMENT

The aeration equipment commonly used in the industrial waste field consists of air-diffusion units, turbine aeration systems in which air is released below the rotating blades of an impeller, and surface aeration units in which oxygen transfer is accomplished by high surface turbulence and liquid sprays. Generic types of aeration equipment are shown in Fig. 5-6.

A manufacturer will generally designate the oxygen-transfer capability of his equipment in terms of the pounds of O_2 transferred per horsepower-hour [kg O_2/(kW · h)] or the oxygen-transfer efficiency or pounds of O_2 transferred per hour (kg O_2/h) per diffusion unit. This is called the SOR or standard oxygen rating in tap water, at 20°C, zero dissolved oxygen at sea level.

The actual oxygen transferred in wastewater (AOR) is computed from

$$N = N_o\left(\frac{\beta C_s - C_L}{9.1}\right)\alpha \times 1.02^{(T-20)} \tag{5.17}$$

in which N = AOR, lb O_2/hp · h [kg O_2/(kW · h)]
N_o = SOR, lb O_2/hp · h [kg O_2/(kW · h)]

Diffused Aeration Equipment

Basically, there are two types of diffused aeration equipment: units producing a small bubble from a porous medium and units using a large orifice or a hydraulic shear device to produce large air bubbles.

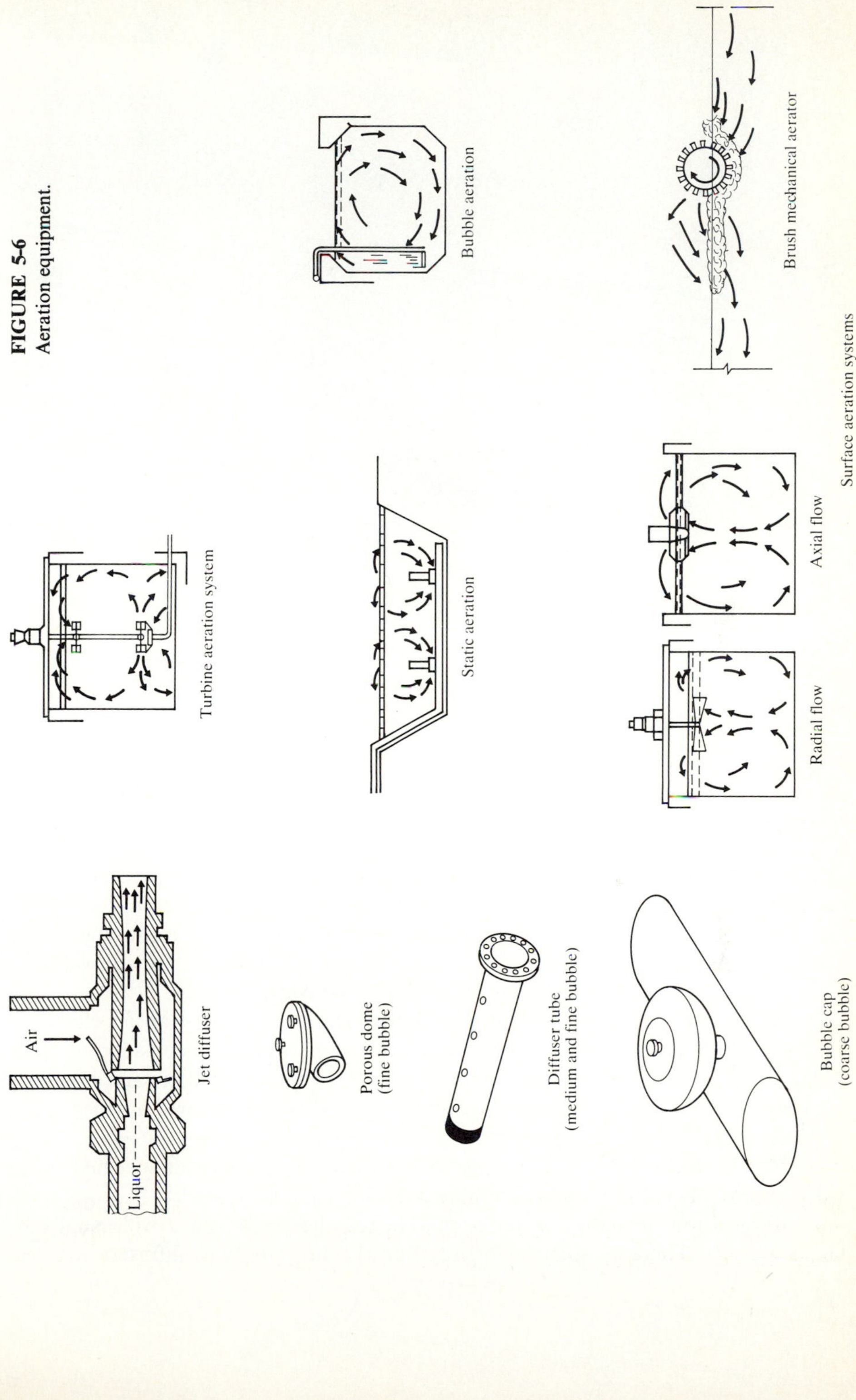

FIGURE 5-6
Aeration equipment.

FIGURE 5-7
Diffused aeration system.

Porous media are either tubes or plates constructed of carborundum or other finely porous media. Tubes are placed at the side wall of the aeration tank, perpendicular to the wall, and generate a rolling motion to maintain mixing or across the bottom of the basin. Maximum spacing is required to maintain solids in suspension; minimum spacing is required to avoid bubble coalescence. A diffused aeration system is shown in Fig. 5-7.

In order for adequate mixing to be maintained for side-wall mounts, the maximum width of the aeration tank is approximately twice its depth. This width can be doubled by placing a line of diffusion units along the centerline of the aeration tank. Figure 5-8 shows the performance of air-diffusion units in water in an aeration tank 15 ft (4.6 m) deep and 24 ft (7.3 m) wide.

Large-bubble air-diffusion units will not yield the oxygen-transfer efficiency of fine-bubble diffusers since the interfacial area for transfer is considerably less. These units have the advantage, however, of not requiring air filters and of generally requiring less maintenance. These units generally operate over a wider range of air flows per unit.

Static aerators consist of vertical cylindrical tubes placed at specified intervals in an aeration basin and containing fixed internal elements.

In aeration of sewage or industrial wastes, the coefficient α will usually be less than 1.0. There is some evidence that the large-bubble diffusers are less

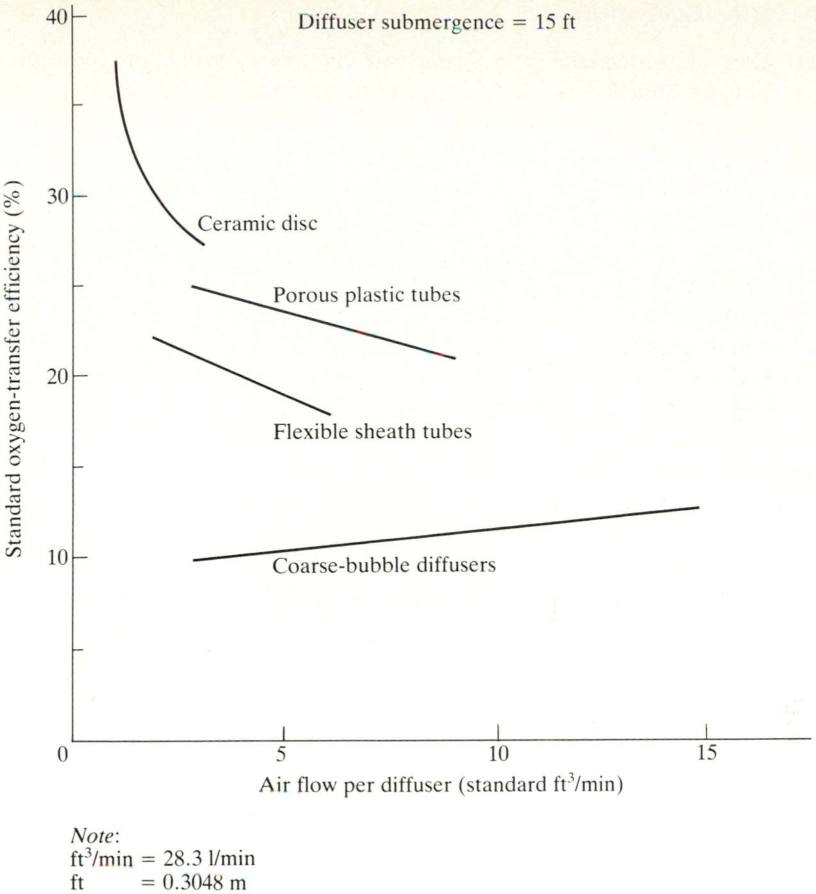

FIGURE 5-8
Effect of air flow rate per diffuser on oxygen-transfer efficiency.

affected by the presence of surface-active materials than are the fine-bubble diffusers.

The performance of all air-diffusion units can be expressed by the relationship, as modified from Eq. (5.15),

$$N = CG_s^n \frac{H^m}{W^p}(C_{s,m} - C_L)1.02^{(T-20)}\alpha \tag{5.18}$$

where N = lb O_2/h transferred per unit [$N \times 0.45$ = kg O_2/(h · unit)]
C = constant characteristic of the oxygen transfer in water at 20°C
$C_{s,m}$ = oxygen saturation in the waste at tank middepth
W = aeration tank width
n, m, and p are exponents characteristic of the aeration device

Turbine Aeration Equipment

Turbine aeration units disperse compressed air by the shearing and pumping action of a rotating impeller. Since the degree of mixing is independently controlled by the power input to the turbine, there are no restrictive limitations to tank geometry. Figure 5-9 shows a typical turbine aeration unit.

The performance of a turbine aeration unit can be expressed by the relationship

$$N = CN_t^x G_s^n d^v (C_{s,m} - C_L) 1.02^{(T-20)} \alpha \tag{5.19}$$

where N_t is the impeller peripheral speed in feet per second (m/s) and d is the impeller diameter in feet (m); x, n, and v are constants.

Air is usually fed to the turbine through a sparge ring located just beneath the impeller blades. The ratio of the turbine diameter to the equivalent tank diameter varies from 0.1 to 0.2. Most aeration applications have employed impeller tip speeds of 10 to 18 ft/s (3.1 to 5.5 m/s).

Quirk[12] showed that Eq. (5.19) can be modified to the form:

$$O_2 \text{ transfer efficiency} = CP_d^n \tag{5.20}$$

where $P_d = \mathrm{hp}_R/\mathrm{hp}_c$ in which hp_R and hp_c are the horsepower (kW) of the turbine and compressor, respectively. The optimum oxygenation efficiency occurs when the power split between turbine and blower horsepower is near a 1 : 1 ratio.

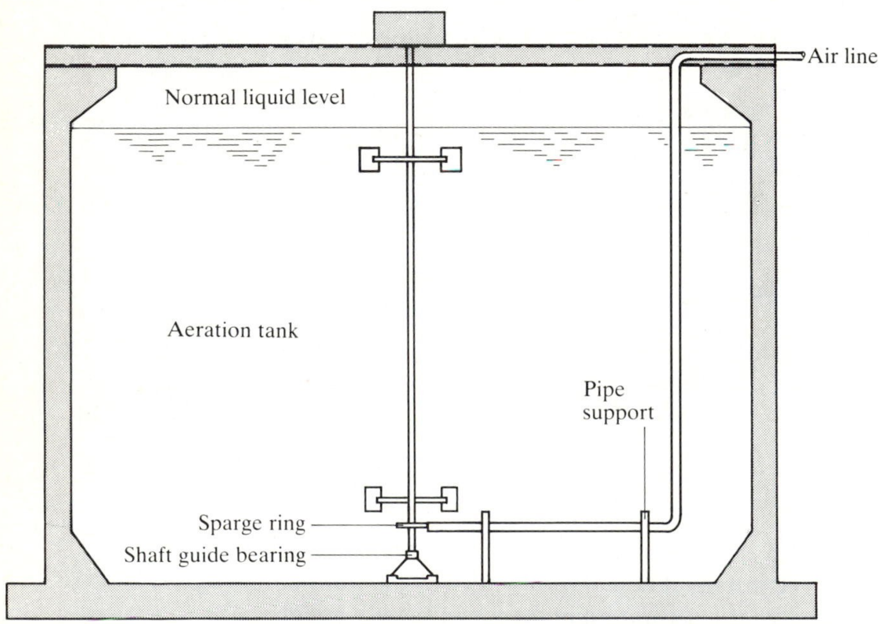

FIGURE 5-9
Typical turbine aerator installation.

The horsepower requirement of a turbine aerator is expressed by the relationship

$$hp = cd^n N_t^m \tag{5.21}$$

where d = impeller diameter, ft (1 ft = 0.305 m)
N_t = impeller speed, r/s

The exponent n generally varies from 4.8 to 5.3; m generally varies from 2.0 to 2.5. The actual power drawn by the impeller will decrease as air is introduced under the impeller owing to the decreased density of the aerating mixture. The transfer efficiency of turbine aeration units will vary from 1.6 to 2.9 lb O_2/hp · h [0.97 to 1.76 kg O_2/(kW · h)], depending on the turbine-blower power split.

To eliminate swirling and vortexing, baffles are normally required. In a round tank, four baffles are placed equally around the circumference of the wall. In a square tank, two baffles on opposite walls are used, while in a rectangular tank with a length/width ratio greater than 1.5 no baffles are required.

Surface-Aeration Equipment

Surface-aeration units are of two types, those employing a draft tube and those with only a surface impeller. In both types, oxygen transfer occurs through a vortexing action and from the surface exposure of large volumes of liquid sprayed

FIGURE 5-10
Low-speed surface aerator.

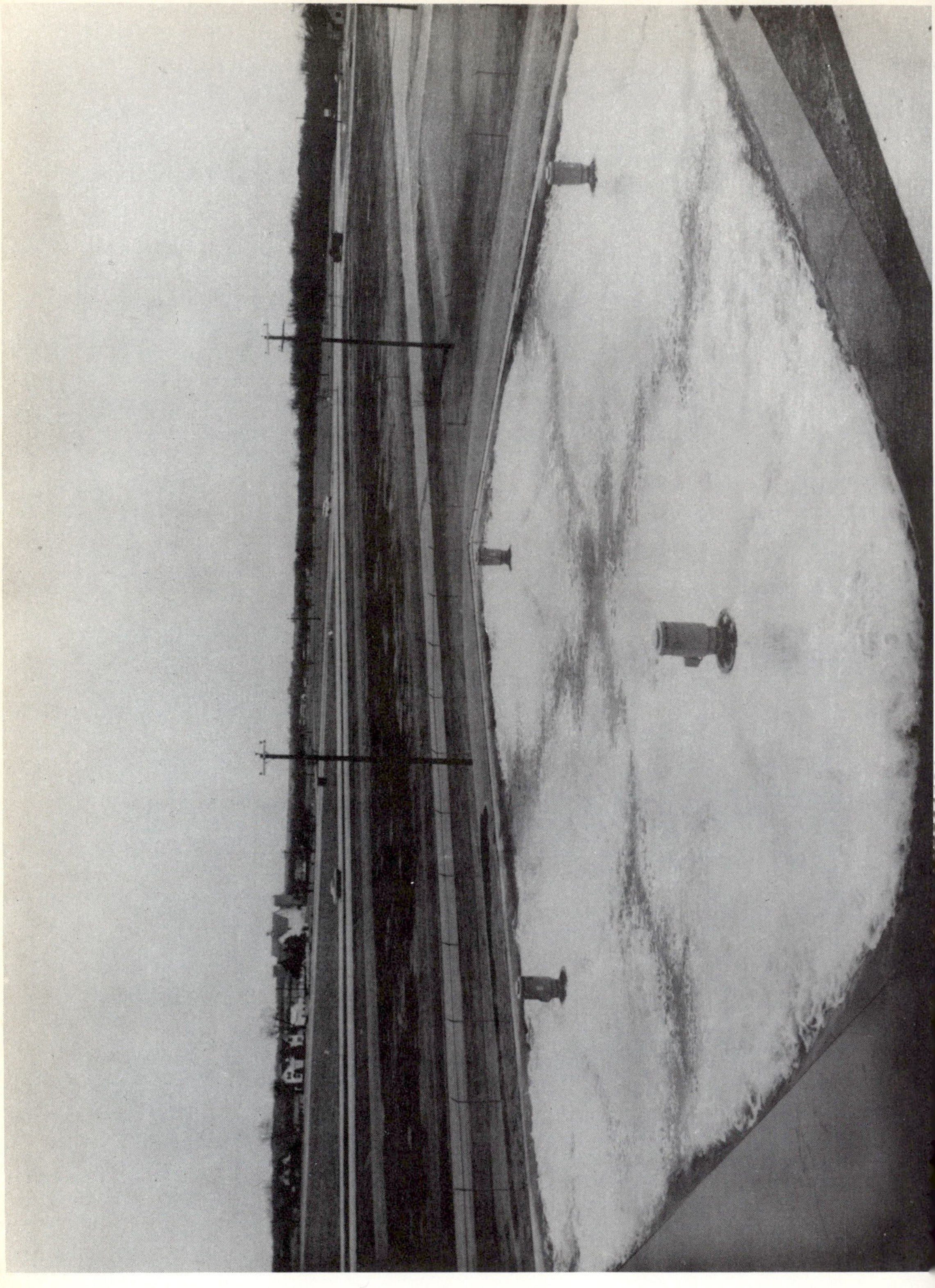

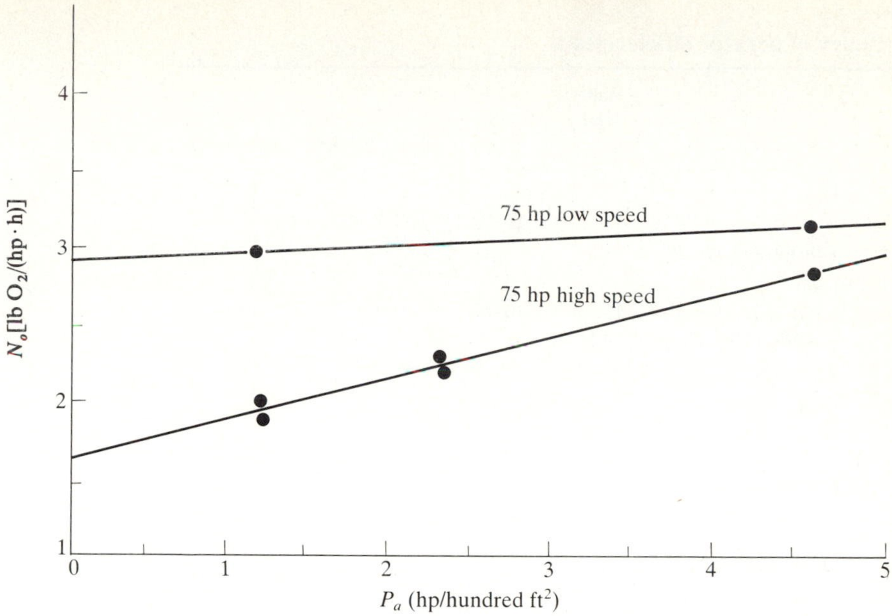

Note:
lb/(hp·h) = 0.608 kg/(kW·h)
hp = 0.75 kW
hp/hundred ft^2 = 80.3 W/m^2

FIGURE 5-12
Comparative effect of surface area on high- and low-speed surface aerators.[13]

over the surface of the aeration tank. Examples of these units are shown in Figs. 5-10 and 5-11. The transfer rate is influenced by the diameter of the impeller and its speed of rotation, and by the submergence level of the rotating element. Under optimum submergence conditions, the transfer rate per unit horsepower (W) remains relatively constant over a wide range of impeller diameters.

The quantity of oxygen transferred at the liquid surface is a function of the power level, and the overall oxygen-transfer rate generally increases with increasing power level. Kormanik *et al.*[13] have shown a correlation between the oxygen-transfer rate and horsepower per unit surface area, as shown in Fig. 5-12. To maintain uniform dissolved oxygen concentrations requires power levels of 6 to 10 hp/million gal (1.2 to 2.0 W/m^3). To maintain biological solids in suspension, a minimum bottom velocity of 0.4 ft/s (12 cm/s) is required for 5000 mg/l MLSS.

To prevent bottom scour minimum depths of 6 to 8 ft (1.8 to 2.4 m) are recommended and to maintain solids in suspension maximum depths of 12 ft (3.7 m) for high-speed aerators and 16 ft (4.9 m) for low-speed aerators.

FIGURE 5-11
High-speed surface aerator.

TABLE 5.3
Summary of aerator efficiencies

Type of aerator	Water depth, ft	OTE, %	lb O_2/hp · h†	Reference
Fine bubble				
Tubes—spiral roll	15	15–20	6.0–8.0	14
Domes—full floor coverage	15	27–31	10.8–12.4	14
Coarse bubble				
Tubes—spiral roll	15	10–13	4.0–5.2	14
Spargers—spiral roll	14.5	8.6	3.4	15
Jet aerators	15	15–24	4.4–4.8	14
Static aerators	15	10–11	4.0–4.4	14
	30	25–30	6.0–7.5	
Turbine	15	10–25	—‡	12
Surface aerator				
Low speed	12	—	5.9–7.5	13
High speed	12	—	3.3–5.0	13

† Wire horsepower must correct for overall blower efficiency.
‡ Horsepower depends on power split.
Note:
ft = 0.3048 m
lb/(hp · h) = 0.608 kg/(kW · h)

The brush aerator, which has been very popular in Europe, uses a high-speed rotating brush that sprays liquid across the tank surface. A circular liquid motion is induced in the aeration tank. The performance of the brush aerator is related to the speed of rotation and submergence of the brush. The efficiency of various aerators is shown in Table 5.3.

Example 5.2. For an oxygen requirement of 1500 lb O_2/h (680 kg O_2/h) in an aeration basin of 3.0 million gal (1.14×10^4 m^3), design the following three aeration systems. Common input to each system is as follows:

$$C_L = 1.50 \text{ mg/l}$$
$$\text{Temperature} = 27.5°\text{C}$$
$$\beta = 0.90$$

Design a sparger system.

Diffused aerators:
$G_s = 7.0$ standard ft^3/(min · unit) [0.198 m^3/(min · unit)]
$\alpha = 0.80$
$E_o = 10.3\%$ (oxygen-transfer efficiency)
$n = 1.02$
$m = 0.72$
$p = 0.35$
$c = 0.0081$

Turbine aerators:

$d/D = 0.12$
$P_d = 0.95$ (optimum power split)
$E_o = 20\%$ (oxygen-transfer efficiency)
$\alpha = 0.85$

High-speed surface aerators:

$N_o = 0.26P_a + 1.65$

where N_o = lb O_2/(hp · h)
P_a = hp/hundred ft^2
$\alpha = 0.90$

Solution

Diffused aerators
Select depth = 16 ft (4.88 m), width = 30 ft (9.14 m).
Calculate $C_{s,m}$:
At 27.5°C:

$$C_s = 8.0 \text{ mg/l}$$

$$O_t = \frac{21 \times (1 - E) \times 100}{79 + 21 \times (1 - E)}$$

$$= \frac{21 \times (1 - 0.103) \times 100}{79 + 21 \times (1 - 0.103)}$$

$$= 19.25\%$$

$$C_{s,m} = \frac{C_s}{2}\left(\frac{p + 0.433 \text{ lb/(in}^2 \cdot \text{ft)} \times h}{p} + \frac{O_t}{20.9}\right)$$

$$= \frac{8.0}{2}\left(\frac{14.7 + 0.433 \times 16}{14.7} + \frac{19.25}{20.9}\right)$$

$$= 4.0(1.47 + 0.92)$$

$$= 9.57 \text{ mg/l}$$

Calculate air diffuser performance [Eq. (5.18)]:

$$N = CG_s^n \frac{H^m}{W^p} (\beta C_{s,m} - C_L) \times 1.02^{(T-20)} \times \alpha$$

$$= 0.0081 \times 7.0^{1.02} \times \frac{16^{0.72}}{30^{0.35}} \times (0.90 \times 9.57 - 1.50) \times 1.02^{(27.5-20)} \times 0.80$$

$$= 0.871 \text{ lb/(h} \cdot \text{unit) [0.395 kg/(h} \cdot \text{unit)]}$$

No. of units required

$$= \frac{1500 \text{ lb}}{0.871 \text{ lb/(h} \cdot \text{unit)}} = 1722$$

Length of aeration basin

$$= \frac{3.0 \times 10^6 \text{ gal}}{16 \times 30 \times 7.48 \text{ gal/ft}^3} = 836 \text{ ft (255 m)}$$

Spacing between units

$$= \frac{836 \text{ ft}}{1722} = 0.486 \text{ ft (0.15 m)}$$

Calculate air flow rate of the blower:

$$G_s = 7.0 \times 1722 = 12{,}054 \text{ standard ft}^3\text{/min (341 m}^3\text{/min)}$$

Calculate horsepower of the blower required:

$$\text{Efficiency} = 70\%, \qquad \Delta P = 8 \text{ lb/in}^2$$

$$\text{Total hp} = \frac{G_s \times \Delta P \times 144 \text{ (in}^2\text{/ft}^2)}{E \times 33{,}000 \text{ (ft} \cdot \text{lbf/min} \cdot \text{hp)}}$$

$$= \frac{12{,}054 \times 8 \times 144}{0.70 \times 33{,}000}$$

$$= 600 \text{ hp (447 kW)}$$

Turbine aerators

$$\text{Select depth} = 16 \text{ ft (4.88 m)}$$

$$O_t = \frac{21 \times (1 - 0.20)}{79 + 21 \times (1 - 0.20)} \times 100$$

$$= 17.54\%$$

$$C_{s,m} = 8.0 \times \tfrac{1}{2}(1.47 + 0.84) = 9.24 \text{ mg/l}$$

Calculate air flow rate. The oxygen-transfer efficiency is defined in Eq. (5.16) and can be calculated by

$$E_o = \frac{R_r \text{ lb/h}}{1.05 G_s \text{ standard ft}^3\text{/min}}$$

$$G_s = \frac{1500}{1.05 \times 0.20}$$

$$= 7140 \text{ standard ft}^3\text{/min (202 m}^3\text{/min)}$$

Calculate required horsepower of the compressor:

$$\text{Efficiency} = 70\%, \qquad \Delta P = 8 \text{ lb/in}^2$$

$$\text{Total hp} = \frac{7140 \times 8 \times 144}{0.70 \times 33{,}000}$$

$$= 356 \text{ hp (266 kW)}$$

Calculate required horsepower of the turbine:

$$\text{hp}_r = P_d \times \text{hp}_c$$
$$= 0.95 \times 356$$
$$= 338 \text{ hp (252 kW)}$$

Calculate the oxygenation capacity:

$$N = \frac{R_r}{\text{hp}_c + \text{hp}_r}$$
$$= \frac{1500}{356 + 338}$$
$$= 2.16 \text{ lb O}_2/\text{hp} \cdot \text{h [1.31 kg/(kW} \cdot \text{h)]}$$

Calculate the oxygenation capacity under standard conditions:

$$C_{s,\,m(20)} = 9.2 \times \frac{9.24}{8.0} = 10.63 \text{ mg/l}$$

$$N_o = \frac{NC_{s,\,m(20)}}{(\beta C_{s,\,m} - C_L)\theta^{(T-20)}a}$$
$$= \frac{2.16 \times 10.63}{(0.90 \times 9.24 - 1.50) \times 1.02^{(27.5-20)} \times 0.85}$$
$$= 3.42 \text{ lb O}_2/(\text{hp} \cdot \text{h}) \text{ [2.08 kg/(kW} \cdot \text{h)]}$$

High-speed surface aerators

$$\text{Select depth} = 10 \text{ ft (3 m)}$$

$$\text{Surface area} = \frac{3.0 \times 10^6 \text{ gal}}{7.48 \text{ g/ft}^3 \times 10 \text{ ft}}$$
$$= 40{,}100 \text{ ft}^2 \text{ (3730 m}^2\text{)}$$

Pick a value of P_a to calculate the required horsepower for mixing and keeping the solids in suspension; use 1.0 hp/100 ft^2:

$$\text{hp}_{\text{mixing}} = \frac{40{,}100 \text{ ft}^2 \times 1 \text{ hp}}{100 \text{ ft}^2} = 400 \text{ hp (300 kW)}$$

Calculate N_o assuming $P_a = 3$ hp/100 ft^2:

$$N_o = 0.26P_a + 1.65 = 0.26 \times 3 + 1.65$$
$$= 2.43 \text{ lb O}_2/(\text{hp} \cdot \text{h}) \text{ [1.48 kg/(kW} \cdot \text{h)]}$$

Calculate N:

$$N = N_o \frac{(0.90 \times 8.0 - 1.50) \times 1.02^{(27.5-20)} \times 0.90}{9.2} = 0.65N_o$$

$$= 0.65 \times 2.43 = 1.60 \text{ lb O}_2/(\text{hp} \cdot \text{h}) \text{ [0.973 kg/(kW} \cdot \text{h)]}$$

Required horsepower for oxygenation:

$$hp_{oxygenation} = \frac{1500 \text{ lb/h}}{1.60 \text{ lb (hp} \cdot \text{h)}} = 940 \text{ hp (700 kW)}$$

Check P_a:

$$P_a = \frac{940 \text{ hp}}{3 \times 10^6 \text{ gal}} \times \frac{7.48 \text{ gal}}{\text{ft}^3} \times 10 \text{ ft} = 2.34 \text{ hp/hundred ft}^2 < 3 \text{ hp/hundred ft}^2$$

Assume $P_a = 2.5$ hp/hundred ft^2

Then

$$N = 0.65 \times (0.26 \times 2.5 + 1.65) = 1.50 \text{ lb O}_2\text{/(hp} \cdot \text{h) [0.91 kg/(kW} \cdot \text{h)]}$$

and

$$hp_{oxygenation} = \frac{1500 \text{ lb/h}}{1.5 \text{ lb O}_2\text{/(hp} \cdot \text{h)}} = 100 \text{ hp (746 kW)}$$

Check P_a:

$$P_a = \frac{1000 \text{ hp}}{3 \times 10^6} \times 7.48 \times 10 = 2.49 \text{ hp/hundred ft}^2 \simeq 2.5 \text{ hp/hundred ft}^2$$

Then

$$hp_{required} = 1000 \text{ hp (746 kW)}$$

Measurement of Oxygen-Transfer Efficiency

Although several procedures have been followed to estimate the transfer efficiency of aeration devices, the non-steady-state aeration procedure has been generally adopted as a standard. This test involves the chemical removal of dissolved oxygen by the addition of sodium sulfite with cobalt added as a catalyst. The cobalt concentration should be 0.05 mg/l to avoid a "cobalt effect" on the magnitude of $K_L a$. The increase in oxygen concentration is measured during aeration under specified conditions, and the overall transfer coefficient calculated from Eq. (5.6). Dissolved oxygen can be measured by the Winkler test or by use of a dissolved oxygen probe. The suggested procedure for non-steady-state aeration is as follows:

1. Remove the dissolved oxygen in the aeration unit by adding sodium sulfite and cobalt chloride. A concentration of 0.05 mg/l should be added, and 8 mg/l Na_2SO_3 per mg/l dissolved oxygen.
2. Thoroughly mix the tank contents. In a diffused aeration unit, aeration for 1 to 2 min will usually be sufficient.
3. Start the aeration unit at the desired operating rate. Sample for dissolved oxygen at selected time intervals. (At least 5 points should be obtained before 90 percent of saturation is reached.)

4. In a large aeration tank multiple sampling points should be selected (both longitudinally and vertically) to compensate for concentration gradients.
5. If a dissolved oxygen probe is used, it can be left in the aeration tank; the values from the probe can be recorded at appropriate time intervals.
6. Record the temperature and measure oxygen saturation. If water is being aerated, it is usually satisfactory to select the saturation value from Table 5.1. For diffused aeration units, saturation should be corrected to the tank mid-depth according to Eq. (5.8).
7. Compute the oxygen-transfer rate in accordance with Eq. (5.6) (see Example 5.1).

Equation (5.6) can be integrated to yield

$$C_s - C_L = (C_s - C_o)e^{-K_L at} \tag{5.22}$$

where C_o = dissolved oxygen at time zero. To evaluate the constants in Eq. (5.22) a nonlinear least-squares program is recommended by ASCE,[5] which provides the best set for $K_L a$, C_o, and C_s.

It is occasionally desirable to measure the oxygen-transfer rate in the presence of activated sludge. Either a steady-state or non-steady-state procedure can be followed for this purpose.

In the aeration of activated sludge, Eq. (5.6) must be modified to account for the effect of the oxygen-utilization rate of the sludge-liquid mixture:

$$\frac{dc}{dt} = K_L a(C_s - C_L) - r_r \tag{5.23}$$

where r_r is the oxygen-utilization rate in milligrams per liter per hour.

Under steady-state operation $dc/dt \rightarrow 0$, and $K_L a$ can be computed from the relationship

$$K_L a = \frac{r_r}{C_s - C_L} \tag{5.24}$$

Aeration is continued until a constant oxygen-uptake rate is obtained in order to insure a steady state.

In the nonsteady procedure, the dissolved oxygen is allowed to approach zero by the microbial respiration. Aeration is then begun and the buildup in dissolved oxygen recorded as in the non-steady-state water procedure. $K_L a$ can be computed from the slope of a plot of dc/dt versus C_L, in accordance with a rearrangement of Eq. (5.23):

$$\frac{dc}{dt} = (K_L aC_s - r_r) - K_L aC_L \tag{5.25}$$

since $K_L a(C_s - r_r)$ will be a constant for any specific operating condition.

Other Measuring Techniques

The off-gas analysis utilizes a hood over a portion of the aeration tank to collect the off-gases for O_2 analysis with a non-membrane-type probe.[14] This technique is applicable for diffused aeration systems with transfer efficiencies above 5 percent.

A radioactive tracer technique has been developed by Neal and Tsivoglou.[15] Their technique uses the stripping of krypton from the aeration tank to measure $K_L a$ under the exact aeration tank mixing and waste conditions. The oxygen-transfer coefficient is then calculated from the known ratio of $(K_L a)_{O_2}/(K_L a)_{\text{krypton}}$.

5.3 AIR STRIPPING OF VOLATILE ORGANICS (VOC)

The physical process of transferring VOC from water into air is called desorption or air stripping. This can be accomplished by injection of water into air via spray systems, spray towers, or packed towers, or by injection of air into water through diffused or mechanical aeration systems. The most common systems in use today are packed towers or aeration systems. (Aeration systems are usually used in conjunction with biological wastewater-treatment processes.) A typical packed tower is shown in Fig. 5-13.

Packing is usually open-structured, of chemically inert material, usually plastic, which is selected to give high surface areas for good contacting while offering a low pressure drop through the tower. Some of the factors that effect removal of VOC are the contact area, the solubility of the contaminant, the diffusivity of the contaminant in the air and water, and the temperature. All these factors except diffusivity and temperature are influenced by the air and water flow rates and the type of packing media. The efficiency of transfer of contaminant from water to air depends on the mass-transfer coefficient and Henry's law constant. The mass-transfer coefficient defines the transfer of contaminant from water to air per unit volume of packing per unit time. The ability of a contaminant to be air-stripped can be estimated from its Henry's law constant. A high Henry's constant indicates that the contaminant has low solubility in water and can therefore be removed by air stripping. In general, Henry's constant increases with increasing temperature and decreases with increasing solubility.

Packed Towers

Packed-tower air strippers can be designed for a wide variety of flow ranges, temperatures, and organics. One of the first steps in applying air stripping is to estimate the maximum possible removal for a given contaminant, based on its Henry's law constant, temperature, and volumetric air/water ratio. The following equation can be used:

$$\frac{C_2}{C_1} = \left(1 + \frac{H_c A_w}{RT}\right)^{-1} \qquad (5.26)$$

where C_2 = final concentration of organic, μg/L or ppb
C_1 = initial concentration of organic, μg/L or ppb
H_c = Henry's law constant, atm · m^3/mol
A_w = volumetric air/water ratio (G/L)
R = universal gas constant, 8.206×10^{-5} atm · m^3/(mol · K)
T = temperature, K

The amount of packing needed to obtain the removal found from Eq. (5.26) is determined from the mass-transfer coefficient for the type of packing used.[16]

While Henry's constant is not accurately defined for many organics, Table 5.4 lists estimated values for organics commonly found in wastewaters. The effect

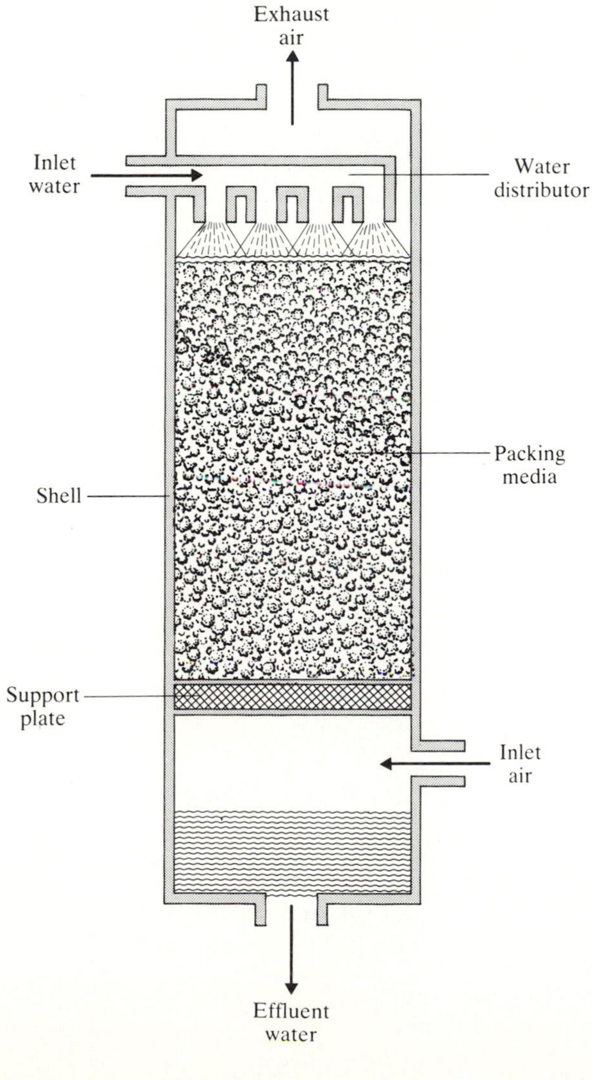

FIGURE 5-13
Typical packed tower.

TABLE 5.4
Estimated Henry's constants for various organics at 20°C

Compound	H_c, atm·m³/mol
Vinyl chloride	6.4
Dichlorofluoromethane	2.1
1,1-Dichloroethylene	1.7×10^{-1}
1,2-Dichloroethylene	1.7×10^{-1}
Carbon tetrachloride	2.5×10^{-2}
Tetrachloroethylene	2.3×10^{-2}
Chloroethane	1.5×10^{-2}
Trichloroethylene	1.0×10^{-2}
Methyl chloride	8.0×10^{-3}
Ethylbenzene	5.7×10^{-3}
Benzene	4.6×10^{-3}
Chlorobenzene	4.0×10^{-3}
1,1,1-Trichloroethane	3.6×10^{-3}
Chloroform	3.4×10^{-3}
Methylene chloride	2.5×10^{-3}
Hexachloroethane	1.1×10^{-3}
1,1,2-Trichloroethane	7.8×10^{-4}
Naphthalene	3.6×10^{-4}
Phenol	2.7×10^{-7}

of temperature is also not well defined. For trichloroethylene (TCE), the following temperature relationship was found for Henry's constant:

$$H_c = \exp\left(9.7 - \frac{4308}{T}\right) \tag{5.27}$$

In order to calculate the amount of packing needed to obtain the removal found from Eq. (5.26), the mass-transfer coefficient ($K_L a$) must be known for the type of packing media used. This must usually be obtained from a pilot plant study. For TCE and 16-mm ($\frac{5}{8}$-in) Pall rings,

$$K_L a = \exp\left(8.52 - \frac{2515}{T}\right) \tag{5.28}$$

where $K_L a$ has units of inverse minutes (min^{-1}). By using the data obtained in Eqs. (5.26) to (5.28) the volume of packing V_p can be estimated:

$$V_p = HA = \frac{L^*}{K_L a} \frac{\ln (C_1/C_2) - RT/(A_w H_c)(C_1/C_2) - 1}{1 - RT/(A_w H_c)} \tag{5.29}$$

where H = height of the packing, m
A = cross-sectional area of the tower, m²
L^* = volumetric water flow rate, m³/min

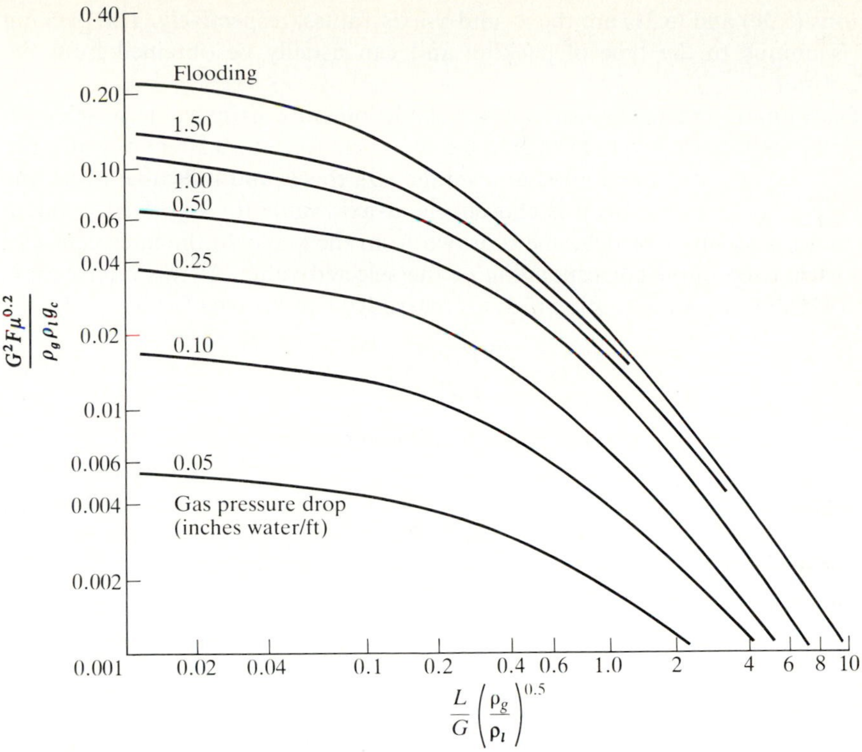

FIGURE 5-14
Generalized flooding and pressure drop curves for packed towers

Once the packing volume has been calculated, the tower diameter can be found from the desired pressure drop calculated in Fig. 5-14, which is a generalized correlation for predicting pressure drop in packed towers, and the following equations:

$$X = \frac{L}{G}\left(\frac{\rho_g}{\rho_l}\right)^{0.5} = \frac{1}{A_w}\left(\frac{\rho_l}{\rho_g}\right)^{0.5} \tag{5.30}$$

where X = x-axis value on Fig. 5-14
L = water loading rate, lb · ft²/s [kg/(s · m²)]
G = air loading rate, lb · ft²/s [kg/(s · m²)]
ρ_g = air density, lb/ft³
ρ_l = water density, lb/ft³

$$Y = \frac{G^2 F \mu^{0.2}}{\rho_g \rho_l g_c} \tag{5.31}$$

where Y = y-axis value on Fig. 5-14
F = packing factor [97 for 16-mm ($\frac{5}{8}$-in) Pall rings]
μ = water viscosity, cP
g_c = gravitational acceleration constant, 32.2 ft²/s

Equations (5.30) and (5.31) are the x- and y-axis values, respectively. The packing factor is unique to the type of packing and can usually be obtained from the manufacturer.

To estimate the tower size, an allowable pressure drop value is selected. Usually a pressure drop in the range of 0.6 to 1.3 cm (0.25 to 0.50 in) of water per foot (30.5 cm) of packed bed gives an average size tower and a flexible operating range. Once the pressure drop is chosen, the x-axis value for Fig. 5-14 is calculated from Eq. (5.30). A vertical line is drawn from the x axis to the intersection of the pressure drop curve corresponding to the selected value. At this intersection, the y-axis value is read. Equation (5.31) is rearranged and solved for G:

$$G = \left(\frac{Y \rho_g \rho_l g_c}{F \mu^{0.2}} \right)^{0.5} \tag{5.32}$$

The tower cross-sectional area is found by dividing the air flow rate by G and the diameter is calculated from the area. The total packed bed depth is calculated by dividing the volume of packing by the cross-sectional area. The total packed bed depth is calculated by dividing the volume of packing by the cross-sectional area. The total pressure drop is the pressure drop per foot of packing times the packed bed depth.

Diffused and Surface Aeration

Models used to estimate stripping with surface or diffused aeration are based on the premise that the transfer rates for volatile solutes are proportional to one another. Because oxygen satisfies the volatility criterion and because a large mass-transfer data base is available for oxygen, oxygen has been selected as the reference compound:[17]

$$(K_L a)_i = \psi (K_L a)_{O_2} \tag{5.33}$$

It has been shown that ψ depends on the liquid-phase diffusivity ratio D_i/D_{O_2} and is approximately constant over a wide range of temperature and mixing intensity. Values of ψ for halogenated aliphatic compounds containing one or two carbon atoms have been reported in the range 0.5 to 0.7. The values of ψ are virtually identical in clean water and wastewater, indicating that the transfer rates of dissolved oxygen and organic solutes are inhibited to the same degree as wastewater constituents. From a mass balance around a CSTR:

$$\frac{QC_{L_1}}{V} - \frac{QC_{L_2}}{V} + K_L a(C_s - C_{L_2}) = 0$$

For oxygen: $$\frac{C_{L_2}}{C_{L_1}} = (1 + K_L a t)^{-1}$$

For volatile organics: $$1 - \frac{C_{L_2}}{C_{L_1}} = 1 - [1 + \psi (K_L a)_{O_2} t]^{-1} \tag{5.34}$$

Surface Aeration

In order to estimate the removal of volatile organics by surface aeration, in a single complete mix basin, $K_L a$ for O_2 transfer is calculated under wastewater conditions. Substitute $(K_L a)_{O_2}$, t, and ψ into Eq. (5.34) and calculate the fractional removal of the VOC.

Diffused Aeration

Under well-mixed conditions, the fractional saturation of the air bubble with respect to an organic compound i is

$$\frac{C_{g,E}}{C_g} = 1 - e^{-\phi} \tag{5.35}$$

in which $C_{g,E}$ = solute concentration in the exiting gas bubble
$C_g = C_L H_C$
= gas phase concentration (g/m^3) in equilibrium with the liquid phase

ϕ = saturation parameter defined as

$$\phi = \frac{(K_L a)_i V}{(H_c)_i Q_g}$$

where V = liquid volume
Q_g = gas flow, m^3/s

The fraction of volatile organic removed from the liquid phase is

$$1 - \frac{C_{L,e}}{C_{L,i}} = 1 - \left[1 + \frac{Q_g}{Q_L}(H_c)_i(1 - e^{-\phi})\right]^{-1} \tag{5.36}$$

TABLE 5.5
Fate of toxic organics in acclimated activated sludge bioreactors operated at 6-day SRTs under steady-state conditions

		Influent, %		
Compound	**Effluent**	**Off-gas**	**Biosorbed**	**Biodegraded**
Nitrobenzene	2	<1	0	98
Benzene	<1	16	0	84
Toluene	<1	17	0	83
Ethylbenzene	<1	22	0	78
o-Xylene	<1	25	0	75
Chlorobenzene	<1	20	0	80
1,2-Dichlorobenzene	6	59	0	35
1,2,4-Trichlorobenzene	10	90	<1	0
Lindane	93	0	7	0

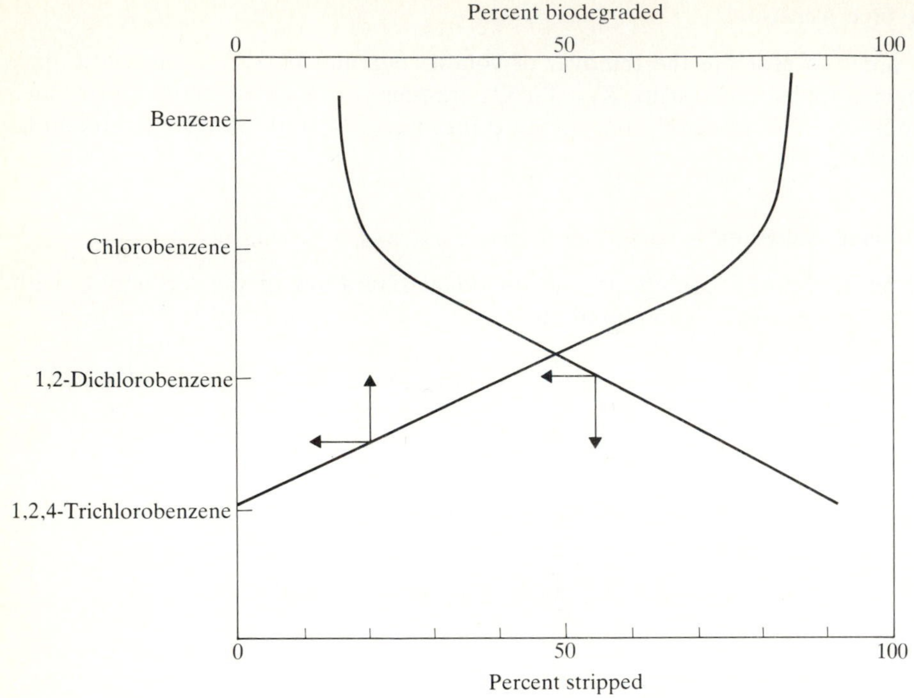

FIGURE 5-15
Stripping and biodegradation of organics in the activated sludge process.[18]

where $C_{L,e}$ = volatile concentration in effluent
$C_{L,i}$ = volatile concentration in inffluent
Q_L = liquid flow, m^3/s

The stripping of chloroform from an aeration basin is shown in Example 5.3. Weber and Jones[18] have evaluated the relative stripping and biodegradation for several organics in the activated sludge process as shown in Table 5.5. It should be noted that in the activated sludge process, adsorption of the organics greatly reduces the amount that is stripped. The relative amounts of chlorinated benzenes stripped and biodegraded are shown in Fig. 5.15.

Example 5.3 Stripping of chloroform in an activated sludge basin. Compute the percentage of chloroform that will be stripped in an aeration basin using fine bubble diffusers.

Parameters:

$$(K_L a)_{O_2} = 3.0/\text{h} = 8.33 \times 10^{-4}/\text{s}$$

$$H = 3.1 \times 10^{-3}\ \text{atm} \cdot \text{m}^3/\text{mol}$$

$$T = 20°\text{C} = 293\ \text{K}$$

$$H_c = \frac{H}{RT} = 41.6H = 0.13$$

$$\psi = 0.6$$

$$D = 18 \text{ ft } (5.50 \text{ m})$$

$$\text{Basin volume} = 1.0 \text{ mg} = 133{,}000 \text{ ft}^3 \ (3785 \text{ m}^3)$$

$$Q_g = 2078 \text{ standard ft}^3/\text{min } (59 \text{ m}^3/\text{min}) \text{ or } 2078 \times \frac{14.7}{14.7 + 9 \times 0.433}$$

$$= 2078 \times \frac{14.7}{18.6}$$

$$= 1642 \text{ ft}^3/\text{min } (47 \text{ m}^3/\text{min}) \text{ at average depth}$$

$$Q_L = \frac{6.8 \times 10^6 \text{ gal/d}}{7.48 \text{ gal/ft}^3 \times 1440 \text{ min/d}}$$

$$= 631 \text{ ft}^3/\text{min } (18 \text{ m}^3/\text{min})$$

The air/liquid ratio is computed:

$$\frac{Q_g}{Q_L} = \frac{1642}{631} = 2.60$$

The transfer coefficient for chloroform is [Eq. (5.33)]

$$(K_L a)_{\text{chloroform}} = 0.6 \times 8.33 \times 10^{-4} = 5.0 \times 10^{-4}/\text{s}$$

The saturation parameter is

$$\phi = \frac{(K_L a) V_L}{H_C Q_g}$$

$$= \frac{5.0 \times 10^{-4}/\text{s} \times 60 \text{ s/min} \times 133{,}000 \text{ ft}^3}{0.13 \times 1642 \text{ ft}^3/\text{min}}$$

$$= 18.7$$

The fraction removal of chloroform by stripping can be calculated from Eq. (5.36) as

$$\frac{C_{L,i}}{C_{L,e}} = 1 + (0.13 \times 2.60)(1 - e^{-18.7})$$

$$= 1.34$$

$$\frac{C_{L,e}}{C_L} = 0.75 \text{ or 25 percent removed}$$

Steam stripping may be applied for the removal of high concentrations of volatile organics. The principal index used to estimate the steam-stripping capability is the boiling point of the organic compound. A compound should exhibit a relatively low boiling point (150°C) and an acceptable Henry's law constant for effective steam stripping. A contaminated condensate is produced that may be recovered or treated further.

PROBLEMS

5.1. An air diffuser yielded a $K_L a$ of 6.5/h at 10°C. The air flow is 50 cm^3/min, the bubble diameter 0.15 cm and the bubble velocity 28 cm/s. The depth of the aeration column is 250 cm with a volume of 4000 cm^3.

(*a*) Compute K_L.

(*b*) Compute $K_L a$ at 25°C.

5.2. An aeration basin has a volume of 2.5 million gal (9463 m^3) with an oxygen requirement of 1000 lb/h (454 kg/h), for the following conditions:

Temperature	25°C
C_L	1.5 mg/1
α diffused air	0.75
α turbine air	0.85
α surface air	0.85
β	0.90

(*a*) Design an aeration system using:

1. Air diffusers; $G_s = 7.0$ standard ft^3/(min · unit), $c = 0.0081$, $n = 1.02$, $m = 0.72$, $p = 0.35$
2. Turbine aerators, $E_0 = 20\%$, $d/D = 0.12$, $P_d = 0.95$
3. High-speed surface aerators

Adjust the basin geometry to the equipment selected.

(*b*) Estimate the horsepower for each case.

5.3. Design a packed stripping tower to reduce TCE from 1000 to 100 μm/l. The following criteria apply:

Water flow rate	500 gal/min (1.93 m^3/min)
Water temperature	15°C (288 K)
Viscosity	1.1 cP
Air temperature	25°C (293K)
Packing media	16-mm ($\frac{5}{8}$-inch) Pall rings

REFERENCES

1. Lewis, W. K., and W. G. Whitman: *Ind. Eng. Chem.*, vol. 16, p. 1215, 1924.
2. Danckwertz, P. V.: *Ind. Eng. Chem.*, vol. 43, p. 6, 1951.
3. Dobbins, W. E.: *Advances in Water Pollution Research*, vol. 2, p. 61, Pergamon Press, 1964.
4. Mueller, J. A., *et al.*: *Proc. 37th Ind. Waste Conf.*, May 1982, Purdue University.
5. ASCE: Draft Report by ASCE Oxygen Transfer Standard Subcommittee, 1981.
6. Carpani, R. E., and J. M. Roxburgh: *Can. J. Chem. Engng*, vol. 36, p. 73, April 1958.
7. Gameson, A. H., and H. B. Robertson: *J. Appl. Chem. Engng*, vol. 5, p. 503, 1955.
8. Eckenfelder, W. W.: *J. Sanit. Engng Div. ASCE*, vol. 85, pp. 88–99, 1959.
9. Pasveer, A.: *Sewage Ind. Wastes*, vol. 27, pt. 10, p. 1130, 1955.
10. Ippen, H. T., and C. E. Carver: MIT Hydrodynamics Lab. Tech. Rep. 14, 1955.
11. Cappock, P. D., and G. T. Micklejohn: *Trans. Inst. Chem. Engrs Lond.*, vol. 29, p. 75, 1951.
12. Quirk, T. P.: personal communication.
13. Kormanik, R., *et al.*: *Proc. 28th Ind. Waste Conf.*, 1973, Purdue University.
14. Redman, D., and W. C. Boyle: Report to Oxygen Transfer Subcommittee of ASCE, 1981.
15. Neal, L. A., and E. C. Tsivoglou: *J. Water Pollut. Control Fed.*, vol. 46, p. 247, 1974.
16. Gross, R. L., and S. G. Termaath: *Env. Prog.*, vol. 4, pt. 2, p. 119, 1985.
17. Roberts, P. V., *et al.*: Environment Protection Agency NTIS P13 84-158,856, 1984.
18. Weber, W. J., and B. E. Jones: Environment Protection Agency NTIS PB86-182,425/AS, 1983.

CHAPTER 6

PRINCIPLES OF BIOLOGICAL OXIDATION

Organics are removed in a biological-treatment process by one or more mechanisms, namely sorption, stripping, or biodegradation. Table 6.1 identifies several organics and the mechanisms responsible for their removal.

6.1 ORGANICS REMOVAL MECHANISMS

Sorption

Limited sorption of nondegradable organics on biological solids occurs for a variety of organics, and this phenomena is not a primary mechanism of organic removal in the majority of cases. An exception is Lindane, as reported by Weber and Jones,[1] who showed that while no biodegradation occurred, there was significant sorption. It is probable that other pesticides will respond in a similar manner in biological wastewater-treatment processes.

While sorption on biomass does not seem to be a significant removal mechanism for toxic organics, sorption on suspended solids in primary treatment may be significant. The importance of this phenomena is the fate of the organics during subsequent sludge-handling operations. In some cases, toxicity to anaerobic digestion may result or land-disposal alternatives may be restricted.

While sorption of organics on biomass is usually not significant, this is not true of heavy metals. Metals will complex with the cell wall and bioaccumulate. While low concentrations of metals in the wastewater are generally not inhibitory

TABLE 6.1
Comparison of removal mechanisms for various priority pollutants in activated sludge

Biodegradation†	Biodegradation and stripping‡	Stripping§	Refractory¶
Nitrobenzene	Ethyl acetate	1,2-Dichloroethane	Bis(2-ethylhexyl) phthalate
2,4-Dichlorobenzene	Benzene	1,1,1-Trichloroethane	
2,4-Dinitrobenzene	1,2-Dichlorobenzene	1,1,2,2-Tetrachloroethane	
Acrolein	Methylene chloride	1,2-Dichloropropane	
Acrylonitrile	Toluene	Trichloroethylene	
Pentachlorophenol	Ethylbenzene		
Phenol	Chloroform		
Phenanthrene	Carbon tetrachloride		
Naphthalene			
Benzidine			

† Compounds that exhibit greater than 95 percent biodegradation.
‡ Compounds that exhibit between 5 and 50 percent removal by stripping and total removal of greater than 95 percent.
§ Compounds that exhibit greater than 95 percent removal by stripping.
¶ Compounds that exhibit less than 80 percent total removal.

to the organic removal efficiency, their accumulation on the sludges can markedly affect subsequent sludge-treatment and disposal operations.

Stripping

Volatile organic carbon (VOC) will air-strip in biological-treatment processes, i.e., trickling filters, activated sludge, aerated lagoons. Depending on the VOC in question, both air stripping and biodegradation may occur, as shown on page 142 and described by Kincannon and Stover.[2] Stripping of VOC in biological-treatment processes is currently receiving considerable attention in the United States as legislation is severely limiting the permissible atmospheric emissions of VOC.

Biodegradation

When organic matter is removed from solution by microorganisms, two basic phenomena occur: oxygen is consumed by the organisms for energy and new cell mass is synthesized. The organisms also undergo progressive autoxidation in their cellular mass. These reactions can be illustrated by the following generation equations:

$$\text{Organics} + a'O_2 + N + P \xrightarrow[K]{\text{cells}} a \text{ new cells} + CO_2 + H_2O + \text{nonbiodegradable soluble residue} \quad (6.1)$$

$$\text{Cells} + b'O_2 \xrightarrow[b]{} CO_2 + H_2O + N + P + \text{nonbiodegradable cellular residue} \quad (6.2)$$

Of primary concern to the engineer in the design and operation of industrial waste-treatment facilities are the rate at which these reactions occur, the amounts of oxygen and nutrient they require, and the quantity of biological sludge they produce.

In Eq. (6.1), K is a rate coefficient and is a function of the biodegradability of the organic or the mixture of organics in the wastewater. The coefficient a' is the fraction of the organics removed that is oxidized to end products for energy and the coefficient a is the fraction of organics removed that is synthesized to cell mass. b is the fraction per day of degradable biomass oxidized and b' the oxygen required for this oxidation.

A small portion of the organics removed in Eq. (6.1) remain as nondegradable byproducts which appear in the effluent as TOC or COD, but not BOD. A portion of the cell mass generated in Eq. (6.1) remains as a nondegradable residue.

For the design or operation of a biological-treatment facility, a primary objective is to balance Eqs. (6.1) and (6.2) for the wastewater in question.

6.2 THE MECHANISMS OF ORGANIC REMOVAL BY BIOOXIDATION

The major organic-removal mechanism for most wastewaters is biooxidation in accordance with Eqs. (6.1) and (6.2).

It should be noted that when treating industrial wastewaters, the sludge must be acclimated to the wastewater in question. For the more complex wastewaters, acclimation may take up to six weeks, as shown in Fig. 6-1 for benzidine. When acclimating sludge the feed concentration of the organic in question must be less than the inhibition level, if one exists.

BOD removal from a wastewater by a biological sludge may be considered as occurring in two phases. An initial high removal of suspended, colloidal, and soluble BOD is followed by a slow progressive removal of remaining soluble BOD. Initial BOD removal is accomplished by one or more mechanisms, depending on the physical and chemical characteristics of the organic matter. These mechanisms are:

1. Removal of suspended matter by enmeshment in the biological floc. This removal is rapid and is dependent upon adequate mixing of the wastewater with the sludge.
2. Removal of colloidal material by physicochemical adsorption on the biological floc.
3. A biosorption of soluble organic matter by the microorganisms. There is some question as to whether this removal is the result of enzymatic complexing or is a surface phenomenon and whether the organic matter is held to the bacterial surface or is within the cell as a storage product or both. The amount of immediate removal of soluble BOD is directly proportional to the concentra-

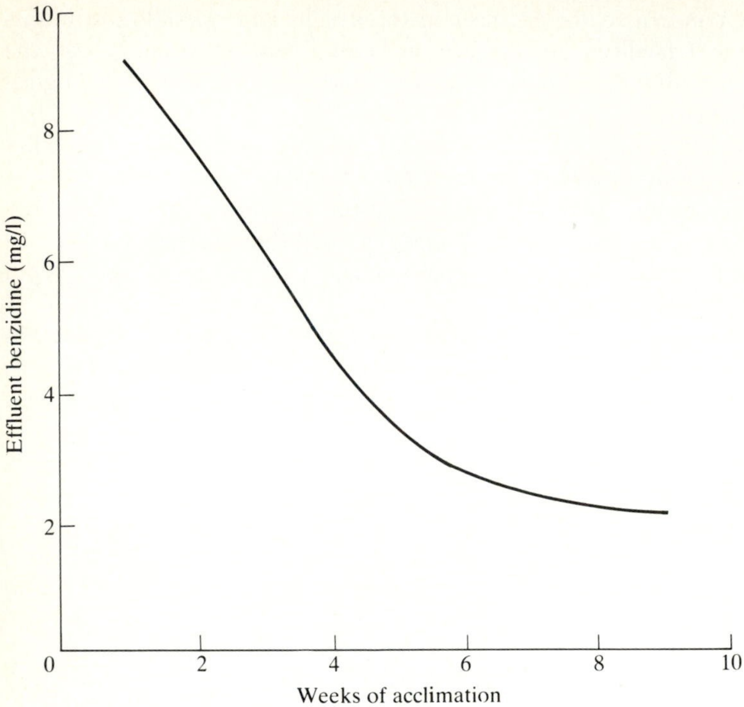

FIGURE 6-1
Acclimation for the degradation of benzidine.

tion of sludge present, the sludge age, and the chemical characteristics of the soluble organic matter.

The biosorption phenomena is related to the floc load in a contact time of 10 to 15 min:

$$\text{Floc load} = \frac{\text{mg BOD applied}}{\text{g VSS}}$$

The relationship between floc load and organic removal by biosorption is shown in Fig. 6-2.

The type of sludge generated markedly affects its sorptive properties. In general, sludge generated from a batch or plug-flow configuration will have better sorptive properties than that generated from a complete mix configuration.

These three reactions occur immediately on contact of sludge with wastewater. The colloidal and suspended material must undergo sequential breakdown to smaller molecules in order that it may be made available to the cell for oxidation and synthesis. The time required for this in an acclimated system is related primarily to the characteristics of the organic matter and to the concentration of

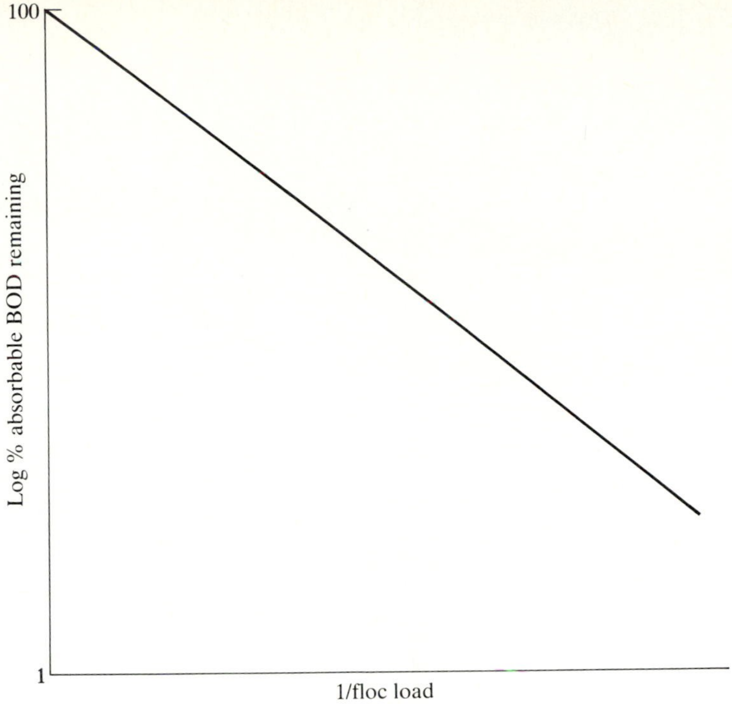

FIGURE 6-2
Biosorption relationship for soluble degradable wastewaters.

active sludge. In a complex waste mixture at high concentrations of BOD, the rate of synthesis is independent of concentration as long as all components remain and, as a result, there is a constant and maximum rate of cellular growth. With continuing aeration, the more readily removable components are depleted and the rate of growth will decrease with decreasing concentration of BOD remaining in solution.

This causes a decrease in cellular mass and cellular carbon accompanied by a corresponding decrease in cellular nitrogen, as shown in Fig. 6-3. This phenomenon has been demonstrated by Gaudy,[3] by Englebrecht and McKinney,[4] and by McWhorter and Heukelekian.[5] Gaudy,[3] in the treatment of pulp-mill wastes, showed a peak in cellular carbohydrate after a 3-h aeration with a corresponding cellular protein peak after a 6-h aeration. The decrease in cellular mass after substantial exhaustion of substrate, shown by Engelbrecht and McKinney,[4] can be attributed to conversion of stored carbohydrate to cellular protoplasm.

In Fig. 6-3 declining removal begins at point *C*. Stored carbohydrate is used by the cell over the time interval *BC*; this results in an increase in cellular nitrogen. The cellular nitrogen is at its peak at point *C*, when the stored carbohydrate is depleted. The cellular mass is shown to increase over the time interval *CD*

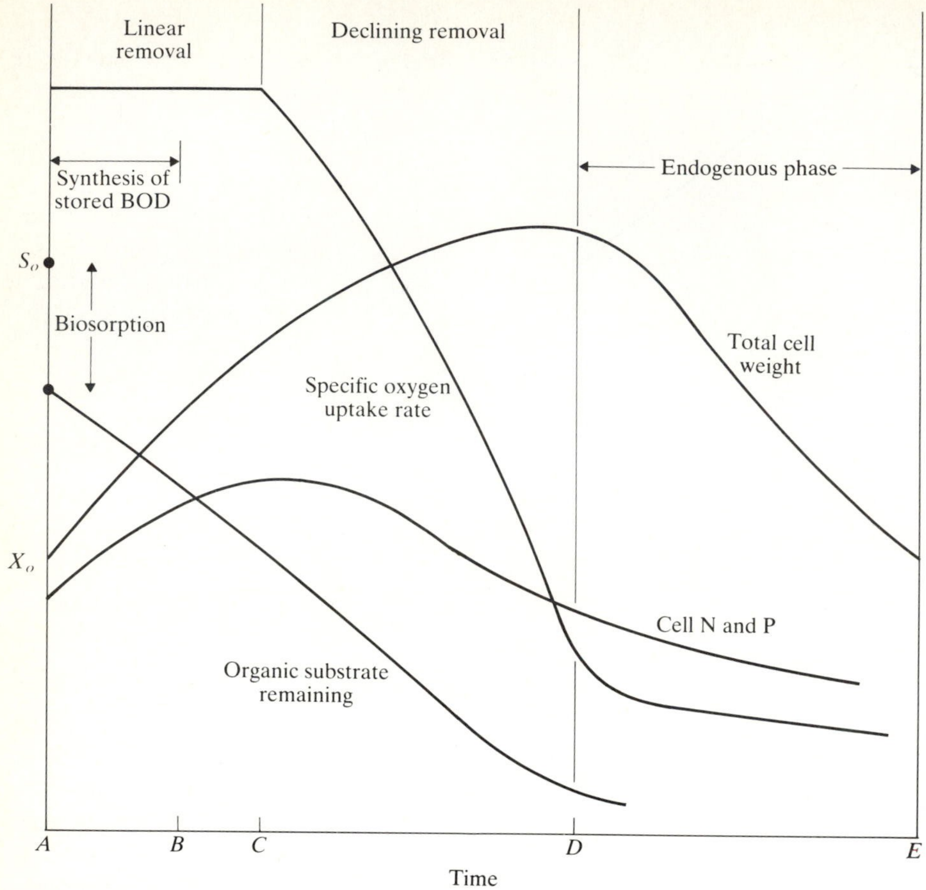

FIGURE 6-3
Reactions occurring during biooxidation.

(declining growth and removal). Depending on the concentration of BOD remaining at point *C* and the rate of removal, the sludge mass may tend to remain constant or even increase while the cellular nitrogen would remain substantially constant. Beyond point *D*, autoxidation results in a decrease in both cell weight and cell nitrogen.

The oxygen utilization per unit of cell mass k_r will remain constant at a maximum rate during the log-growth phase, since substrate is not limiting the rate of synthesis.

The oxygen utilization will continue at a maximum rate until depletion of the sorbed BOD; after this it will decrease as the rate of BOD removal decreases. In wastes containing suspended and colloidal matter the oxygen-uptake rate will also reflect the rate of solubilization and subsequent synthesis of colloidal and suspended BOD.

Mathematical Relationships of BOD Removal

Several mathematical models have been offered to explain the mechanism of BOD removal by biological oxidation processes. All these models have shown that at high BOD levels the rate of BOD removal per unit of cells will remain constant to a limiting BOD concentration below which the rate will become concentration-dependent and decrease. Wuhrmann[6] and Tischler and Eckenfelder[7] have shown that single substances are removed by a zero-order reaction to very low substrate levels. Some reactions are shown in Fig. 6-4. In a mixture of substances being removed at different rates, a constant maximum removal rate will prevail until one of the substances is completely removed. As other substances are progressively removed, the overall rate will decrease. As Gaudy, Komolrit, and Bhatla[8] have shown, sequential substrate removal will also yield a decreasing overall rate.

For specific organics, these kinetics can be formulated in accordance with the Monod relationship:

$$-\frac{1}{X_v}\frac{dS}{dt} = \frac{\mu_m}{a}\frac{S}{K_s + S} \tag{6.3}$$

where S = substrate concentration
μ_m = maximum specific growth rate of organisms
X_v = volatile suspended solids
K_s = Monod's constant, which is defined as the substrate concentration when the rate is one-half the maximum rate
a = biomass yield coefficient

It is assumed that the volatile suspended solids is proportional to the cell mass.

In the case of a multicomponent wastewater where concurrent zero-order reactions occur, the overall removal can be formulated as shown in Fig. 6-5. The zero-order removal for three components is shown in Fig. 6-5*a*. When considering the total removal of all components as BOD, COD, or TOC, the overall rate will remain constant until time t_1, when component A is substantially removed. The overall rate will then decrease to reflect components B and C. At time t_2, component B is substantially removed and the rate will decrease to reflect only component C. For a wastewater consisting of many components, the breaks in Fig. 6-5*b* are not apparent and a curve results. In most cases, this curve can be linearized as shown in Fig. 6-5*c*. Including variations in initial concentration, a batch or plug-flow oxidation of a multicomponent wastewater can be expressed as[9]

$$\frac{S}{S_o} = e^{-K_b X_v t/S_o} \tag{6.4}$$

where S_o = initial substrate concentration
K_b = batch or plug-flow kinetic rate coefficient

In a complete mix basin, the overall removal rate will decrease as the concentration of organics remaining in solution decreases, since the more readily degradable organics will be removed first. The kinetics of the oxidation can therefore be expressed as[10]

$$\frac{S_o - S}{X_v t} = K\frac{S}{S_o} \tag{6.5}$$

The relative biodegradability of various organic compounds is shown in Table 6.2. The rate coefficient K for various industrial wastewaters is shown in Table 6.3.

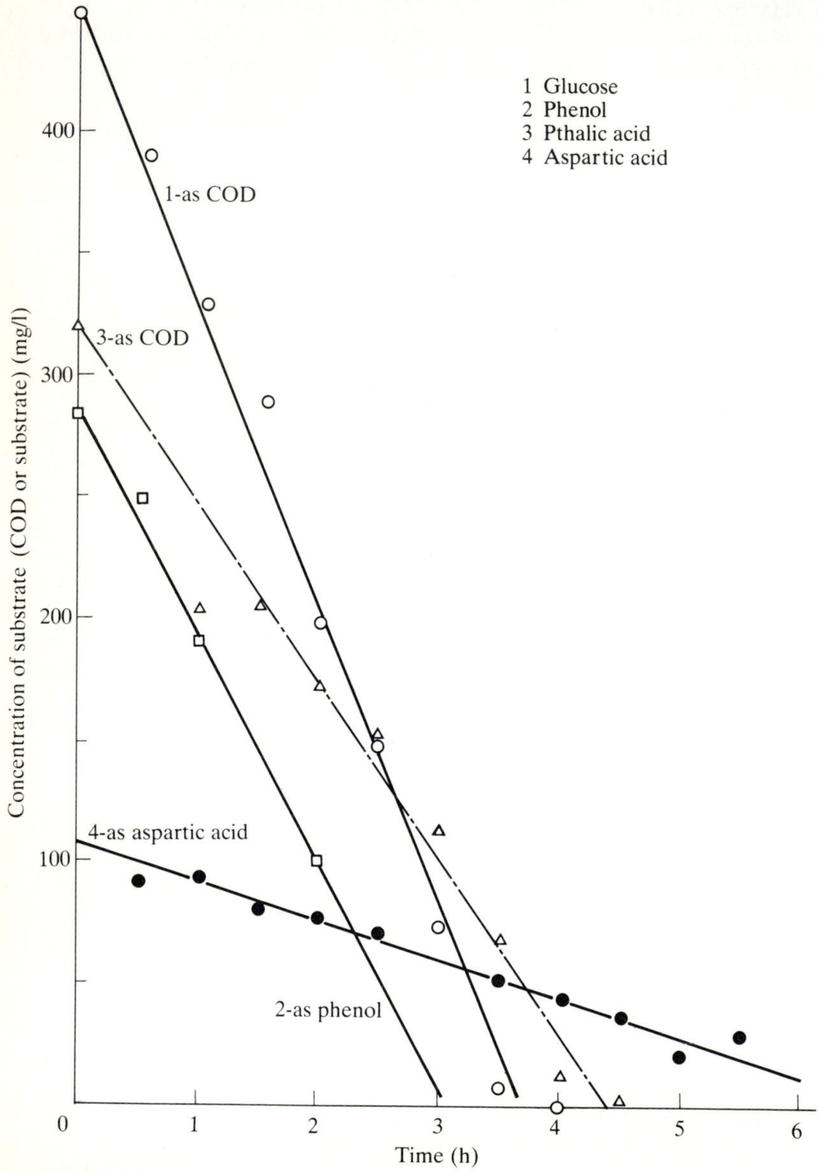

FIGURE 6-4
Zero-order removal rates for specific substrates.

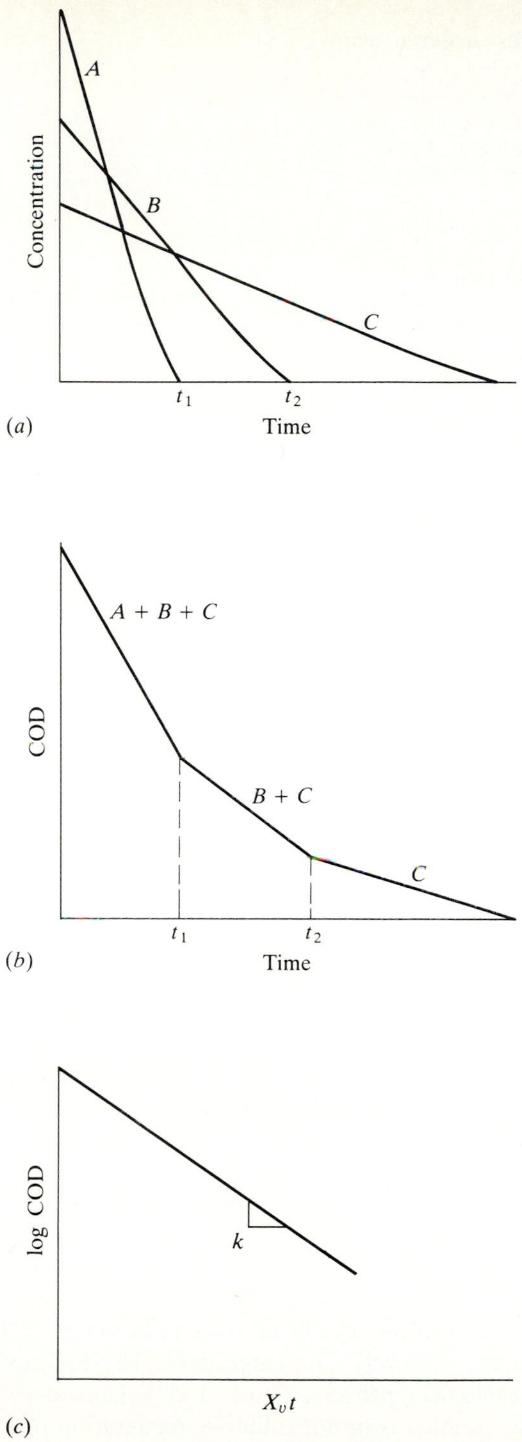

FIGURE 6-5
Schematic representation of multi-component substrate removal.

TABLE 6.2
Relative biodegradability of certain organic compounds

Biodegradable organic compounds†	Compounds generally resistant to biological degradation
Acrylic acid	Ethers
Aliphatic acids	Ethylene chlorohydrin
Aliphatic alcohols	Isoprene
(normal, iso, secondary)	Methyl vinyl ketone
Aliphatic aldehydes	Morpholine
Aliphatic esters	Oil
Alkyl benzene sulfonates with	Polymeric compounds
exception of porpylene-based	Polypropylene benzene
benzaldehyde	sulfonates
Aromatic amines	Selected hydrocarbons
Dichlorophenols	Aliphatics
Ethanolamines	Aromatics
Glycols	Alkyl-aryl groups
Ketones	Tertiary aliphatic alcohols
Methacrylic acid	Tertiary aliphatic sulfonates
Methyl methacrylate	Trichlorophenols
Monochlorophenols	
Nitriles	
Phenols	
Primary aliphatic amines	
Styrene	
Vinyl acetate	

† Some compounds can be degraded biologically only after extended periods of seed acclimation.

As shown in Eq. (6.1), nonbiodegradable byproducts result from the oxidation of organics and from the endogenous oxidation in Eq. (6.2). It has been shown that many of these byproducts are high-molecular-weight organics and, at least in some cases, are toxic to aquatic life. Data from the oxidation of glucose and from a plastics additive wastewater are shown in Table 6.4.

Since the coefficient K is related to the wastewater composition, changes in composition will result in a change in K. For example, as shown in Fig. 6-6, the K for a dairy waste will remain constant since the composition does not change. By contrast, the K for a multiproduct chemicals plant will change every time the product mix changes, resulting in a statistical variability as shown in Fig. 6-6.

Oxygen Utilization and Sludge Yield

As shown in Eq. (6.2), endogenous respiration results in a degradation of cell mass. A portion of the volatile cell mass, however, is nondegradable, i.e., does not degrade in the time frame of the biological process. Quirk and Eckenfelder[11] showed that a portion of the volatile biomass is nondegradable. As aeration pro-

TABLE 6.3
Reaction rate coefficients for organic wastewaters

Wastewater	K, d^{-1}	Temperature, °C
Potato processing	36.0	20
Peptone	4.03	22
Sulfite paper mill	5.0	18
Vinyl acetate monamer	5.3	20
Polyester fiber	14.0	21
Formaldehyde, propanol, methanol	19.0	20
Cellulose acetate	2.6	20
AZO dyes, epoxy, optical brighteners	2.2	18
Petroleum refinery	9.1	20
Vegetable tannery	1.2	20
Organic phosphates	5.0	21
High nitrogen organics	22.2	22
Organic intermediates	20.6	26
	5.8	8
Viscose rayon and nylon	8.2	19
	6.7	11
Soluble fraction of domestic sewage	8.0	20

ceeds the degradable portion of the biomass is oxidized, resulting in a decrease in the degradable fraction. Through kinetic and mass balances, the degradable fraction can be related to the endogenous rate coefficient and the sludge age:

$$X_d = \frac{X'_d}{1 + bX'_n\theta_c} \tag{6.6}$$

where X_d = degradable fraction of the VSS
X'_d = degradable fraction of the VSS at generation, that is, Eq. (6.1) average of 0.8
X'_n = nondegradable fraction of the VSS at generation, that is, Eq. (6.1) average of 0.2
b = endogenous rate coefficient, d^{-1}
θ_c = sludge age, d

TABLE 6.4
Molecular weight distribution of biological effluents

Molecular weight	PA and dyestuffs		Glucose Chudoba[12] COD, %
	Influent TOC, %	Bioeffluent TOC, %	
> 10,000	—	11.5	45
500–10,000	—	14.5	16
< 500	100	74.0	39

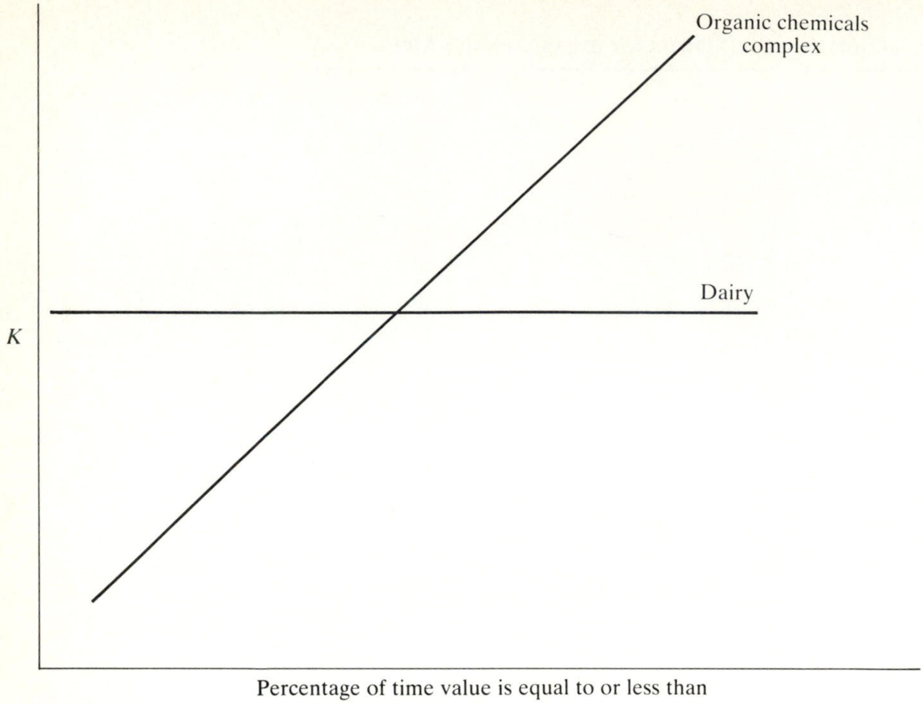

FIGURE 6-6
Variability in K as related to wastewater composition.

The degradable fraction is related to the viable mass. The degradable mass can be related to oxygen-uptake rates, ATP, dehydrogenase enzyme content, or by plate count measurement, as shown in Fig. 6-7. It is significant to note that while volatile suspended solids are traditionally used as a measure of biomass only the active mass is responsive in the process. In plant operation, some measure of active mass is necessary for process control in order to detect toxic shocks, shock loads, etc. Oxygen-uptake rate is most commonly used for this purpose.

Sludge age is defined as the average length of time the microorganisms are under aeration. In a flow-through system, sludge age is the reciprocal of the dilution rate (V/Q). In order for growth to occur and to effect BOD removal, the growth rate becomes

$$\frac{1}{\theta_c} = R_h \tag{6.7}$$

where θ_c is the sludge age and R_h the dilution ratio.

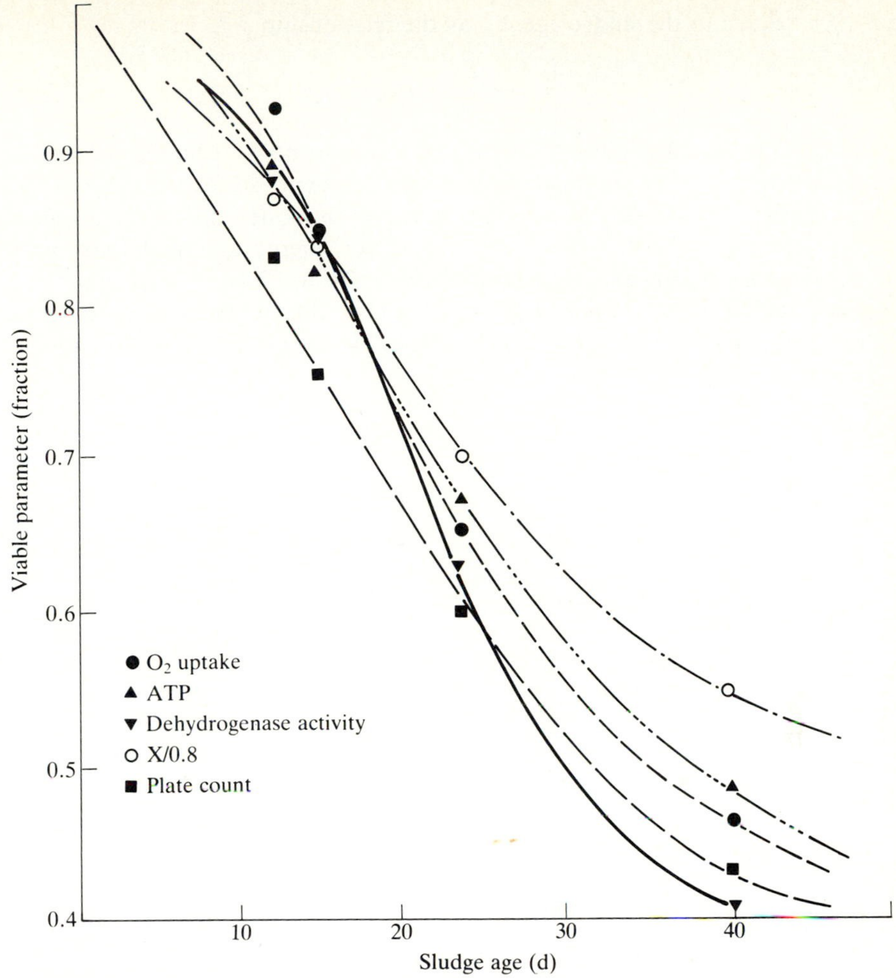

FIGURE 6-7
Viable fraction of activated sludge as related to sludge age.

In a recycle system such as an activated sludge plant, the sludge age is defined:

$$\theta_c = \frac{X_v t}{\Delta X_v} \tag{6.7a}$$

in which ΔX_v is the sludge wasted per day in milligrams per liter.

Process performance can also be related to the organic loading to the process as defined by the food/microorganism ratio F/M:

$$\frac{F}{M} = \frac{S_o}{X_v t} \tag{6.8}$$

The F/M is related to the sludge age, θ_c, by the relationship

$$\frac{1}{\theta_c} = a\frac{F}{M} - bX_d \tag{6.9}$$

Several investigators have shown that a constant mass of biological cells is synthesized from a given weight of organic matter removed (expressed as total oxygen demand, COD). McKinney[13] has indicated that one-third of the ultimate oxygen demand (COD) of a substrate is used for energy and that two-thirds results in synthesis. Using an oxygen-to-cellular-volatile-solids conversion of 0.7 g VSS/g O_2, 0.47 g VSS (volatile suspended solids) will be synthesized for each gram of COD removed. Variations in this value have been attributed to endoge-

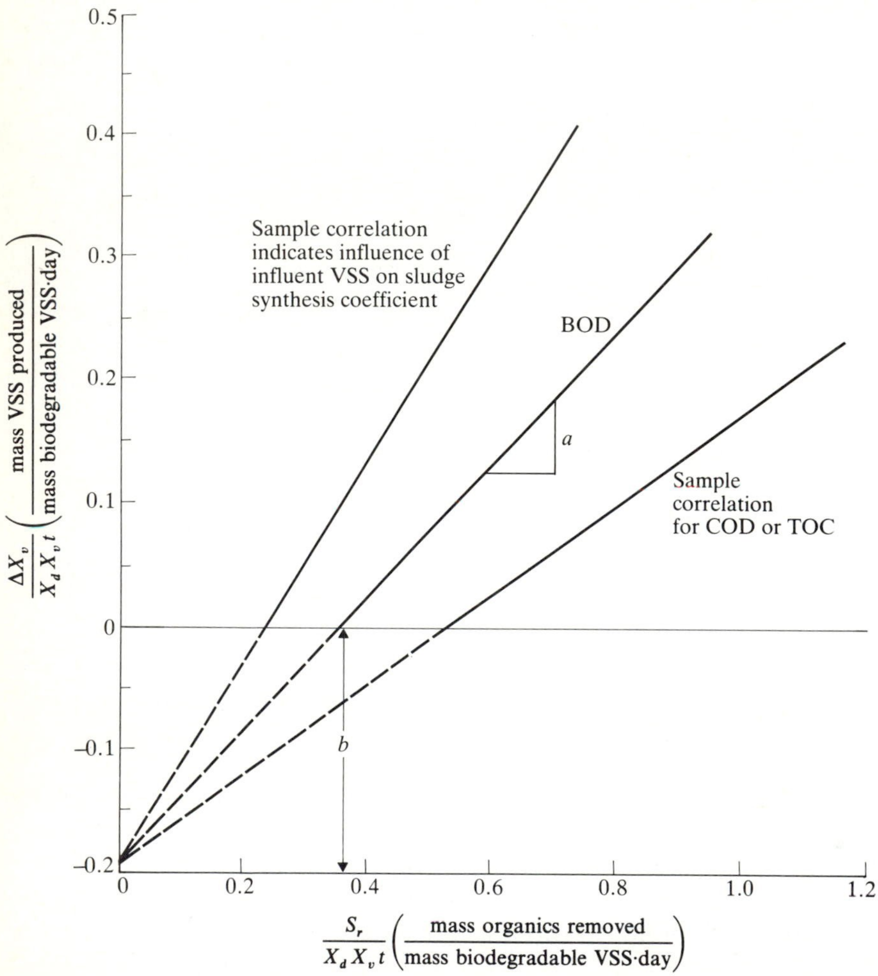

FIGURE 6-8
Determination of sludge production coefficients.

nous respiration effects. Sawyer[14] and Gellman and Heukelekian[15] have shown that for sewage and several industrial wastes 0.5 g VSS is synthesized per gram of BOD_5 removed. Busch and Myrick[16] showed total synthesis from glucose to be 0.44 g cells/g COD. Porges and Hoover[17] similarly showed a synthesis from lactose to be 0.43 g VSS/g COD removed. Using NO_3^- as a nitrogen source, McWhorter and Heukelekian[5] found an average synthesis of 0.315 g VSS/g COD from glucose. As shown by Pipes,[18] the reduction of nitrate to amino nitrogen requires energy such that some of the COD is consumed for this reduction, resulting in a lower yield than when ammonia is used as a nitrogen source.

When the nutrient nitrogen level is lowered below the optimum value, cell yield tends to increase as more substrate is shunted to the buildup of insoluble cell polymers.

Sludge generation from the biological oxidation of various wastewaters has been summarized by Eckenfelder:[19]

$$\Delta X_v = aS_r - bX_d X_v t \tag{6.10}$$

where S_r represents the soluble substrate removed.

In a system with sludge recycle and wastage of excess sludge, sludge age can be computed from Eq. (6.7*a*) and is defined as

$$\theta_c = \frac{X_v t}{aS_r - bX_d X_v t} \tag{6.11}$$

When volatile suspended solids are present in the wastewater, they will contribute to the sludge yield and Eq. (6.10) is modified:

$$\Delta X_v = aS_r + a_x(1 - f)X_i - bX_d X_v t + fX_i \tag{6.12}$$

in which X_i = influent volatile suspended solids
f = fraction not degraded
a_x = g VSS produced/g VSS removed

In cases where both θ_c and X_d are unknown, θ_c can be computed by combining Eqs. (6.6) and (6.11) for soluble substrates to yield

$$\theta_c = \frac{-(aS_r - bX_v t) + \sqrt{(aS_r - bX_v t)^2 + 4(abX'_n S_r)(X_v t)}}{2abX'_n S_r} \tag{6.13}$$

When θ_c is computed, X_d can be calculated from Eq. (6.6).

The correlations for sludge yield are shown in Fig. 6-8.

Data obtained from the treatment of a soluble pharmaceutical waste are shown in Fig. 6-9. Although the synthesis coefficient a is 0.645 on a BOD_5 basis, the coefficient that results from plotting on a COD basis is 0.37.

It has been previously shown that the total oxygen requirements in a system are related to the oxygen consumed to supply energy for synthesis and the oxygen consumed for endogenous respiration [Eqs. (6.1) and (6.2)], in which a' is the fraction of organics consumed to supply energy for synthesis and b' the autoxidation rate of the sludge. In a similar manner, the oxygen requirements can

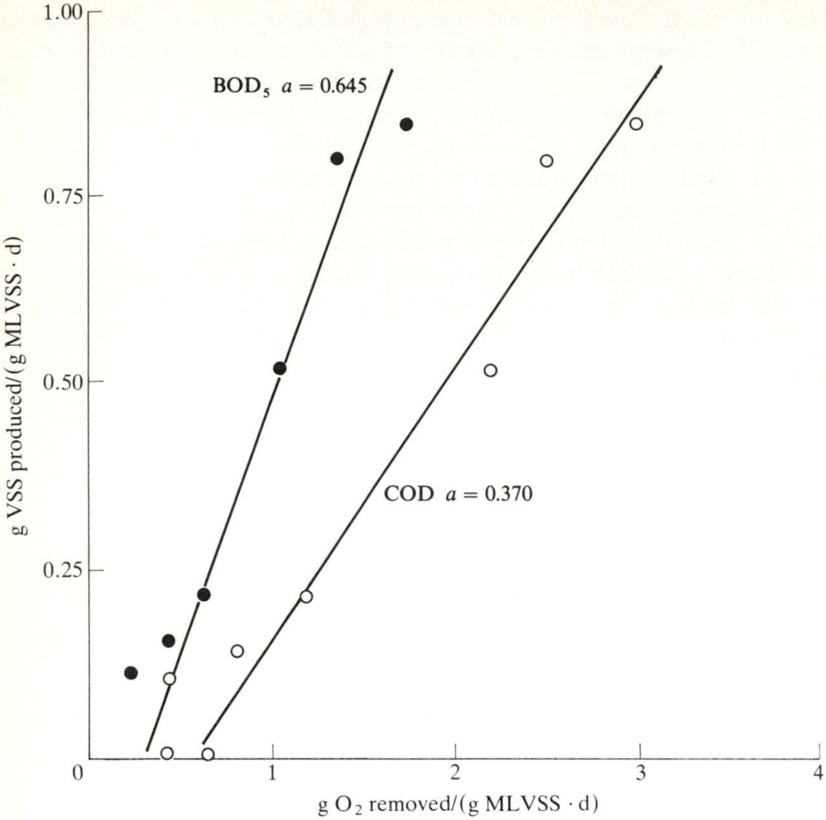

FIGURE 6-9
Cell synthesis from a soluble pharmaceutical wastewater.

be calculated:

$$r_r = a'S_r + b'X_d X_v t \tag{6.14}$$

in which r_r is the oxygen uptake rate in milligrams per liter per hour.

During the log-growth phase, the unit oxygen-uptake rate $k_r(r_r/X_v)$ is constant, and hence r_r will increase with increasing synthesis of new cells. As the substrate concentration decreases the oxygen-uptake rate decreases. When the available substrate is exhausted, the oxygen-uptake rate decreases to the endogenous rate. The calculation of a, a', and b' from process operating data for a chemicals wastewater is shown in Example 6.1.

Example 6.1. Soluble TOD removed across aeration = 12,000 lb/d
Sludge reaeration (endogenous)

$$\text{OUR}_e = 0.45 \text{ mg/(l} \cdot \text{min)}$$

$$\text{VSS}_e = 9000 \text{ mg/l}$$

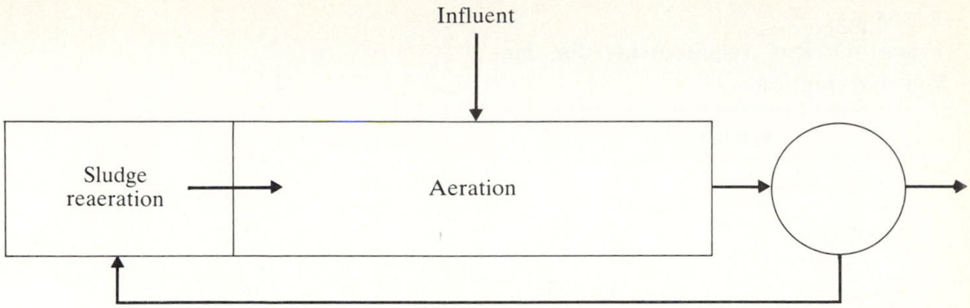

FIGURE 6-10

Aeration

$$\mathrm{OUR}_a = 1.73\ \mathrm{mg/(l\cdot min)}$$

$$\mathrm{VSS}_a = 6000\ \mathrm{mg/l}$$

$$\mathrm{Volume}_a = 0.26\ \text{million gal}\ (984\ \mathrm{m^3})$$

$$b'X_d = \frac{\mathrm{OUR}_e}{\mathrm{VSS}_e} = \frac{0.45\ \mathrm{mg/(l\cdot min)} \times 1440\ \mathrm{min/d}}{9000\ \mathrm{mg/l}} = 0.07/\mathrm{d}$$

$$a' = \frac{8.34V_a[\mathrm{OUR}_a \times 1400 - (\mathrm{OUR}_e \times 1440 \times \mathrm{VSS}_a)/\mathrm{VSS}_e]}{\text{soluble TOD removed}}$$

$$a' = \frac{8.34[(\mathrm{lb/million\ gal})/(\mathrm{mg/c})] \times 0.26\ \text{million gal} \times 1440\ \mathrm{min/d}}{12{,}000\ \mathrm{lb/d}}$$

$$\times [1.73\ \mathrm{mg/(l\cdot min)} - (0.45\ \mathrm{mg/(l\cdot min)} \times 6000\ \mathrm{mg/l})/9000\ \mathrm{mg/l}]$$

$$= 0.37$$

$$a = \frac{1-a'}{1.4} = \frac{1-0.37}{1.4} = 0.45$$

Nutrient Requirements

Several mineral elements are essential for the metabolism of organic matter by microorganisms. All but nitrogen and phosphorus are usually present in sufficient quantity in the carrier water. An exception is process wastewater generated from deionized water. Iron and other trace nutrients may be deficient in this case. Trace nutrient requirements are shown in Table 6.5.

Sewage provides a balanced microbial diet, but many industrial wastes (cannery, pulp and paper, etc.) do not contain sufficient nitrogen and phosphorus and require their addition as a supplement.

The quantity of nitrogen required for effective BOD removal and microbial synthesis has been the subject of much research. Early work by Helmers *et al.*[20] indicates a nitrogen requirement of 4.3 lb N/100 lb $\mathrm{BOD_{rem}}$ (4.3 kg N/100 kg $\mathrm{BOD_{rem}}$) and a phosphorus requirement of 0.6 lb P/100 lb $\mathrm{BOD_{rem}}$ (0.6 kg P/100 kg

TABLE 6.5
Trace nutrient requirements for biological oxidation

	mg/mg BOD
Mn	10×10^{-5}
Cu	14.6×10^{-5}
Zn	16×10^{-5}
Mo	43×10^{-5}
Se	14×10^{-10}
Mg	30×10^{-4}
Co	13×10^{-5}
Ca	62×10^{-4}
Na	5×10^{-5}
K	45×10^{-4}
Fe	12×10^{-3}
CO_3	27×10^{-4}

BOD_{rem}). These represent average values derived from the treatment of several nitrogen-supplemented industrial wastes. When insufficient nitrogen is present, the amount of cellular material synthesized per unit of organic matter removed increases as an accumulation of polysaccharide. At some point, nitrogen-limiting conditions restrict the rate of BOD removal. Nutrient limiting conditions will also stimulate filamentous growth as discussed on page 171.

The nitrogen content of sludge as generated in the process has been shown to average 12.3 percent based on the VSS. The nitrogen content of the sludge will decline during the endogenous phase. The nitrogen content of the nondegradable cellular mass has been shown to average 7 percent. The decrease in nitrogen content of an activated sludge with a soluble substrate is shown in Fig. 6-11. The phosphorus content of sludge at generation has been found to average 2.6 percent with the nondegradable cellular mass having a phosphorus content of 1 percent. The nitrogen and phosphorus requirements can be calculated considering the nitrogen and phosphorus content of the sludge wasted from the process:

$$N = 0.123 \frac{X_d}{0.8} \Delta X_v + 0.07 \frac{0.8 - X_d}{0.8} \Delta X_v \tag{6.15}$$

$$P = 0.026 \frac{X_d}{0.8} \Delta X_v + 0.01 \frac{0.8 - X_d}{0.8} \Delta X_v \tag{6.16}$$

Not all organic nitrogen compounds are available for synthesis. Ammonia is the most readily available form, and other nitrogen compounds must be converted to ammonia. Nitrite, nitrate, and about 75 percent of organic nitrogen compounds are also available.

Phosphorus may be fed as phosphoric acid in larger plants and ammonia as anhydrous or aqua ammonia. In small plants, nutrients may be fed as diammonium phosphate. In many cases, in aerated lagoons treating pulp and paper-mill

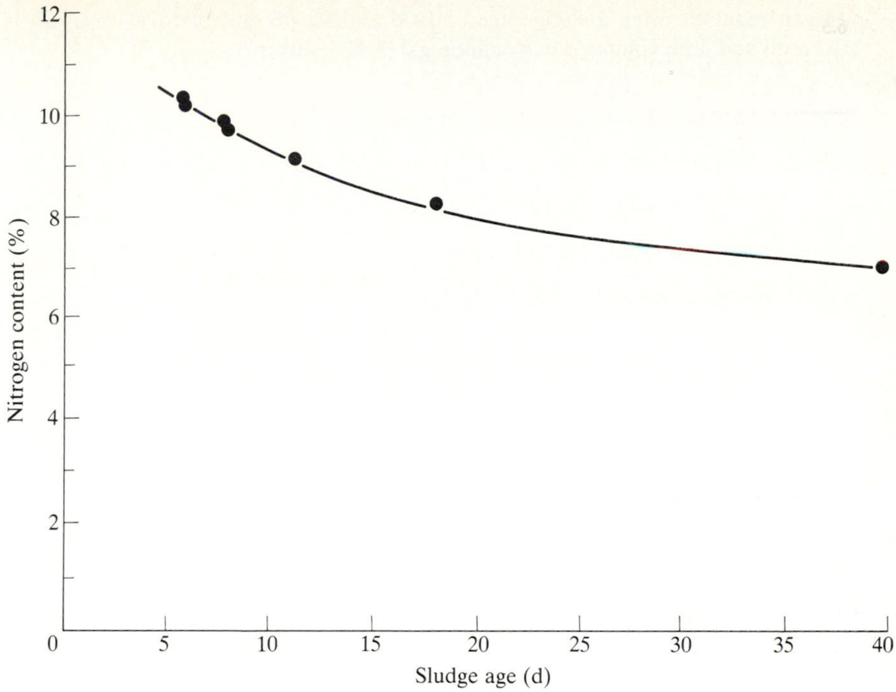

FIGURE 6-11
Nitrogen content in biological volatile suspended solids.

wastewaters, nitrogen and phosphorus have not been added, but rather the retention time has been increased. Calculated rate coefficients with and without the addition of nutrients is shown in Table 6.6. Nutrient requirements are shown in Example 6.2.

Example 6.2. Three million gallons per day (1.14×10^4 m^3/d) of an industrial wastewater with a BOD_5 of 2300 mg/l and a nitrogen content of 1.5 mg/l is to be treated in the activated sludge process with 5 million gal/d (1.89×10^4 m^3/d) of domestic sewage with a BOD_5 of 190 mg/l and a nitrogen content of 35 mg/l. Assuming a BOD k_{10} of 0.17/d, a of 0.52, and b of 0.1/d, compute the additional

TABLE 6.6

	K, d^{-1}	
Waste	**Without nutrients**	**With nutrients**
Kraft paper	0.35	1.33
Board mill	0.70	3.20
Hardboard	0.34	1.66

nitrogen required using a sludge age of 10 d and 92 percent removal. MLVSS is 2500 mg/l and tank volume is 9.04 million gal (3.42×10^4 m^3).

Solution. The total BOD removal is computed:

$$\text{BOD}_{\text{in}} \approx 2300 \text{ mg/l} \times 3 \text{ million gal/d} \times 8.34 + 190 \text{ mg/l} \times 5 \text{ million gal/d} \times 8.34$$

$$= 65{,}468 \text{ lb/d } (29{,}697 \text{ kg/d})$$

$$65{,}468 \times 0.92 = 60{,}230 \text{ lb/d } (27{,}321 \text{ kg/d})$$

The degradable fraction of the active sludge is [Eq. (6.6)]

$$X_d = \frac{0.8}{1 + 0.2 \times 0.1 \times 10}$$

$$= 0.66$$

The waste activated sludge is computed [Eq. (6.10)]:

$$\Delta X_v = 0.52 \times 60{,}230 \text{ lb/d} - 0.1 \times 0.66 \times 9.04 \text{ million gal} \times 2500 \text{ mg/l} \times 8.34$$

$$= 18{,}879 \text{ lb/d } (8564 \text{ kg/d})$$

The nitrogen available in the wastewater is

$$3 \text{ million gal/d} \times 1.5 \text{ mg/l} \times 8.34 + 35 \text{ mg/l} \times 5 \times 8.34 = 1495 \text{ lb/d } (678 \text{ kg/d})$$

The nitrogen lost in the waste activated sludge is [Eq. (6.15)]

$$N = 0.123 \times \frac{0.66}{0.8} \times 18{,}879 + 0.07 \times \frac{0.8 - 0.66}{0.8} \times 18{,}879$$

$$= 2146 \text{ lb/d } (973 \text{ kg/d})$$

The additional nitrogen required is

$$\text{N} = 2146 - 1495 = 651 \text{ lb/d } (295 \text{ kg/d})$$

Effect of Temperature

Variations in temperature affect all biological processes. There are three temperature regimes: the mesophilic over a temperature range of 4 to 39°C, the thermophilic which peaks at a temperature of 55°C, and the psychrophilic which operates at temperatures below 4°C. For economic and geographical reasons, most aerobic biological-treatment processes operate in the mesophilic range. In the mesophilic range, the rate of the biological reaction will increase with temperature to a maximum value at approximately 31°C for most aerobic waste systems. A temperature above 39°C will result in a decrease in rate for mesophilic organisms.

At temperatures above 96°F (35.5°C) there is a deterioration in the biological floc. Protozoa have been observed to disappear at 104°F (40°C) and a dispersed floc with filaments to dominate at 110°F (43.3°C). This causes a sharp

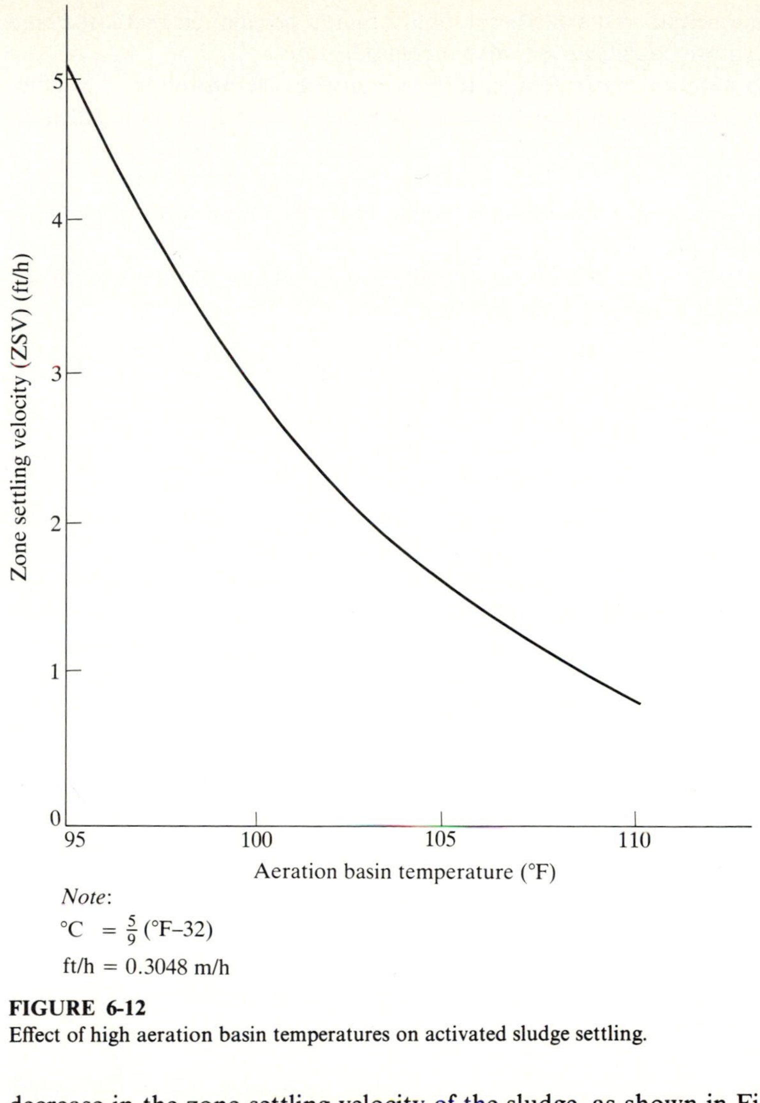

Note:
°C = $\frac{5}{9}$ (°F–32)
ft/h = 0.3048 m/h

FIGURE 6-12
Effect of high aeration basin temperatures on activated sludge settling.

decrease in the zone-settling velocity of the sludge, as shown in Fig. 6-12. It is for this reason that a maximum aeration basin temperature of 96°F (35.5°C) is recommended. Over a temperature range of 4 to 31°C, the K rate is corrected:

$$K_T = K_{20°C}\,\theta^{(T-20)} \tag{6.17}$$

The coefficient θ has been found to vary from 1.03 to 1.10 for a variety of industrial wastewaters.

It should be noted that the temperature coefficient, θ, for municipal wastewater has a value of 1.015, showing relatively little effect. This is because a major portion of the BOD is present as colloidal or suspended organics which are biocoagulated with very little temperature effect. When industrial wastewaters are discharged to municipal systems, the soluble organic content is increased and hence the temperature coefficient θ is also increased.

Decreasing aeration basin temperature will also cause an increase in effluent suspended solids. The solids are of a dispersed nature and are nonsettleable. For example, the Union Carbide plant at South Charleston, West Virginia, had an effluent suspended solids of 42 mg/l during the summer and 104 mg/l during the winter. Removal of these suspended solids requires the addition of coagulating chemicals (see page 228).

The endogenous coefficient b is also affected by temperature through the relationship

$$b_T = b_{20°C} \times 1.04^{(T-20)} \tag{6.18}$$

Effect of pH

A relatively narrow effective pH range will exist for most biooxidation systems. For most processes, this covers a range of pH 5 to 9 with optimum rates occurring over the range pH 6.5 to 8.5. It is significant to note that this relates to the pH of the mixed liquor in contact with the biological growths and not the pH of the waste entering the system. The influent waste is diluted by the aeration tank contents and is neutralized by reaction with the CO_2 produced by microbial respiration. In the case of both caustic and acidic wastes, the end product is bicarbonate (HCO_3^-) which effectively buffers the aeration system at near pH 8.0. Application of the complete mixing concept is essential in order to take advantage of these reactions. The amount of caustic that can be present in a wastestream is related to the BOD removal, which in turn will determine the CO_2 produced for reaction with the caustic. If we consider 0.9 lb or kg of CO_2 produced per lb or kg of COD removed (assuming conventional loadings) and that 70 percent of this is reactive with caustic alkalinity present, then 0.63 lb or kg of caustic alkalinity (as $CaCO_3$) will be neutralized per pound or kilogram of COD removed.

Since the oxidation of organic acids results in the production of CO_2, the permissible concentration of organic acids in the wastestream is related to the reaction rate of acid degradation to CO_2.

As long as the buffering capacity of the process is maintained, the pH of the aeration tank contents should remain near pH 8.0, even under conditions of fluctuating caustic or acidic loads.

Toxicity

Toxicity in biological oxidation systems may be due to one of several causes:

1. An organic substance, such as phenol, which is toxic in high concentrations, but biodegradable in low concentrations.
2. Substances, such as heavy metals, which have a toxic threshold depending on the operating conditions.

TABLE 6.7
Heavy metal removal in the activated sludge process

	Bio plant influent, mg/l	Bio plant effluent, mg/l
Cr	2.2	0.9
Cu	0.5	0.1
Zn	0.7	0.4

3. Inorganic salts and ammonia which exhibit a retardation at high concentrations.

The toxic effects of organics can be minimized by employing the complete mixing system in which the influent is diluted by the aeration tank contents, and the microorganisms are only in contact with the effluent concentration. In this way wastes with concentrations many times the toxic threshold can be successfully treated.

Heavy metals exhibit a toxicity in low concentrations to biological sludges. Acclimation of the sludge to the metal, however, will increase the toxic threshold considerably.

While an acclimated biological process is tolerant of the presence of heavy metals, the metal will concentrate in the sludge by complexing with the cell wall. Metal concentrations in the sludge as high as 4 percent have been reported. Heavy metal removal from petroleum refinery wastewater is shown in Table 6.7. This in turn creates problems relative to ultimate sludge disposal. The accumulation of copper on activated sludge with increasing sludge age is shown in Fig. 6-13.

High concentrations of inorganic salts are not toxic in the conventional sense, but rather exhibit progressive inhibition and a decrease in rate kinetics. Biological sludges, however, can be acclimated to high concentrations of salt. Processes are successfully operating with as high as 5 percent salt by weight. Frequently, high salt content will increase the effluent suspended solids.

6.3 SLUDGE-QUALITY CONSIDERATIONS

One of the factors essential to the performance of the activated sludge process is effective flocculation of the sludge, with subsequent rapid settling and compaction. McKinney[13] related flocculation to the food/microorganism ratio and showed that certain organisms normally present in activated sludge flocculate rapidly under starvation conditions. More recently, it has been shown that flocculation results from the production of a sticky polysaccharide slime layer to which organisms adhere. Flagellettes are also entrapped in this slimy material.

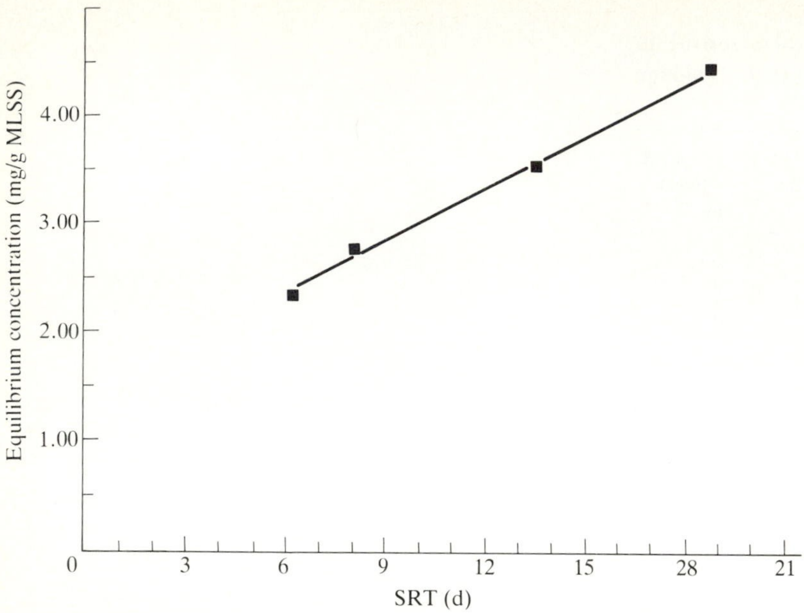

FIGURE 6-13
Copper accumulation in the activated sludge process.

Filamentous organisms are present in most activated sludge (exceptions are found in the chemical and petrochemical industry). Palm, Jenkins, and Parker[21] have identified three generic types of activated sludge, as shown in Fig. 6-14. Nonbulking sludge will result from plug flow or selector plant configuration or from complex organic wastewaters. Bulking sludges result from degradable wastewaters treated in a complete mix process or from oxygen or nutrient deficiency.

Pin floc usually results from low F/M (long sludge age) operation. Recently a number of filamentous organisms have been identified in activated sludge treating municipal and industrial wastewaters.[22] Depending on process operating conditions one or more of these organisms may dominate in the process, as shown in Table 6.8. Identification and control of filaments has recently been reported.[22] Proper process design and operation should not permit the filaments to overgrow the floc formers.

Filamentous overgrowth is affected by:

1. Wastewater composition. Wastewaters containing glucoselike saccharides (glucose, saccharose, lactose, maltose, etc.) promote filamentous growth while laundry, textile, and chemical wastewaters inhibit filamentous growth in a completely mixed system. In general, the more readily degradable the substrate, the more prone the system is to filamentous bulking. A comparison of bulking tendency for several substrates is shown in Table 6.9.[23]

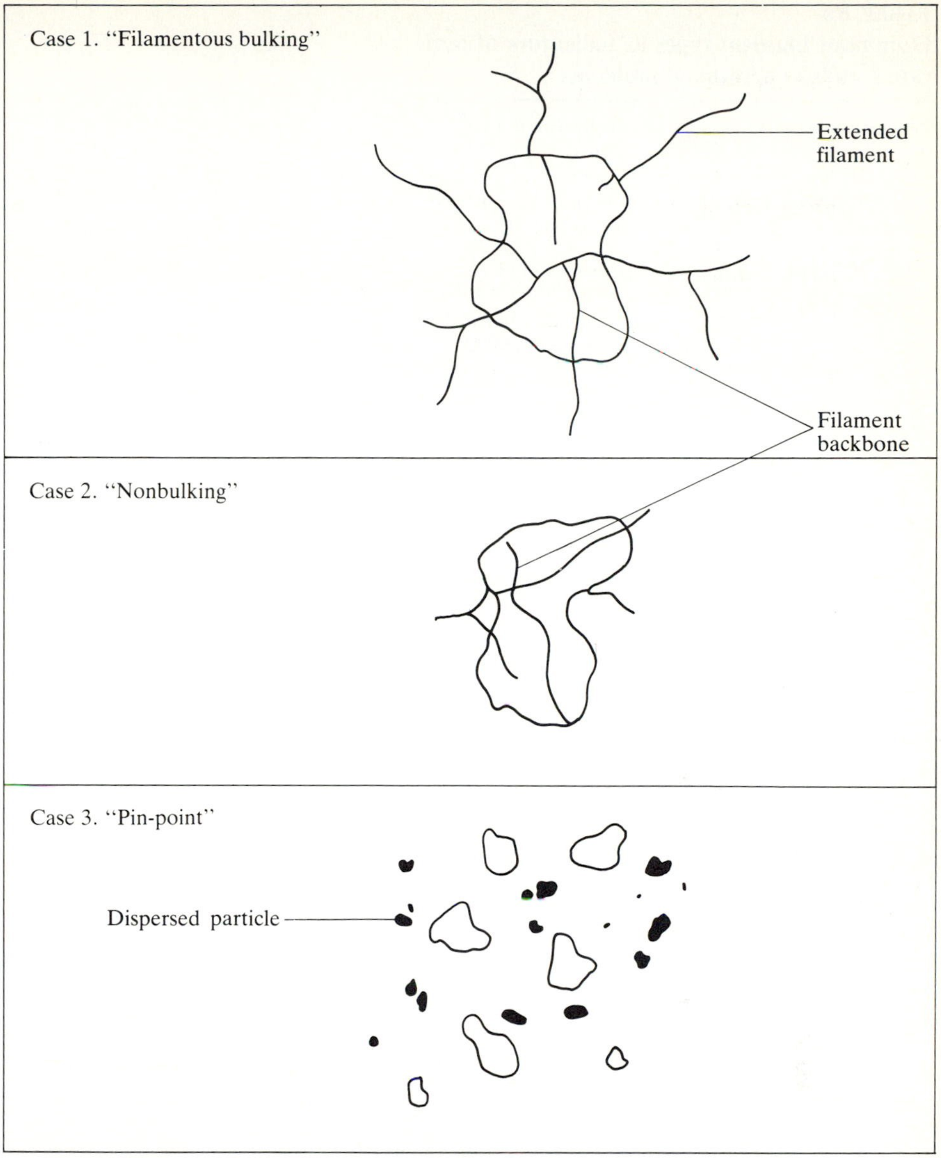

FIGURE 6-14
Activated sludge types.

2. Dissolved oxygen concentration. Oxygen must diffuse into the flocs in order to be available to the organisms within the floc. The depth of oxygen penetration within the floc depends on the bulk concentration of oxygen in the surrounding liquid and on the oxygen-utilization rate of the floc. The oxygen-utilization rate is proportional to the organic loading (F/M). Hence, as the organic

TABLE 6.8
Dominant filament types as indicators of activated sludge operational problems[22]

Suggested causative conditions	Indicative filament types
Low dissolved oxygen (D.O.)	Type 1701, *S. natans*, *H. hydrossis*
Low food/mass ratio (F/M)	*M. parvicella*, *H. hydrossis*, *Nocardia* spp., types 021N, 0041, 0675, 0092, 0581, 0961, and 0803
Septic wastewater/sulfides	*Thiothrix* spp., *Beggiatoa* spp., and type 021N
Nutrient deficiences	*Thiothrix* spp., *S. natans*, type 021N; and possibly *H. hydrossis* and types 0041 and 0675
Low pH	Fungi

loading increases, the dissolved oxygen necessary to maintain a fully aerobic floc also increases. The thin filaments (1 to 4 μm) can readily obtain oxygen at concentrations less than 0.1 mg/l. The relationship between dissolved oxygen and F/M developed from the data of Palm, Jenkins, and Parker[21] is shown in Fig. 6-15.

3. Substrate concentration. Chudoba *et al.*[24] have shown that the growth rate of specific organisms can be defined by the Monod equation (6.3).

With degradable substrates at low concentrations, growth is favored by the filaments, as shown in Fig. 6-16.[25] This explains why complete mix systems with low mixed liquor substrate concentrations favor filamentous growth. At high

TABLE 6.9
Effect of substrate composition on filamentous bulking

Substrate	SVI in CMAS
Glucose	700
Citric acid	700
Phenol	296 ± 196
Galactose	130 ± 25

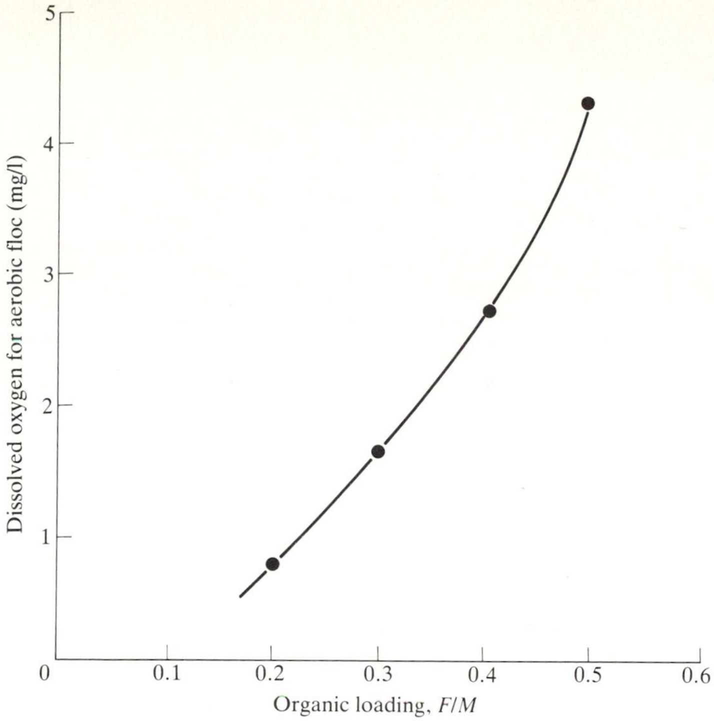

FIGURE 6-15
Relationship between dissolved oxygen and F/M for an aerobic floc.

substrate concentrations, the floc formers remove substrate at a greater rate than the filaments so they will dominate the system. In order to achieve high substrate gradients, a plug-flow configuration or the use of a selector or contactor must be employed. In a kraft pulp and paper mill, plug-flow operation never generated a filamentous sludge while complete mix operation on the same wastewater resulted in a filamentous sludge.

One of the more common causes of filamentous bulking in industrial wastewaters is inadequate nitrogen or phosphorus. There are numerous examples, particularly in the pulp and paper industry, in which severe filamentous bulking resulted from inadequate nitrogen. Restoration of adequate nitrogen restored a flocculated sludge within three sludge ages. Studies on a Wisconsin pulp mill indicated a minimum concentration of NH_3–N in the effluent of 1.5 mg/l to favor zoogleal growth. Other studies indicate that higher ammonia concentrations may be required in some cases. Minimum soluble phosphorus concentrations in the effluent of 0.5 mg/l have been reported as required for optimal zoogleal growth.

Filamentous Bulking Control

Many filamentous organisms are aerobic and can be destroyed by prolonged periods of anaerobiosis. Most of the bacteria, on the other hand, are facultative

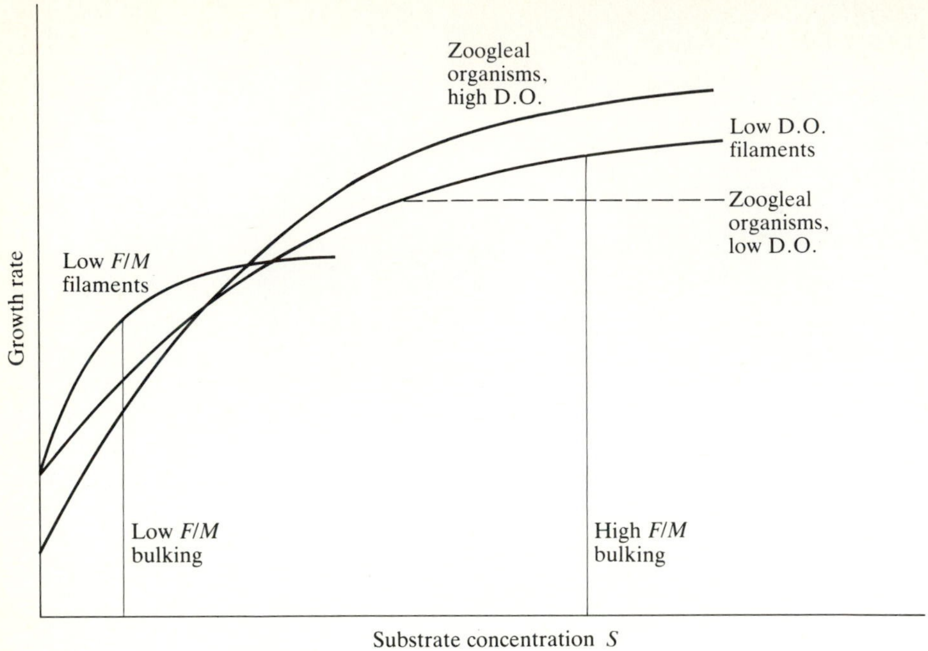

FIGURE 6-16
Microbial growth characteristics as related to substrate concentration.

and can exist for extended periods without oxygen. Although available data are somewhat contradictory, it would appear that anaerobic or anoxic conditions maintained within the process will restrict the growth of these filaments.

The use of an anoxic zone was studied at an organic chemicals plant. Three systems were run in parallel: a conventional air system, an oxygen system, and an air system with an anoxic zone. The settling properties of the respective sludges indicated that the oxygen system yielded the best settling sludge, while the air system with insufficient dissolved oxygen demonstrated severe filamentous bulking. The anoxic-aerobic system was almost devoid of filaments, but exhibited poorer flocculation and settling than the oxygen system.

Chlorine or hydrogen peroxide can be added to the return sludge or the aeration basin to reduce bulking. Hydrogen peroxide is selective for some filament types. Hydrogen peroxide dosages are in the order of 20 to 50 mg/l. Chlorine dosages may vary from 9 to 10 lb of Cl_2 per day per thousand pound of MLSS [9 to 10 kg Cl_2/(d · thousand kg MLSS)] in cases of severe bulking and 1 to 2 lb Cl_2/(d · thousand lb MLSS) [1 to 2 kg Cl_2/(d · thousand kg MLSS)] in cases of moderate bulking. Because of the growth rate of the filaments, if the hydraulic retention time exceeds 8 h, chlorination must be applied directly to the aeration basin. Since the filaments exhibit a high negative zeta potential they can be flocculated by the addition of cationic polyelectrolytes. This treatment is expensive and the filaments are not destroyed in the process.

6.4 NITRIFICATION AND DENITRIFICATION

Nitrification

Nitrogen changes in a biological-treatment process are shown in Fig. 6-17. Nitrification is the biological oxidation of ammonia to nitrate with nitrite formation as an intermediate. The microorganisms involved are the autotrophic species *Nitrosomonas* and *Nitrobacter*, which carry out the reaction in two steps:

$$2NH_4^+ + 3O_2 \rightarrow 2NO_2^- + 4H^+ + 2H_2O \text{ (\textit{Nitrosomonas})}$$

$$2NO_2^- + O_2 \rightarrow 2NO_3^- \text{ (\textit{Nitrobacter})}$$

The cell yield for *Nitrosomonas* has been reported as 0.05 to 0.29 mg VSS/mg NH_3–N and for *Nitrobacter* 0.02 to 0.08 mg VSS/mg NH_3–N. A value of 0.15 mg VSS/mg NH_3–N is usually used for design purposes. It is generally accepted that the specific growth rate of *Nitrobacter* is higher than the growth rate of *Nitrosomonas* and hence there is no accumulation of nitrite in the process and the growth rate of *Nitrosomonas* will control the overall reaction.

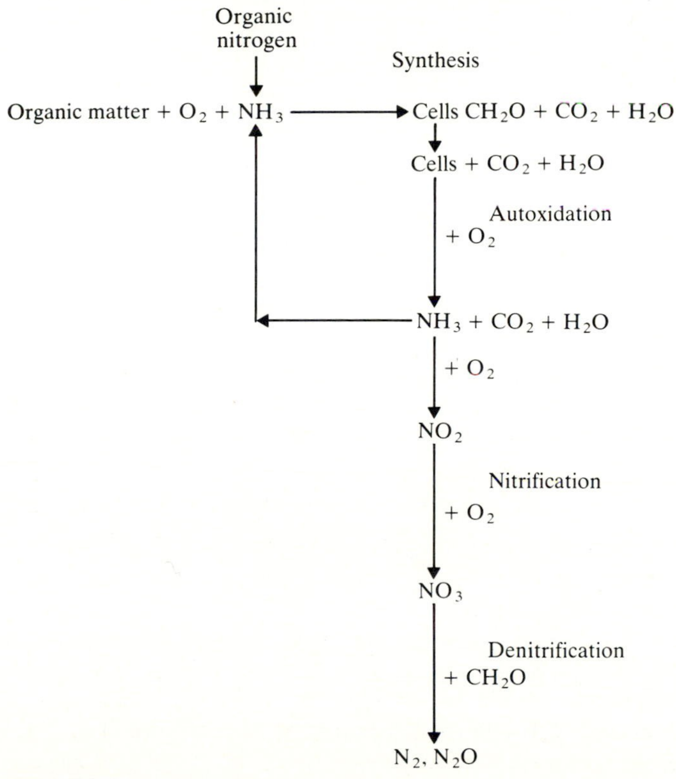

FIGURE 6-17
The nitrogen cycle in biological-oxidation processes.

For effective nitrification to occur, the sludge age must be greater than the reciprocal of the growth rate of *Nitrosomonas*. Shorter sludge ages will result in a washout of these organisms. Several investigators have reported nitrification to occur over a temperature range of 5 to 45°C with the optimum range being 25 to 35°C. The critical sludge age can be computed from the relationship:

$$\theta_c = 2.13e^{0.098(15-T)} \tag{6.19}$$

The U.S. EPA recommends a safety factor in the order of 2.5[26] to the θ_c calculated in Eq. (6.19) for design purposes to ensure that ammonia breakthrough does not occur during diurnal peaks in loads. The required sludge age can be achieved by varying either the MLVSS or the hydraulic detention time, as follows:

$$\theta_c = \frac{X_v t}{aS_r - bX_d X_v t} \tag{6.11}$$

In order to achieve nitrification at variable influent concentrations, the nitrification rate must be considered.

The fraction of nitrifiers in the sludge can be estimated from the relationship:

$$f_N = \frac{a_N N_r}{a_N N_r + aS_r} \tag{6.20}$$

where a_N = nitrifiers yield coefficient
N_r = nitrogen oxidized
S_r = substrate removed

The rate of nitrification has been reported by Wong-Chong and Loehr[27] as essentially constant with biomass concentrations (as VSS) up to 1500 mg/l and decreases above that level. From these data, the nitrification rate, at 20°C, can be estimated as 1.04 mg NH_3–N oxidized per milligram of nitrifying microbial mass per day. The overall rate of nitrification can then be expressed at 20°C as

$$R_N = 1.04 f_N X_v \tag{6.21}$$

in which R_N is the nitrification rate in milligrams per liter per day; f_N is the fraction of nitrifiers in the mixed liquor. Other sources have shown nitrification rates as high as 2.5 mg NH_3–N/(mg VSS · d).

It should be noted that the nitrogen oxidized is the nitrogen in the influent wastewater less that which is assimilated (see page 162):

$$N_r = TKN - SON - N_{synthesized}$$

The TKN is the total Kjeldhal nitrogen in the wastewater. The SON is the residual soluble organic nitrogen which is nonbiodegradable. In municipal wastewater, this is in the order of 1.0 mg/l. In industrial wastewaters, this can be highly variable depending on the composition of the wastewater. For complex wastewaters, this nitrogen fraction should be determined by a laboratory study. The $N_{synthesized}$ is related to sludge age, as shown in Fig. 6.11.

Like all biochemical reactions, the rate of nitrification is temperature-dependent with θ in the range of 1.03 to 1.15.

The oxygen requirements for nitrification will be 4.33 mg O_2/mg NH_3–N oxidized. Since nitrification generates acidity, 7.14 mg alkalinity is required per milligram of NH_3–N oxidized.

Nitrifying organisms are subject to inhibition by various organic and inorganic compounds. Hockenbury and Grady[28] have summarized inhibition data for selected organic compounds as shown in Tables 6.10 and 6.11. Anthoisen[29] has shown toxicity due to ammonia or nitrite. Since only ammonia and nitrite in the un-ionized form are toxic, these effects are a function of pH, as shown in Fig. 6-18.

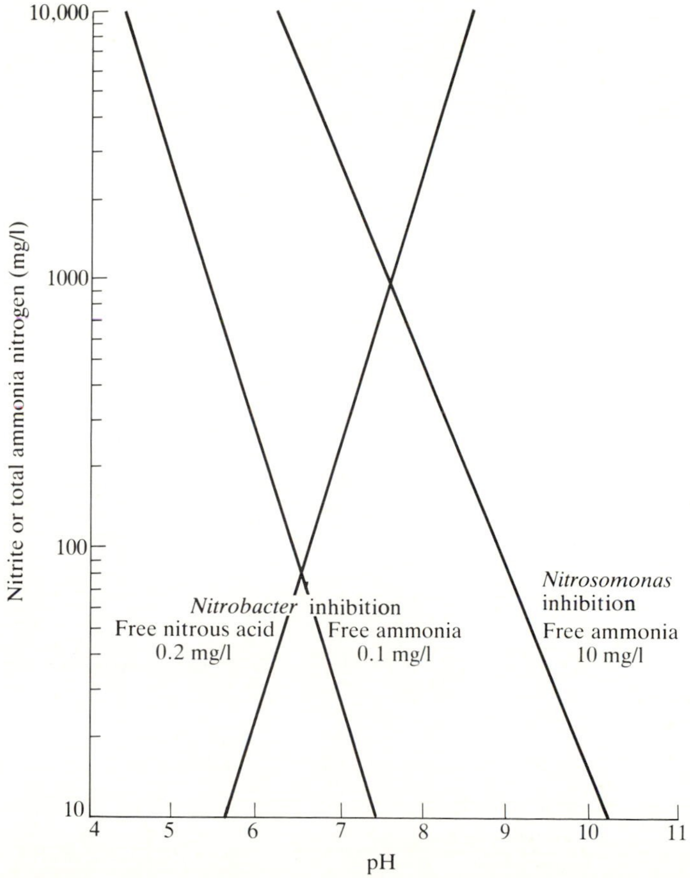

FIGURE 6-18
Ammonia and nitrite inhibition to nitrification.

TABLE 6.10
Organic compounds that inhibit activated sludge nitrification

Compound	Concentration,† mg/l
Acetone‡	2000.000
Allyl alcohol	19.500
Allyl chloride	180.000
Allyl isothiocyanate	1.900
Benzothiazole disulfide	38.000
Carbon disulfide‡	35.000
Chloroform‡	18.000
o-Cresol	12.800
Diallyl ether	100.000
Dicyandiamide	250.000
Diguanide	50.000
2,4-Dinitrophenol	460.000
Dithiooxamide	1.100
Ethanol‡	2400.000
Guanidine carbonate	16.500
Hydrazine	58.000
8-Hydroxyquinoline	72.500
Mercaptobenzothiazole	3.00
Methylamine hydrochloride	1550.000
Methyl isothiocyanate	0.800
Methyl thiuronium sulfate	6.500
Phenol‡	5.600
Potassium thiocyanate	300.000
Skatol	7.000
Sodium dimethyl dithiocarbamate	13.600
Sodium methyl dithiocarbamate	0.900
Tetramethyl thiuram disulfide	30.000
Thioacetamide	0.530
Thiosemicarbazide	0.180
Thiourea	0.076
Trimethylamine	118.000

† Concentration giving approximately 75% inhibition.

‡ In the list of industrially significant chemicals.

The optimal pH for nitrification varies between 6 and 7.5, depending on the formation of free ammonia and free nitrous acid. Low dissolved oxygen levels will also affect the rate of nitrification. Hence, a dissolved oxygen level in excess of 2.0 mg/l is recommended where maximum nitrification rates are desired.

Denitrification

Biological denitrification is achieved under anoxic (absence of molecular oxygen) conditions by heterotrophic microorganisms that utilize nitrate as a hydrogen

TABLE 6.11
Effects of concentrations of organic compounds on degree of inhibition of ammonia oxidation exhibited

Compound	Degree of inhibition at concentration indicated, %				Estimated concentration giving 50% inhibition, mg/l
	100 mg/l	50 mg/l	10 mg/l	As noted, mg/l	
Dodecylamine	96	95	—	66‡	<1
Aniline†	86	—	—	76,§ 89¶	<1
n-Methylaniline	90	83	71	—	<1
1-Naphthylamine	81	81	45	—	15
Ethylenediamine†	73	—	41	61††	17
Napthylethylenediamine diHCl	93	79	29	—	23
2,2′-Bipyridine	91	81	23	—	23
p-Nitroaniline	64	52	46	—	31
p-Aminopropiophenone	80	56	22	—	43
Benzidine diHCl	84	56	12	—	45
p-Phenylazoaniline	54	47	0	—	72
Hexamethylene diamine†	52	45	27	—	85
p-Nitrobenzaldehyde	76	32	29	—	87
Triethylamine	35	—	—	63‡‡	127
Ninhydrin	30	26	31	—	>100
Benzocaine	30	27	0	—	>100
Dimethylgloxime	30	9	—	56‡‡	140
Benzylamine	26	10	0	—	>100
Tannic acid	20	7	—	22‡‡	>150
Monoethanolamine†	16	—	—	20§§	>200

† From the list of industrially significant chemicals.
‡ 1 mg/l.
§ 2.5 mg/l.
¶ 5 mg/l.
†† 30 mg/l.
‡‡ 150 mg/l.
§§ 200 mg/l.

acceptor when an organic energy source is available. Denitrification will also occur under conditions of endogenous respiration, although at a much slower rate:

$$NO_3^- + \text{substrate} \rightarrow N_2 + CO_2 + H_2O + OH^- + \text{cells}$$

Nitrate nitrogen is converted to reduced forms such as N_2, N_2O, and NO. The breakdown of carbonaceous organics in the denitrification reactions is similar to that in the aerobic process. Thermodynamic data show a higher energy yield from the aerobic metabolism of organic carbon compared to anoxic denitrification, a factor that indicates the need for strict anoxic conditions in a denitrifying system. Many activated sludge systems are operated so that aerobic processes

TABLE 6.12
Denitrification rates with various carbon sources

Carbon source	lb NO_3–N/(lb VSS·d)	Temperature, °C
Brewery waste	0.22–0.25	20
Methanol	0.18	19–24
Volatile acids	0.36	20
Molasses	0.10	10
	0.036	16

Note: lb/(lb·d) = kg/(kg·d)

and denitrification can occur simultaneously due to the presence of a dissolved oxygen gradient across bacterial flocs.

The denitrification rate can be formulated as

$$R_{DN} = K_{DN} X_v \times 1.09^{(T-20)}(1 - \text{D.O.}) \tag{6.22}$$

in which R_{DN} = overall denitrification rate
K_{DN} = specific denitrification rate

The denitrification rate under aerobic conditions will depend on the anoxic fraction of the biological floc and the availability of the carbon substrate. The D.O. term in Eq. (6.22) indicates that the denitrification rate decreases linearly to zero when the dissolved oxygen concentration reaches 1.0 mg/l.

Denitrification has been studied using a number of different carbon sources. Specific denitrification rates with various carbon sources are shown in Table 6.12.

In the anoxic reaction, the formation of 3.57 mg alkalinity as $CaCO_3$ for each milligram of NO_3–N denitrified is predicted. This amounts to one-half of the alkalinity that is consumed during nitrification. Most of this alkalinity, however, is lost by reaction with the CO_2 generated.

Approximately 3 mg/l of BOD is consumed for each mg/l of NO_3–N reduced. A number of process alternatives are available which achieve nitrification-denitrification using internal carbon sources in which some form of aerobic-anoxic sequencing is provided. The simplest configuration enables carbonaceous oxidation, nitrification, and denitrification to occur in a single basin. This can be accomplished by careful positioning of aeration equipment to maintain defined aerobic-anoxic zones in different basin sections. A more simple system utilizes a single basin for both aeration and sedimentation by providing intermittent aeration and nonaeration cycles to yield aerobic and anoxic time sequences. No matter what system is utilized, it is necessary to provide a sufficiently long solids retention time to insure the growth of nitrifying organisms: a sufficient active fraction of biomass at a dissolved oxygen level adequate for nitrification to occur and an anaerobic or anoxic phase of sufficient duration to permit dissimilatory nitrate reduction. An example of nitrification and denitrification is shown in Examples 6.3 and 6.4.

Example 6.3. A 1.0 million gal/d (3.79×10^3 m^3/d) wastewater treatment plant is to achieve nitrification at 15°C. The BOD is 850 mg/l and the NH_3–N is 150 mg/l. The alkalinity is 380 mg/l. Design the facility for the following conditions:

Effluent BOD = 30 mg/l
MLVSS = 3500 mg/l
$a = 0.55$
$b = 0.1\ \text{day}^{-1}$
$X_d = 0.60$
$a_N = 0.15$
$\theta = 1.05$

1. The critical sludge age at 15°C is computed by Eq. (6.19) as

$$\theta_c = 2.13\ \text{d}$$

Using a safety factor of 2.5 to account for variability, the minimum design sludge age is $2.13 \times 2.5 = 5.33$ d. For a MLVSS level of 3500 mg/l the required retention time for organic removal is [from Eq. (6.11)]

$$t = \frac{\theta_c a(S_o - S)}{X_v(1 + bX_d\theta_c)}$$

$$= \frac{5.33 \times 0.55(850 - 30)}{3500(1 + 0.1 \times 0.6 \times 5.33)}$$

$$= 0.52\ \text{d or } 12.5\ \text{h}$$

2. Check nitrification rate.

(*a*) Calculate sludge yield by organic removal [Eq. (6.10)]:

$$\Delta X_v = aS_r - bX_dX_vt$$

$$= 0.55 \times 820\ \text{mg/l} - 0.1/\text{d} \times 0.60 \times 3500\ \text{mg/l} \times 0.52\ \text{d} = 342\ \text{mg/l}$$

$$\Delta X_v = 342\ \text{mg/l} \times 1.0\ \text{million gal/d} \times 8.34\ \frac{\text{lb/million gal}}{\text{mg/l}}$$

$$= 2852\ \text{lb/d}\ (1294\ \text{kg/d})$$

(*b*) Calculate NH_3–N removal by synthesis assuming the waste activated sludge contains 10 percent nitrogen:

$$NH_3\text{–N synthesized} = 0.1 \times \Delta X_v$$

$$= 0.1 \times 342\ \text{mg/l}$$

$$= 34\ \text{mg/l}$$

(*c*) Calculate NH_3–N removal by nitrification:

$$NH_3\text{–N to be oxidized} = 150 - 34$$

$$= 116\ \text{mg/l}$$

(*d*) NH_3–N oxidation rate at 20°C is 1.04 mg NH_3–N oxidized/(mg VSS · d). The nitrifier fraction in the mixed liquor can be estimated as

$$f_N = \frac{a_N N_r}{a_N N_r + aS_r}$$

$$= \frac{0.15 \times 116}{0.15 \times 116 + 0.55 \times 820}$$

$$= 0.037$$

$$R_N = 1.04\theta^{(T-20)} f_N X_v$$

$$= 1.04(1.05)^{(15-20)} \times 0.037 \times 3500$$

$$= 106 \text{ mg/(l} \cdot \text{d)}$$

(*e*) Calculate the required detention time for nitrification:

$$t = \frac{116 \text{ mg/l}}{106 \text{ mg/(l} \cdot \text{d)}}$$

$$= 1.09 \text{ d or } 26.3 \text{ h} > 0.52 \text{ d or } 12.5 \text{ h}$$

Therefore, it is nitrification rate-controlling.

3. Oxygen requirement:

$$116 \text{ mg } NH_3\text{–}N_{ox}\text{/l} \times 4.33 \text{ mg } O_2\text{/mg } NH_3\text{–}N_{ox} = 502 \text{ mg } O_2\text{/l}$$

$$O_2 \text{ for nitrification} = 502 \text{ mg } O_2\text{/l} \times 1.0 \text{ million gal/d} \times 8.34 \frac{\text{lb/million gal}}{\text{mg/l}}$$

$$O_2 = 4189 \text{ lb/d (1900 kg/d)}$$

4. The alkalinity required for nitrification is

$$116 \text{ mg } NH_3\text{–}N_{ox}\text{/l} \times 7.14 \text{ mg alk/mg } NH_3\text{–}N_{ox} = 828 \text{ mg alk/l}$$

Alkalinity to be supplied:

$$828 \text{ mg/l} - 380 \text{ mg/l} = 448 \text{ mg/l}$$

Example 6.4. Design an aerobic-anoxic process for nitrification-denitrification of sewage with the characteristics given in Example 6.3 and the following conditions:

$$\text{Denitrification rate} = 0.06 \text{ lb } NO_3\text{–N/(lb VSS} \cdot \text{d) at 20°C}$$

$$\text{MLVSS} = 3500 \text{ mg/l}$$

$$\theta = 1.10$$

$$K = 6.4 \text{ d}^{-1} \text{ at 15°C}$$

1. Calculate the anoxic volume requirement for denitrification:

$$R_{DN} = 0.06 \text{ mg NO}_3\text{–N/(mg VSS} \cdot \text{d)} \times 1.10^{(15-20)} \times 3500 \text{ mg VSS/l}$$

$$= 0.06(1.10)^{(15-20)} \times 3500$$

$$= 130 \text{ mg NO}_3\text{–N/(l} \cdot \text{d)}$$

$$t = \frac{116 \text{ mg NO}_3\text{–N}_{den}\text{/l}}{130 \text{ mg NO}_3\text{–N}_{den}\text{/(l} \cdot \text{d)}}$$

$$= 0.89 \text{ d} = 21 \text{ h}$$

$$V_{anoxic} = Qt$$

$$= 1.0 \text{ million gal} \times 0.89$$

$$= 0.89 \text{ million gal } (3.37 \times 10^3 \text{ m}^3)$$

2. Calculate BOD removal in the anoxic zone:

$$\text{BOD}_{removal} = \frac{3.0 \text{ mg BOD}_{rem}}{\text{mg NO}_3\text{–N}_{den}} \times 116 \text{ mg NO}_3\text{–N/l} = 348 \text{ mg/l}$$

3. Calculate BOD removal in the aerobic zone (assume $t_e = t$ of Example 6.3) but considering 50 percent reduction in nitrification rate; then $t_e = 0.5 \times 1.09 = 0.55$ d:

$$\text{BOD}_{in} = 850 - 348 = 502 \text{ mg/l}$$

$$\text{BOD}_{effluent} = \frac{S_o^2}{S_o + KX_v t_e}$$

$$= \frac{(502)^2}{502 + 6.4/\text{d} \times 3500 \times 0.55 \text{ d}} = 20 \text{ mg/l}$$

$$\text{BOD}_{removal} = \frac{502 - 20}{502} \times 100 = 96\%$$

6.5 DEVELOPMENT OF PROCESS DESIGN PARAMETERS

A protocol has been developed to generate cost-effective design criteria for biological wastewater-treatment processes. The first step is a partition of the organic content as shown on page 20. Wastewaters with a high sorbable and high biodegradable fraction are usually treatable by biological processes to acceptably low effluent organic content. The sorbable fraction is determined by contacting sludge and wastewater for 15 min and determining the soluble organic removal (page 149). These wastewaters, however, will usually generate filamentous organisms so that process configuration must be considered in both the experimental program and the process design. Wastewaters with a low sorbable and biodegradable fraction will require more detailed analysis and frequently additional physical-

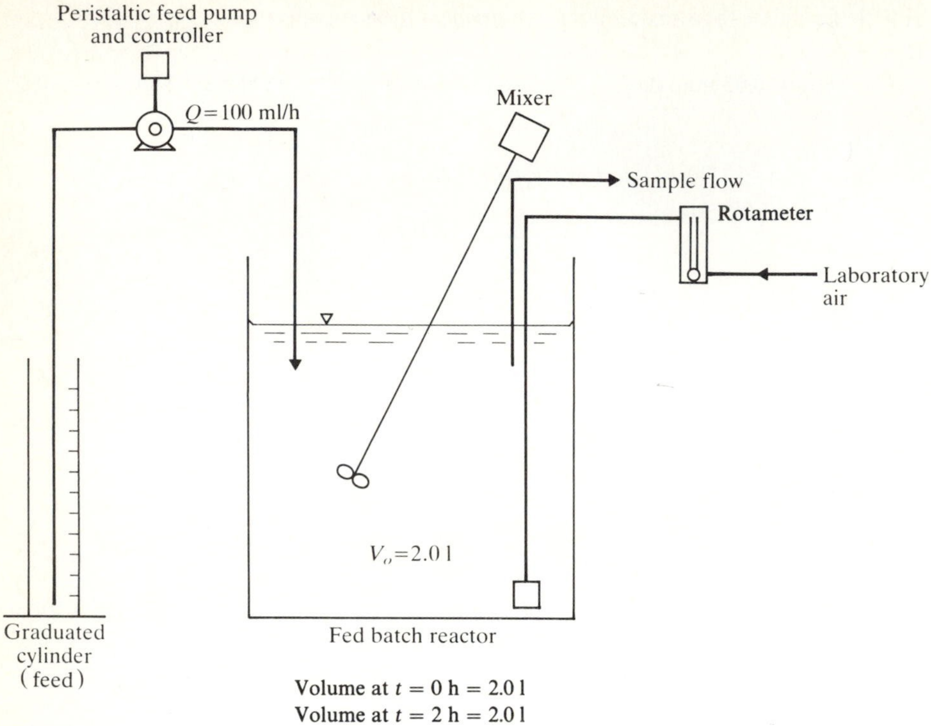

FIGURE 6-19
Fed batch reactor configuration.

chemical technology to meet current effluent criteria. The fed batch reactor (FBR) is used as a screening tool to define both the biodegradability of a wastewater and the inhibition level if one exists. In the FBR procedure wastewater is added to the reactor at a constant rate, as shown in Fig. 6-19. The organic concentration of the wastewater must be sufficiently high to yield a concentration increase in the reactor and to identify inhibition at concentration levels that could be experienced in the full-scale installation. The sludge used in the test is acclimated to a variety of organics representing those compounds most commonly found in industrial wastewaters.

The maximum specific growth rate of degradation, μ_m in Eq. (6.3), can be computed as the difference in the slopes of the feed rate and the substrate concentration buildup as shown in Fig. 6-20. Inhibition is identified from a change in SOUR[mg O_2/(g VSS · h)] and by a change in slope of the substrate buildup curve (Fig. 6-20).

Continuous reactors are used to develop the process design criteria. The activated sludge process selection will depend on the nature of the wastewater. Readily degradable wastewaters will use a plug-flow or selector configuration or an intermittent process (SBR). If the wastewater is sorbable and readily degradable, a sequencing batch reactor (SBR) should be used to generate a nonfilamentous sludge (Fig. 6-21). For a more refractory wastewater, a complete mix reactor as shown in Fig. 6-22 should be used. This reactor configuration is also used for powdered activated carbon (PAC) studies (see page 284).

Prior to generating design data, the sludge should be acclimated to the wastewater. Acclimation is usually achieved by running the unit at an F/M ratio

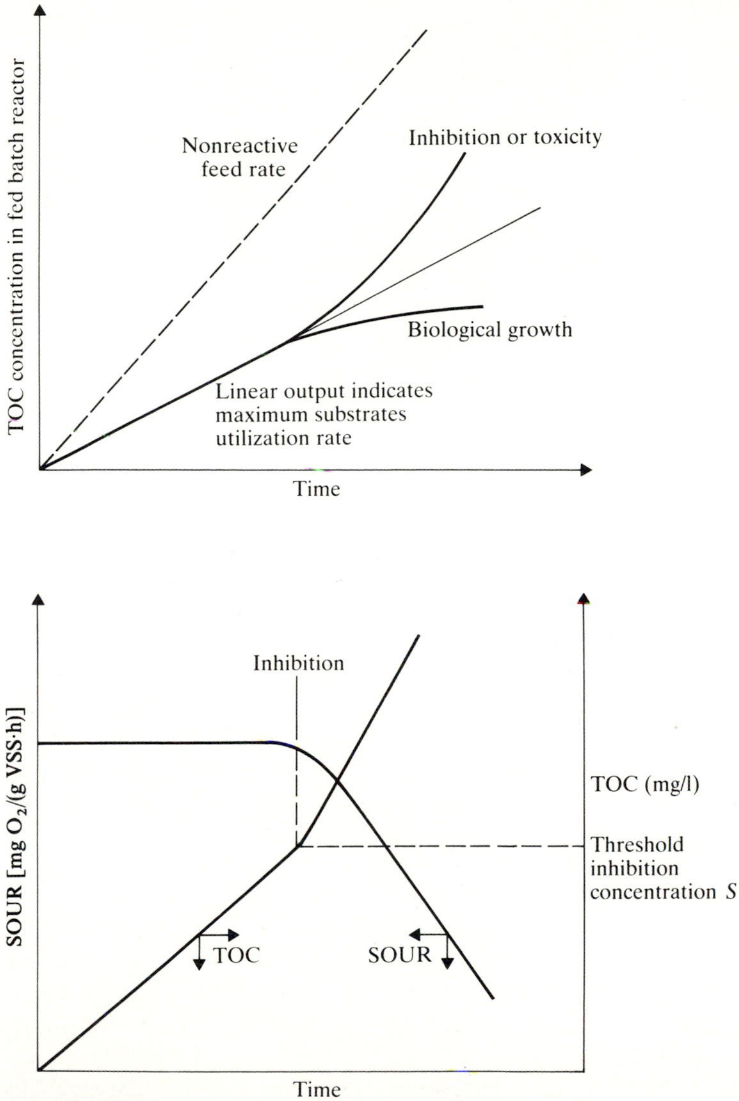

FIGURE 6-20
Theoretical fed batch reactor output with influent substrate mass flow rate greater than $\mu_m X_v$ and inhibition effects.

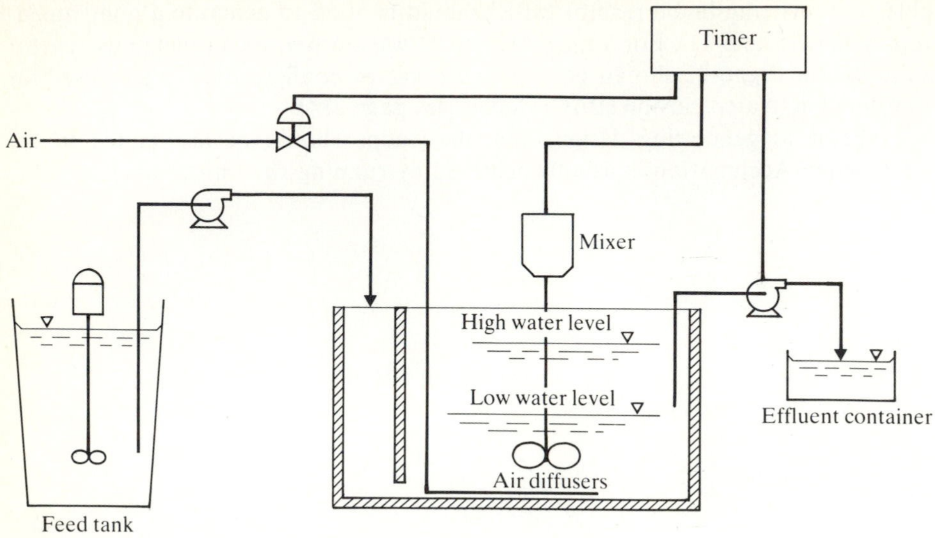

FIGURE 6-21
Pilot plant activated sludge system (SBR modification).

TABLE 6.13
Biological treatment process design study sampling and analysis schedule

Analysis	Frequency
BOD	3/week
COD or TOC	Daily
O_2 uptake rate	Daily
MLVSS	Daily
Dissolved oxygen	Daily
pH	Daily
Temperature	Daily
Nitrogen†	2/week
Phosphorus	2/week
Bioassay‡	Weekly
Specific pollutants	Weekly

† The case of a nitrogen-deficient wastewater. If nitrification is desired, TKN, NH_3–N, and NO_3–N should be run 3 times per week.

‡ The test species will depend on permit conditions.

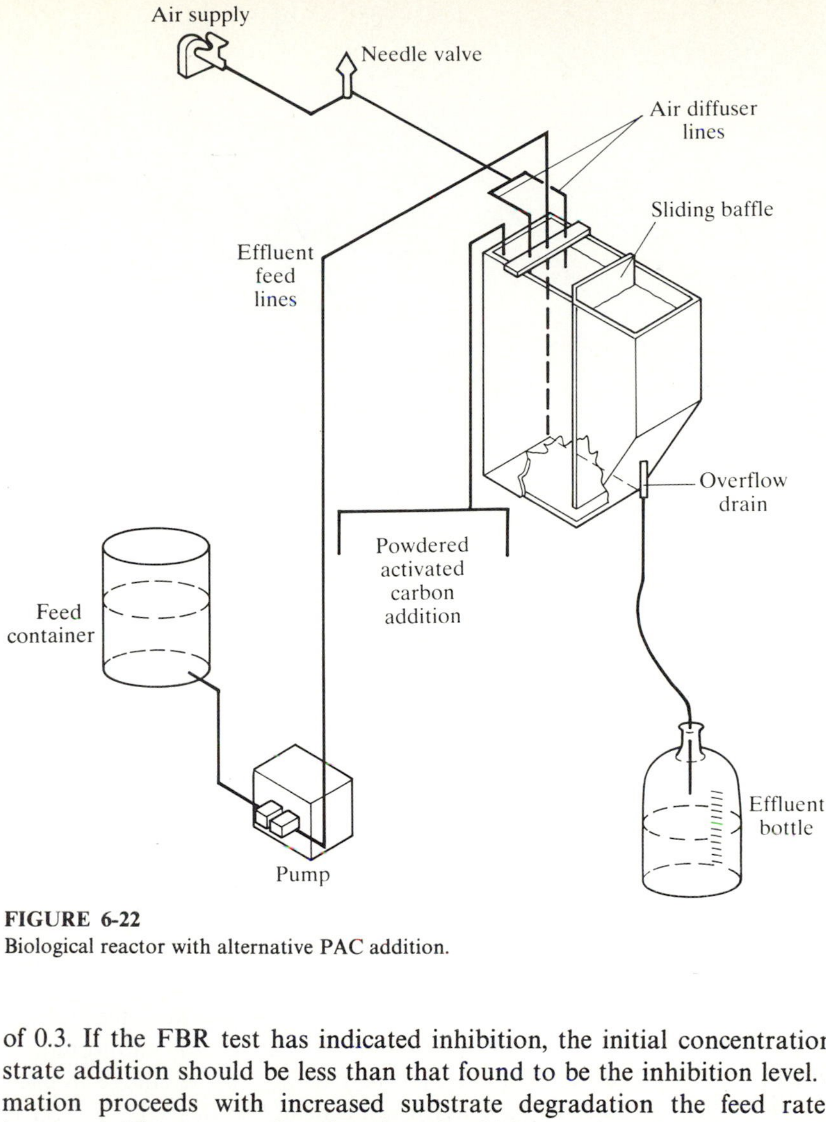

FIGURE 6-22
Biological reactor with alternative PAC addition.

of 0.3. If the FBR test has indicated inhibition, the initial concentration of substrate addition should be less than that found to be the inhibition level. As acclimation proceeds with increased substrate degradation the feed rate can be increased. The course of acclimation should be monitored as shown on page 148 until an equilibrium substrate level is achieved.

Continuous-treatment runs with acclimated sludge are made over a range of sludge ages. This range is selected based on the biodegradability of the wastewater and the effluent quality desired. For example, a readily degradable wastewater may operate over a range of θ_c of 3 to 12 days while a more refractory wastewater may cover a range of 10 to 30 days. The sludge age is maintained by wasting an appropriate reactor volume daily. For example, for a sludge age of 10 d, 10 percent of the reactor is wasted each day. A correction should be made for the suspended solids lost in the effluent. The analyses to be conducted and a

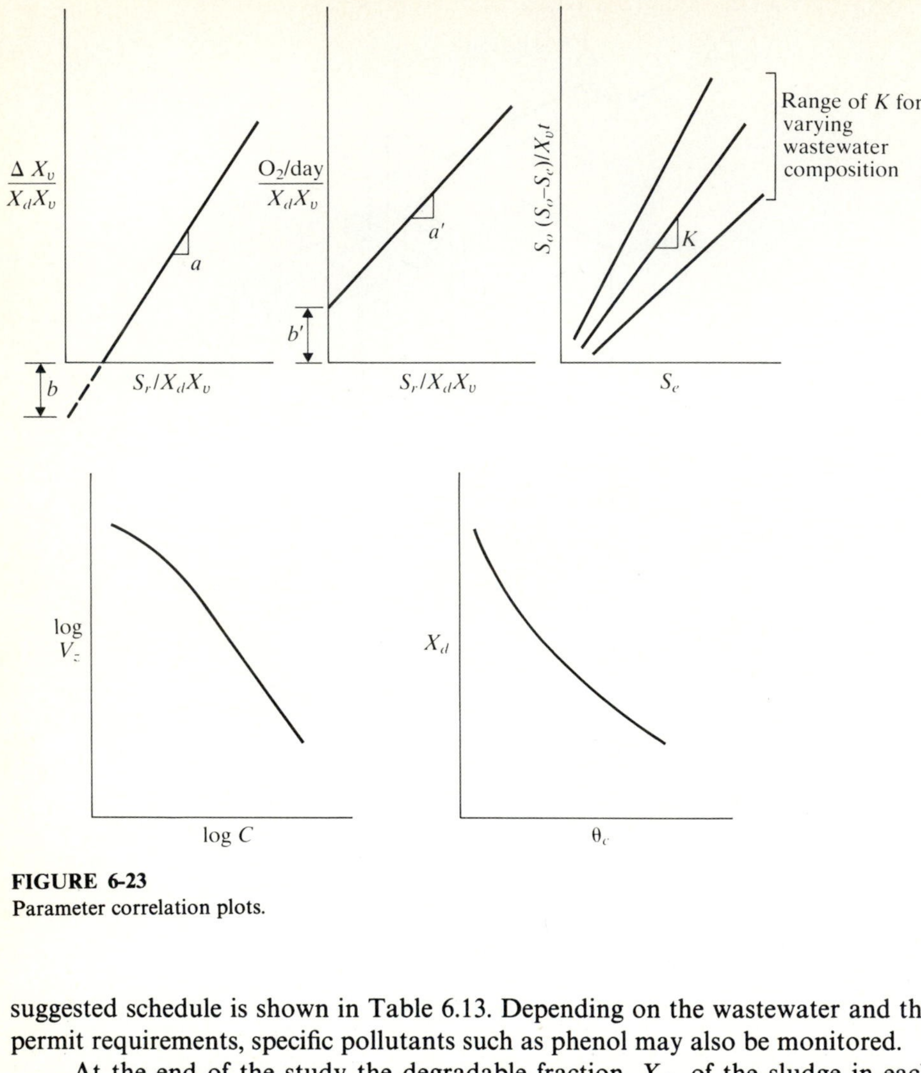

FIGURE 6-23
Parameter correlation plots.

suggested schedule is shown in Table 6.13. Depending on the wastewater and the permit requirements, specific pollutants such as phenol may also be monitored.

At the end of the study the degradable fraction, X_d, of the sludge in each reactor is determined by aerating the sludge until all of the degradable VSS is oxidized. This usually takes 15 to 25 days.

The data obtained can be correlated as shown in Fig. 6-23 and the process design parameters determined from these plots.

PROBLEMS

6.1. The following data were developed from a pilot plant study at 20°C:

(*a*) Determine the K for this wastewater.
(*b*) If $\theta = 1.06$ what is the K at 10°C?

F/M, d^{-1}	X_v, mg/l	S_o, mg/l	S_e, mg/l
0.3	2000	1640	26
0.48	1980	1640	44
0.72	2000	1640	70
0.96	2056	1640	100
1.18	2050	1640	167
2.68	2100	1640	333

6.2. A complete mix activated sludge plant treating an organic chemicals wastewater yielded the following performance data:

$$\text{TOD}_{\text{influent}} = 830 \text{ mg/l}$$
$$\text{TOD}_{\text{effluent}} = 50 \text{ mg/l}$$
$$\text{Oxygen uptake rate (OUR)} = 0.49 \text{ mg/(l} \cdot \text{min)}$$
$$\text{MLVSS} = 3200 \text{ mg/l}$$
$$Q = 3.2 \text{ million gal/d } (12{,}100 \text{ m}^3\text{/d})$$
$$t = 0.8 \text{ d}$$

A sample of mixed liquor was aerated in the laboratory to determine the endogenous oxygen uptake rate where

$$\text{OUR} = 0.3 \text{ mg/(l} \cdot \text{min)}$$
$$\text{MLVSS} = 4700 \text{ mg/l}$$

Calculate the endogenous coefficient b' and the coefficients a and a'.

6.3. Given the following data:

$$Q = 4 \text{ million gal/d}$$
$$S_o = 610 \text{ mg/l}$$
$$S_e = 40 \text{ mg/l}$$
$$\text{Eff SS} = 40 \text{ mg/l}$$
$$\text{mg BOD/mg SS} = 0.3$$
$$K = 3.0\text{/day at } 20°\text{C}$$
$$b = 0.1\text{/day at } 20°\text{C}$$
$$X_v = 3000 \text{ mg/l}$$
$$a = 0.55$$
$$a' = 0.50$$
$$\theta_b = 1.04$$
$$\theta_K = 1.065$$

(*a*) Compute the sludge age and F/M to meet this effluent quality.
(*b*) Compute the MLVSS at 10°C to meet the same effluent quality.
(*c*) Compute the oxygen requirements at 30°C.
(*d*) Compute the excess sludge at 10°C.
(*e*) Compute the N and P at 20°C.

6.4. Develop a relationship between the NH_3–N/BOD ratio and the aerobic volume percentage for a nitrification-denitrification channel. Consider a BOD of 200 mg/l and a range of NH_3–N of 10 to 70 mg/l. Compute the oxygen requirements for each case.

Use the following parameters:

$$R_N = \text{half the maximum rate}$$
$$K_{DN} = 0.06 \text{ mg NO}_3\text{–N/(mg VSS} \cdot \text{day) at } 20°\text{C}$$
$$X_v = 3000 \text{ mg/l}$$
$$a' = 0.55$$
$$K_{\text{BOD, aerobic}} = 8.0/\text{day at } 20°\text{C}$$
$$\text{BOD/NO}_3\text{–N} = 3.0$$
$$a = 0.5$$
$$b = 0.1/\text{day}$$
$$X_d = 0.45$$
$$S_e = 10 \text{ mg/l}$$

REFERENCES

1. Weber, W. J., and B. E. Jones: EPA NTIS PB86-182425/AS, 1983.
2. Kincannon, D. F., and E. L. Stover: EPA Report CR-806843-01-02, 1982.
3. Gaudy, A. F.: *J. Water Pollut. Control Fed.*, vol. 34, pt. 2, p. 124, February 1962.
4. Englebrecht, R. S., and R. E. McKinney: *Sewage Ind. Wastes*, vol. 29, pt. 12(1), p. 350, December 1957.
5. McWhorter, T. R., and H. Heukelekian: *Advances in Water Pollution Research*, vol. 2, Pergamon, New York, 1964.
6. Wuhrmann, K.: *Biological Treatment of Sewage and Industrial Wastes*, vol. 1, Reinhold, New York, 1956.
7. Tischler, L. F., and W. W. Eckenfelder: *Advances in Water Pollution Research*, vol. 2, Pergamon, Oxford, England, 1969.
8. Gaudy, A. F., K. Komolrit, and M. N. Bhatla: *J. Water Pollut. Control Fed.*, vol. 35, pt. 7, p. 903, July 1963.
9. Grau, P.: *Water Res.*, vol. 9, p. 637, 1975.
10. Adams, C. E., W. W. Eckenfelder, and J. Hovious: *Water Res.*, vol. 9, p. 37, 1975.
11. Quirk, T., and W. W. Eckenfelder: *J. Water Pollut. Control Fed.*, vol. 58, pt. 9, p. 932, 1986.
12. Chudoba, J.: *Water Res.*, vol. 192, p. 197, 1985.
13. McKinney, R. E.: *Advances in Biological Waste Treatment*, Pergamon, New York, 1963.
14. Sawyer, C. N.: *Biological Treatment of Sewage and Industrial Wastes*, vol. 1, Reinhold, New York, 1956.
15. Gellman, I., and H. Heukelekian: *Sewage Ind. Wastes*, vol. 25, pt. 10(1), p. 196, 1953.
16. Busch, A. W., and N. Myrick: *Proc. 15th Ind. Waste Conf.*, 1960, Purdue University.
17. Porges, N., and S. Hoover: *Biological Treatment of Sewage and Industrial Wastes*, vol. 1, Reinhold, New York, 1956.
18. Pipes, W.: *Proc. 18th Ind. Waste Conf.*, 1963, Purdue University.
19. Eckenfelder, W. W.: *Principles of Water Quality Management*, CBI, Boston, Mass., 1980.
20. Helmers, E. N., J. D. Frame, A. F. Greenberg, and C. N. Sawyer: *Sewage Ind. Wastes*, vol. 23, pt. 7, p. 834, 1951.
21. Palm, J. C., D. Jenkins, and P. S. Parker: *J. Water Pollut. Control Fed.*, vol. 52, pt. 2, p. 484, 1980.
22. Eickleboom, D. H.: *Prog. Water Technol.*, vol. 8, p. 153, 1977.
23. Jenkins, D., M. Richard, and G. Daigger: *Manual on the Causes and Control of Activated Sludge Bulking and Foaming Water*, Research Committee, Pretoria, S.A., 1984.
24. Chudoba, J., *et al.*: *Water Res.*, vol. 19, pt. 2, p. 191, 1985.
25. Chiesa, C. S., and R. L. Irvine: *Water Res.*, vol. 18, pt. 4, p. 47, 1985.
26. U.S. EPA: *Process Design for Nitrogen Control*, Technology Transfer, U.S. EPA, October 1975.
27. Wong-Chong, G. M., and R. C. Loehr: *Water Res.*, vol. 9, p. 1099, 1975.
28. Hockenbury, M. R., and C. P. L. Grady: *J. Water Pollut. Control Fed.*, vol. 49, p. 768, 1977.
29. Anthoisen, A. C.: *J. Water Pollut. Control Fed.*, vol. 48, p. 835, 1976.

CHAPTER 7

BIOLOGICAL WASTEWATER-TREATMENT PROCESSES

7.1 LAGOONS AND STABILIZATION BASINS

Stabilization basins are a common method of organic wastewater treatment where sufficient land area is available.

Stabilization basins can be divided into two classifications: the impounding and absorption lagoon, and the flow-through lagoon. In the impounding-absorption lagoon either there is no overflow or there is intermittent discharge during periods of high stream flow. The volumetric capacity of the basin is equal to the total waste flow less losses by evaporation and percolation. If there is intermittent discharge, the required capacity is related to the stream flow characteristics. In view of the large area requirements, impounding lagoons are usually limited to industries discharging low daily volumes of waste or to seasonal operations such as the canning industry.

The flow-through lagoon can be considered in three classifications, based on the nature of biological activity.

Type I. Facultative ponds

The facultative pond is divided by loading and thermal stratification into an aerobic surface and an anaerobic bottom. The aerobic surface layer will have a

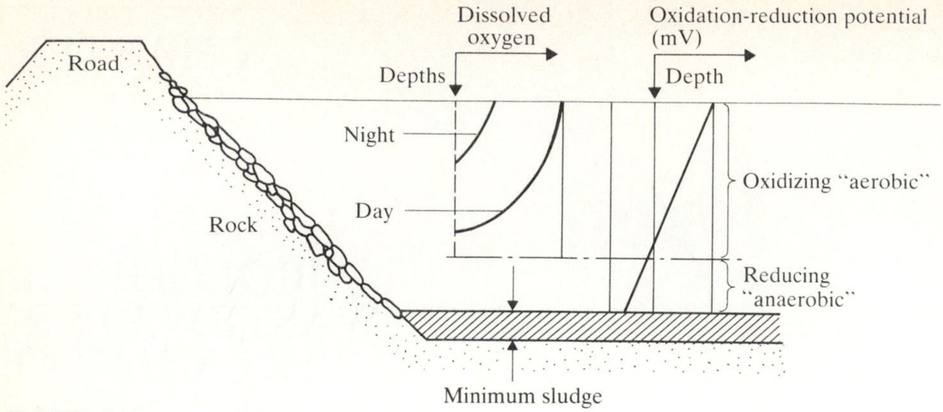

FIGURE 7-1
Waste stabilization pond—facultative type.[1]

diurnal variation, increasing in oxygen content during the daylight hours due to algal photosynthesis and decreasing during the night, as shown in Fig. 7-1. Sludge deposited on the bottom will undergo anaerobic decomposition, producing methane and other gases. Odors will be produced if an aerobic layer is not maintained. Depths will vary from 3 to 6 ft (0.9 to 1.8 m).

Since oxygen generation by photosynthesis depends on light penetration, highly colored wastewaters such as textile and pulp and paper cannot be treated by this technology.

Type II. Anaerobic ponds

Anaerobic ponds are loaded to such an extent that anaerobic conditions exist throughout the liquid volume. The biological process is the same as that occurring in anaerobic digestion tanks, being primarily organic acid formation followed by methane fermentation. The depth of anaerobic ponds is selected to give a minimum surface area/volume ratio and thereby provide maximum heat retention.

Type III. Aerated lagoons

These range from a few days to 2 weeks detention depending on the BOD removal efficiency desired. Oxygen is supplied by diffused or mechanical aeration systems, which also cause sufficient mixing to induce a significant amount of surface aeration. Depths from 6 to 15 ft (1.8 to 4.6 m) are common.

Lagoon Applications

In some industrial waste applications, aerobic ponds have been used after anaerobic ponds to provide high-degree treatment. Stabilization basins are also used to polish effluents from biological-treatment systems such as trickling filters and activated sludge.

In aerobic ponds, the amount of oxygen produced by photosynthesis can be estimated from

$$O_2 = CfS \tag{7.1}$$

where O_2 = oxygen production, lb/(acre · d) or kg/(m^2 · d)
C = 0.25 if O_2 is in lb (acre · d) or 2.8×10^{-5} if O_2 is in kg (m^2 · d)
f = light conversion efficiency, %
S = light intensity, cal/(cm^2 · d)

If the light conversion efficiency is estimated as 4 percent, then $O_2 = S$. S is a function of latitude and the month of the year and may be expected to vary from 100 to 300 calories/(cm^2 · d) during winter and summer for latitude 30°. This, in turn, would imply maximum loadings of 100 to 300 lb BOD_u/(acre · d) [0.011 to 0.034 kg/s (m^2 · d)] in order to maintain any aerobic activity in the pond.

The depth of oxygen penetration in a facultative pond has been estimated as a function of surface loadings, as shown in Fig. 7-2. It should be noted that the data for Fig. 7-2 was developed from oxidation ponds treating domestic wastewater in California. Appropriate adjustments to this curve would have to be made for other types of wastewater being treated in other climatic conditions.

The typical green algae in waste stabilization basins are *Chlamydomonas*, *Chlorella*, and *Euglena*. Common blue-green algae are *Oscillatoria*, *Phormidium*, *Anacystic*, and *Anabaena*. Algae types in a pond will vary seasonally.

In the treatment of highly colored or turbid wastewaters such as kraft pulp and paper mills, light penetration will be minimal and oxygen input will generate primarily from surface reaeration. Gellman and Berger[3] estimated the oxygen input from reaeration at 45 lb O_2/(acre · d) [0.005 kg/(m^2 · d)]. The performance of stabilization basins in the pulp and paper industry from their data is shown in Fig. 7-3.

Several concepts have been employed for the design of facultative and anaerobic ponds.[4,5] Functional equations for anaerobic and facultative ponds have been developed as follows:

For a single pond:

$$\frac{S}{S_o} = \frac{1}{1 + kt} \tag{7.2}$$

For multiple ponds:

$$\frac{S}{S_o} = \frac{1}{(1 + k_1 t_1)(1 + k_2 t_2) \cdots (1 + k_n t_n)} \tag{7.3}$$

For an infinite number of ponds:

$$\frac{S}{S_o} = e^{-k_n t_n} \tag{7.4}$$

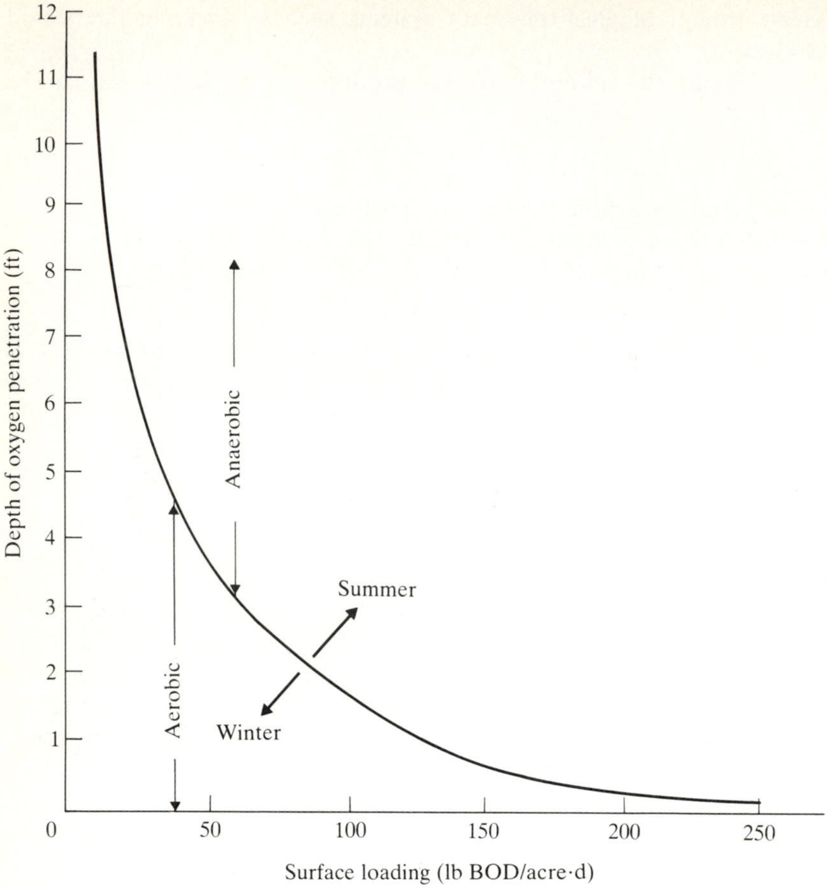

Note:
lb/(acre·d) = 1.121×10^{-4} kg (m^2·d)
ft = 0.3048 m

FIGURE 7-2
Depth of oxygen penetration in facultative ponds.[2]

While variable influent concentrations are considered Eq. (7.2) can be modified to

$$\frac{S}{S_o} = \frac{1}{1 + kt/S_o} \tag{7.5}$$

The equation is functionally similar to that employed for aerated lagoons and activated sludge, except that the rate coefficient k includes the effect of biomass concentration. This is because it is generally impractical or impossible to effectively measure the biomass concentration (VSS) in waste stabilization basins.

When multiple ponds are to be considered [Eq. (7.3)], k is assumed to be the same for all ponds. For complex wastewaters, this is probably not true since

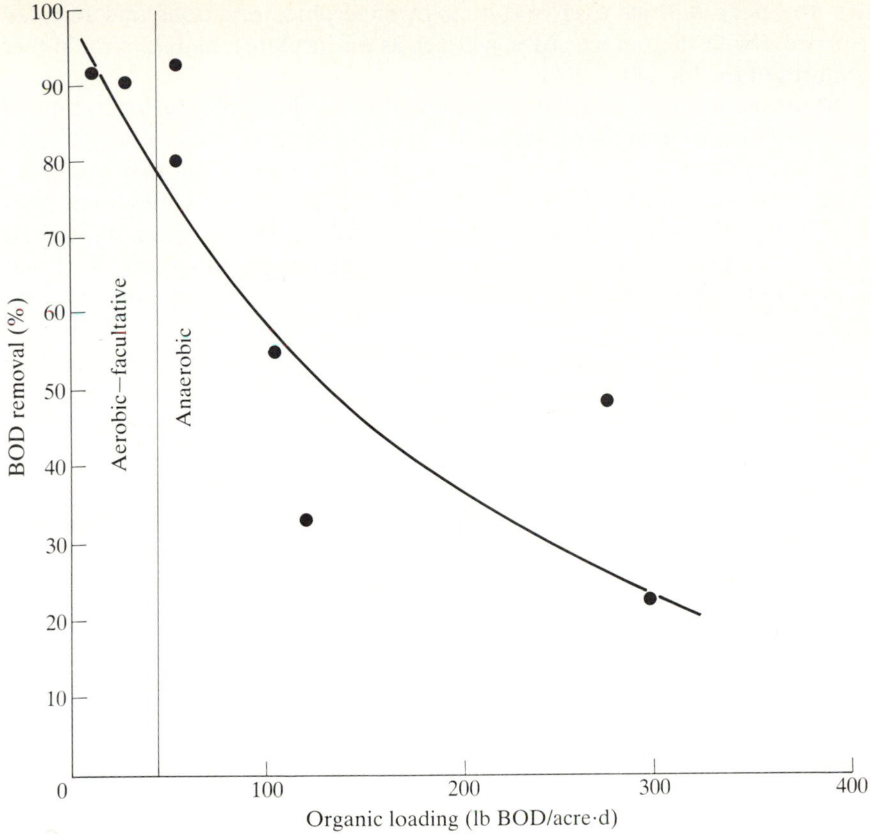

Note: lb/(acre·d) = 1.121×10^{-4} kg (m²·d)

FIGURE 7-3
Waste stabilization pond performance in the pulp and paper industry.

the more readily degradable compounds will be removed in the initial ponds. For these cases, an experimental study would need to be conducted for the wastewater in question to define the change in k.

Data from an organic chemicals plant showed k to be 0.05/d under anaerobic conditions at 20°C and 0.5/d under aerobic conditions.

In a stabilization basin, the average k_m can be computed:

$$k_m = \frac{k_{\text{aerobic}} \times D_{\text{aerobic}} + k_{\text{anaerobic}} \times D_{\text{anaerobic}}}{D_{\text{total}}} \tag{7.5a}$$

As in all biological processes, the biological activity in the pond will be a function of temperature, and the rate coefficient k can be corrected through the application of Eq. (6.16). Evaluation of the 30-d average performance for a kraft pulp and paper mill showed a θ value of 1.053. During winter operation in colder climates, the pond will ice over, resulting in anaerobic conditions and reduced performance. (Note that an ice cover will act as an insulator, maintaining higher temperatures in the liquid.)

Certain design considerations are important to the successful operation of stabilization basins. These have been discussed by Hermann and Gloyna[4] and Marais.[5] Embankments should be constructed of impervious material with maximum-minimum slopes of 3 to 4:1 and 6:1, respectively. A minimum free-board of 3 ft (0.91 m) should be maintained in the basin. Provision should be made for bank protection from erosion. Wind action is important for pond mixing and is effective with a fetch of 650 ft (198 m) in a pond with a depth of 3 ft (0.91 m).

Meat waste has been treated in shallow aerobic ponds 18 in (0.66 m) deep at loadings of 214 lb BOD/(acre · d) [0.024 kg/(m^2 · d)].[6] The waste was presettled and had a concentration of 175 mg/l BOD. BOD reductions of 96 percent in the summer and 70 percent in the winter were obtained.

Use of an aerobic pond treating meat-packing waste after the anaerobic contact process yielded 80 percent BOD removal at a loading of 410 lb BOD/(acre · day) [0.046 kg/(m^2 · d)]. The concentration of BOD entering the pond was 129 mg/l. Removals at various loadings have been summarized by Steffen.[6]

Loadings to anaerobic ponds 8 to 17 ft (2.4 to 5.2 m) deep have been reported to vary from 0.011 to 0.015 lb BOD/(ft^3 · d) [0.176 to 0.240 kg/(m^3 · d)], with high BOD removals on packing-house waste. Treatment of corn waste with an initial BOD of 2936 mg/l in a lagoon with a 9.6-d detention yielded 59 percent BOD reduction. The organic loading was 0.184 lb BOD/(ft^3 · d) [2.95 kg/(m^3 · d)].

In order to improve the efficiency of lagoon operation, nitrogen and phosphorus must be added to nutrient-deficient wastes. The requirements for aerobic and anaerobic processes are discussed on pages 162 and 250, respectively.

Lagoons treating wastes in which aerobic conditions are not maintained frequently emit odors and provide a breeding ground for insects. The odor problems can frequently be eliminated by the addition of sodium nitrate at a dosage equal to 20 percent of the applied oxygen demand. Surface sprays can be used to reduce the fly and insect nuisance and in some cases the odor.

High efficiencies can frequently be attained by the use of anaerobic ponds followed by aerobic ponds.

An anaerobic pond with a 6-d detention and 14-ft (4.3 m) depth loaded at 0.014 lb BOD/(ft^3 · d) [0.224 kg/(m^3 · d)], followed by a 3-ft (0.9 m) deep aerobic pond with a 19-d detention loaded at 50 lb BOD/(acre · day) [0.0056 kg/(m^2 · d)], yielded an overall reduction in BOD of 1100 to 67 mg/l.[7]

Performance data for aerobic, facultative, and anaerobic ponds are summarized for various industrial wastes in Table 7.1.

7.2 AERATED LAGOONS

An aerated lagoon is a basin of significant depth 8 to 16 ft (2.4 to 4.9 m) in which oxygenation is accomplished by mechanical or diffused aeration units and through induced surface aeration.

TABLE 7.1

Performance of lagoon systems

Summary of average data from aerobic and facultative ponds

Industry	Area, acres	Depth, ft	Detention, d	Loading, lb/(acre·d)	BOD removal, %
Meat and poultry	1.3	3.0	7.0	72	80
Canning	6.9	5.8	37.5	139	98
Chemical	31	5.0	10	157	87
Paper	84	5.0	30	105	80
Petroleum	15.5	5.0	25	28	76
Wine	7	1.5	24	221	
Dairy	7.5	5.0	98	22	95
Textile	3.1	4.0	14	165	45
Sugar	20	1.5	2	86	67
Rendering	2.2	4.2	4.8	36	76
Hog feeding	0.6	3.0	8	356	
Laundry	0.2	3.0	94	52	
Miscellaneous	15	4.0	88	56	95
Potato	25.3	5.0	105	111	

Summary of average data from anaerobic ponds

Industry	Area, acres	Depth, ft	Detention, d	Loading, lb/(acre·d)	BOD removal, %
Canning	2.5	6.0	15	392	51
Meat and poultry	1.0	7.3	16	1260	80
Chemical	0.14	3.5	65	54	89
Paper	71	6.0	18.4	347	50
Textile	2.2	5.8	3.5	1433	44
Sugar	35	7.0	50	240	61
Wine	3.7	4.0	8.8		
Rendering	1.0	6.0	245	160	37
Leather	2.6	4.2	6.2	3000	68
Potato	10	4.0	3.9		

Summary of average data from combined aerobic-anaerobic ponds

Industry	Area, acres	Depth, ft	Detention, d	Loading, lb/(acre·d)	BOD removal, %
Canning	5.5	5.0	22	617	91
Meat and poultry	0.8	4.0	43	267	94
Paper	2520	5.5	136	28	94
Leather	4.6	4.0	152	50	92
Miscellaneous industrial wastes	140	4.1	66	128	

Note:

ft = 0.3048 m

lb/(acre · d) = 1.121×10^{-4} kg/(m^2 · d)

acre = 4.0469×10^3 m^2

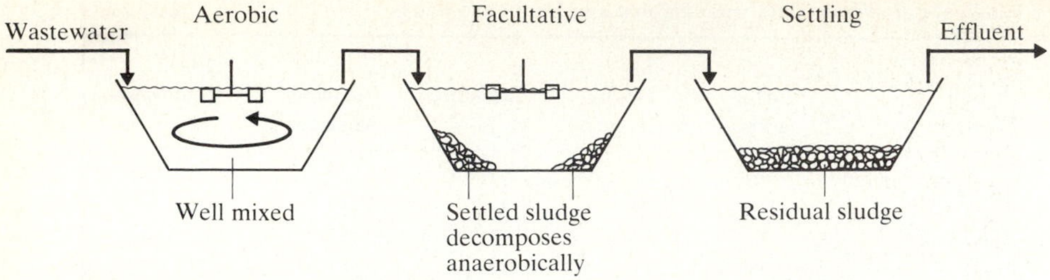

FIGURE 7-4
Aerated lagoon types.

There are two types of aerated lagoons:

1. The aerobic lagoon in which dissolved oxygen and suspended solids are maintained uniformly throughout the basin.
2. The aerobic-anaerobic or facultative lagoon in which oxygen is maintained in the upper liquid layers of the basin, but only a portion of the suspended solids is maintained in suspension. These basin types are shown in Fig. 7-4. A typical aerated lagoon is shown in Fig. 7-5.

In the aerobic lagoon, all solids are maintained in suspension and this system may be thought of as a "flow-through" activated sludge system without solids recycle. Thus, the effluent suspended solids concentration will be equal to the aeration basin solids concentration.

In the facultative lagoon, a portion of the suspended solids settle to the bottom of the basin where they undergo anaerobic decomposition. The anaerobic byproducts are subsequently oxidized in the upper aerobic layers of the basin. The facultative lagoon can also be modified to yield a more highly clarified effluent by the inclusion of a separate postsettling pond or a baffled settling compartment.

Aerobic and facultative lagoons are primarily differentiated by the power level employed in the basin. In the aerobic lagoon, the power level is sufficiently high to maintain all solids in suspension and may vary from 14 to 20 hp/million gal (2.8 to 3.9 W/m^3) of basin volume, depending on the nature of the suspended solids in the influent wastewater. Field data demonstrated that 14 hp/million gal (2.8 W/m^3) was generally sufficient to maintain solids in suspension in pulp and paper-mill effluents, while 20 hp/million gal (3.9 W/m^3) was required for domestic wastewater treatment.

In the facultative lagoon, the power level employed is only sufficient to maintain dispersion and mixing of the dissolved oxygen. Experience in the pulp and paper industry showed that a minimum power level employing low-speed mechanical surface aerators is 4 hp/million gal (0.79 W/m^3). The use of other kinds of aeration equipment might require different power levels to maintain uniform dissolved oxygen in the basin.

FIGURE 7-5
Aerated lagoon.

Aerobic Lagoons

At a constant basin detention time, the equilibrium biological solids concentration and the overall rate of organic removal can be expected to increase as the influent organic concentration increases. For a soluble industrial wastewater, the equilibrium biological solids concentration, X_v, can be predicted from the relationship:

$$X_v = \frac{aS_r}{1 + bt} \tag{7.6}$$

When nondegradable volatile suspended solids are present in the wastewater, Eq. (7.6) becomes

$$X_v = \frac{aS_r}{1 + bt} + X_i \tag{7.7}$$

in which X_i = influent volatile suspended solids not degraded in the lagoon. Combining Eq. (7.6) with the kinetic relationship [Eq. (6.5)], the effluent soluble organic concentration can be computed:

$$\frac{S}{S_o} = \frac{1 + bt}{aKt} \tag{7.8}$$

From Eq. (7.8) it can be concluded that the fraction of effluent soluble organic concentration remaining is independent of the influent organic concentration. For lagoons with fixed detention times this conclusion is justified, because higher influent organic concentrations result in higher equilibrium biological solids levels and, therefore, higher overall BOD removal rates.

The specific organic reaction rate coefficient K is temperature-dependent, and can be corrected for temperature employing Eq. (6.17). The oxygen requirements for an aerobic lagoon are computed using the same relationship as employed for activated sludge [Eq. (6.17)]. Aerobic lagoons are employed for pretreatment of high-strength industrial wastewaters prior to discharge to a joint or municipal treatment system, or as the first basin in a two-basin aerated lagoon series, followed by a facultative basin. It should be noted that while in an aerobic lagoon the soluble organic content is reduced, there is an increase in the effluent suspended solids through synthesis. The relationship obtained for an aerobic lagoon pretreating a brewery wastewater is shown in Fig. 7-6.

Power requirements should generally be designed for summer operation, since the rates of organic removal will be the greatest during this period of operation. The design of an aerobic lagoon is shown in Example 7.1.

Example 7.1. A brewery wastewater is to be pretreated in an aerobic lagoon with a retention period of one day. Estimate the effluent BOD and suspended solids.

$BOD_5 = 2500$ mg/l
$K = 40$/d
$a = 0.5$
Inf SS = 500 mg/l (nondegradable)
$b = 0.1$/d

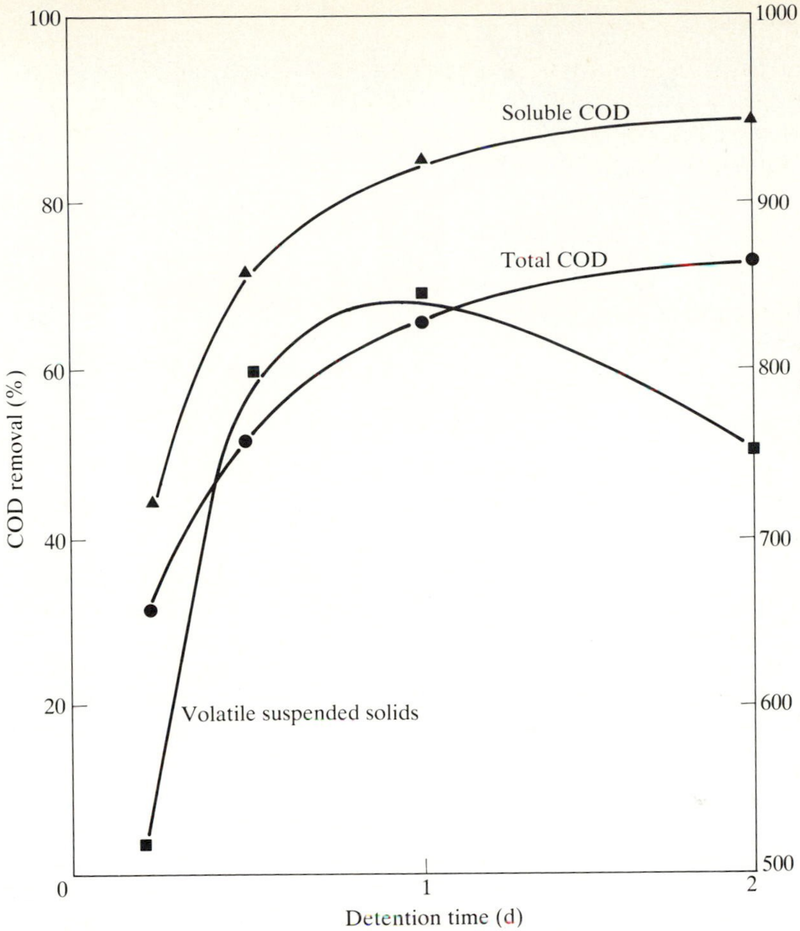

FIGURE 7-6
COD removal from brewery wastewater through an aerobic lagoon.

Solution

$$\frac{S_e}{S_o} = \frac{1 + bt}{aKt}$$

$$S_e = \frac{1 + 0.1 \times 1}{0.5 \times 40 \times 1} \times 2500 \text{ mg/l}$$

$$= 138 \text{ mg/l as BOD}_5$$

$$X_v = \frac{aS_r}{1 + bt} = \frac{(0.5)(2500 - 138)}{1 + 0.1 \times 1} = 1074 \text{ mg/l as biological VSS}$$

$$\text{Effluent BOD} = 138 + 0.3 \frac{\text{mg BOD}}{\text{mg VSS}} (1074 \text{ mg/l VSS}) = 460 \text{ mg/l BOD}$$

$$\text{Effluent SS} = \frac{1074}{0.8} + 500 = 1843 \text{ mg/l (assume 80 percent volatile)}$$

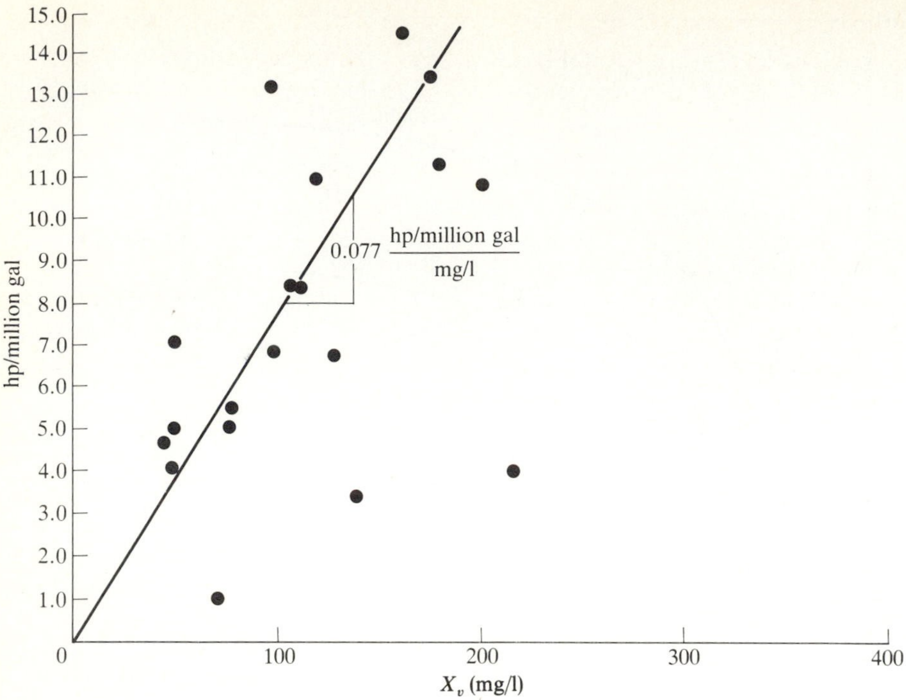

Note: hp/million gal = 0.1970 W/m^3

FIGURE 7-7
Correlation for mixing power input versus MLSS concentrations.

Facultative Lagoons

In a facultative lagoon, the biological solids level maintained in suspension is a function of the power level employed in the basin. Results from the pulp and paper industry are shown in Fig. 7-7. Solids deposited in the bottom of the facultative lagoon will undergo anaerobic degradation, which results in a feedback of soluble organics to the upper aerobic layers. Under these conditions Eq. (6.5) should be modified to

$$\frac{S}{FS_o} = \frac{FS_o}{FS_o + KX_v t} \tag{7.9}$$

in which F is a coefficient that accounts for organic feedback due to anaerobic activity in the deposited sludge layers. The degree of anaerobic activity is highly temperature-dependent and the coefficient F may be expected to vary from 1.0 to 1.4 under winter and summer conditions, respectively, depending on the geographical location of the plant.

In a facultative lagoon biological solids are maintained at a lower level than in an aerobic lagoon, and soluble organics are fed back to the liquid as anaerobic degradation products. It is, therefore, not possible to directly compute the oxygen requirements employing Eq. (6.14). In this case, the oxygen requirements can be related empirically to organic removal and estimated from

$$R_r = F'S_r \tag{7.10}$$

in which F' is an overall oxygen-utilization coefficient for facultative lagoons. Results obtained for various industrial wastewaters would indicate that the coefficient F' is a function of the degree of organic feedback, which in turn is a function of temperature. In general, depending on the geographic location of the plant, F' can be estimated to vary from 0.8 to 1.1 during winter operation when anaerobic activity in the basin is low, and from 1.1 to 1.5 during summer operation when anaerobic activity in the bottom of the basin is at a maximum. The value selected will depend on the geographical location of the plant.

Nutrient requirements in aerated lagoons are computed in a similar way to the activated sludge process. In the case of a facultative lagoon, however, anaerobic decomposition of sludge deposited in the bottom of the basin will feed back nitrogen and phosphorus. This is usually sufficient for organic removal occurring in this basin, and no additional nitrogen and phosphorus need to be added. The relationships for the design of aerated lagoon systems are summarized in Table 7.2.

Temperature Effects in Aerated Lagoons

The performance of aerated lagoons is significantly influenced by changes in basin temperature. In turn, the basin temperatures are influenced by the tem-

TABLE 7.2
Aerated lagoon design relationships

	Aerobic	Facultative	Settling
Relationship	hp/million gal 14 to 20	hp/million gal >4 <10	
Kinetics	$\frac{S_1}{S_o} = \frac{1 + bt}{aKt}$	$S_2 = \frac{F^2 S_1^2 †}{K X_v t § + F S_1}$	$t < 2$ d‡
Oxygen requirements	$O_2/d = a'S_r + b'X_v t$	$O_2/d = \begin{cases} 0.8 - 1.1\ S_r \text{ (winter)}¶ \\ 1.1 - 1.4\ S_r \text{ (summer)} \end{cases}$	
Nutrient requirements	$N = 0.11 \Delta X_v$ $P = 0.02 \Delta X_v$	None required††	
Sludge yield	$X_v = \frac{aS_r}{1 + bt}$	$0.1\ aS_r$ + inerts‡‡ 3 to 7 percent weight concentration	

† BOD feedback from benthal decomposition varies from 1.0 in winter to 1.4 in summer.

‡ To minimize algae growth.

§ Depends on power level in basin (see Fig. 7-7).

¶ Factors employed depend on geographical location.

†† Benthal decomposition feeds back N and P.

‡‡ Most of biovolatile solids will break down through benthal decomposition.

Note: hp/million gal = 1.98×10^{-2} kW/m^3

perature of the influent wastewater and the ambient air temperature. Although heat is lost through evaporation, convection, and radiation, it is gained by solar radiation. While several formulas have been developed to estimate the temperature in aerated lagoons, the following equation will usually give a reasonable estimate for engineering design purposes:

$$\frac{t}{D} = \frac{T_i - T_w}{f(T_w - T_a)} \tag{7.11}$$

where t = basin detention time, day
D = basin depth, ft
T_i = influent wastewater temperature, °F
T_a = mean air temperature, °F (usually taken as the mean weekly temperature)
T_w = basin temperature, °F

Equation (7.11) was developed for surface aeration and hence cannot be used for subsurface aeration.

The coefficient f is a proportionality factor containing the heat-transfer coefficients, the surface area increase from aeration equipment, and wind and humidity effects. f has an approximate value of 90 for most aerated lagoons employing surface aeration equipment.

A general temperature model has been developed by Argamon and Adams,[8] which considers an overall heat balance in the basin including heat gained from solar radiation, mechanical energy input, biochemical reaction, and heat lost by long-wave radiation, evaporation from the basin surface, conduction from the basin surface, evaporation and conduction from the aerator spray, and conduction through the basin walls. Their final equation is

$$\begin{aligned} T_w = T_a + \Bigg[& \frac{Q}{A}(T_i - T_a) + 10^{-6}(1 - 0.0071C_c^2)H_{s,o} + 6.95(\beta - 1) \\ & + 0.102(\beta - 1)T_a - e^{0.0604T_a}\left(1 - \frac{f_a}{100}\right)1.145A^{-0.05}V_w \\ & + \frac{126NFV_w}{A} + \frac{10^{-6}H_m}{A} + \frac{1.8S_r}{A}\Bigg] \\ \Bigg/ \Bigg[& \frac{Q}{A} + 0.102 + (0.068e^{0.0604T_a} + 0.118)A^{-0.05}V_w \\ & + \frac{4.32NFV_w}{A}(3.0 + 1.75e^{0.0604T_a}) + \frac{10^{-6}UA_w}{A}\Bigg] \end{aligned} \tag{7.12}$$

where T_w = basin water temperature, °C
T_a = air temperature, °C
T_i = influent waste temperature, °C
Q = flow rate, m^3/d

A = surface area, m^2
C_c = average cloud cover, tenths
$H_{s,o}$ = average daily absorbed solar radiation under clear sky conditions, $cal/(m^2 \cdot d)$
U = heat-transfer coefficient, $cal/(m^2 \cdot d \cdot °C)$
β = atmospheric radiation factor
f_a = relative humidity, percent
N = number of aerators
F = aerator spray vertical cross-section area, m^2
V_w = wind speed at tree top, m/s
$H_m = 15.2 \times 10^6 p$, where p = aeration power, hp
S_r = organic removal rate, kg COD removed/d
A_w = effective wall area, m^2

Equation (7.12) can be used to predict the temperature of diffused air systems by substituting

$$NFV_w = 2Q_A$$

where Q_A = air flow, m^3/s

Aerated Lagoon Systems

Multiple basins may be most effectively employed in aerated lagoon systems under proper conditions. There may be little advantage to series basin operation at a given temperature in terms of effluent organic concentration, but there are two considerations which favor series operation. First, series operation may be desired where land availability is a concern. When considering a thermal balance, a minimum total basin volume can be obtained by employing two basins in series. The first basin volume is minimized to maintain a high temperature, a high biological solids level, and a resulting high BOD reaction rate in an aerobic lagoon. The second basin is a facultative basin at lower power (mixing) levels that permit solids settling and decomposition in the bottom of the basin. An optimization procedure[9] can be employed to determine the smallest total basin volume and the lowest aeration horsepower for a specified effluent quality. Where low effluent suspended solids are desired, a final settling basin can be employed. The settling basin should have:

1. A sufficiently long detention period to effect the desired suspended solids removal.
2. Adequate volume for sludge storage.
3. Minimal algal growth.
4. Minimal odors from anaerobic activity.

Unfortunately, these design objectives are not always compatible. Frequently short retention times are required to inhibit algal growth, which are too short for proper settling. Also, adequate volume must remain above the sludge deposits at all times to prevent the escape of the odorous gases of decomposition.

In order to achieve these objectives, a minimum detention time of one day is usually required to settle the majority of the settleable suspended solids. Where algal growths pose potential problems, a maximum detention time of 3 to 4 d is recommended. For odor control, a minimum water level of 3 ft (0.9 m) should be maintained above the sludge deposits at all times. The design of an aerated lagoon system is shown in Example 7.2.

Example 7.2. Design a two-stage aerated lagoon system of 12 ft (3.66 m) depth to treat an 8.5 million gal/d (32,170 m^3/d) industrial wastewater with the following characteristics:

Influent: $BOD_5 = 425$ mg/l
Temperature = 85°F (29°C)
SS = 0 mg/l

Ambient temperature: summer = 70°F (21°C)
winter = 34°F (1°C)

Kinetic variables: $K = 6.3$/d at 20°C
$a = 0.5$
$b = 0.2$/d at 20°C
$a' = 0.52$
$b' = 0.28$/d at 20°C
$\theta = 1.035$ for BOD reaction rate
$\theta = 1.024$ for oxygen-transfer efficiency
$F = 1.0$ (winter)
$= 1.4$ (summer)
$F' = 1.5$ (summer)
$N_o = 3.2$ lb O_2/(hp · h) (1.95 kg/kW · h)
$\alpha = 0.85$
$\beta = 0.90$
$C_L = 1.0$ mg/l

The final effluent should have a maximum soluble BOD_5 of 20 mg/l for the summer and 30 mg/l for the winter.

Solution. In general, the required detention time of the lagoon system for reaching a terminal BOD_5 is controlled by winter temperatures while the oxygen requirements and power level will usually be controlled by summer conditions.

(*a*) *Design of basin based on minimum detention time*

1. Compute basin volume by assuming a detention time.
For $t = 2$ day:

$$V = Qt$$

$$= 8.5 \times 2$$

$$= 17 \text{ million gal } (64{,}350 \text{ m}^3)$$

2. Calculate winter water temperature in basin:

$$\frac{t}{D} = \frac{T_i - T_w}{f(T_w - T_a)}$$

or

$$T_w = \frac{DT_i + ftT_a}{D + ft}$$

$$= \frac{12 \times 85 + 1.6 \times 2 \times 34}{12 + 1.6 \times 2}$$

$$= 74.3°\text{F or } 23.5°\text{C}$$

3. Correct the BOD reaction rate for winter conditions:

$$K_{(T_2)} = K_{(T_1)}\theta^{T_2 - T_1}$$

$$K_{23.5} = 6.3 \times 1.035^{23.5-20}$$

$$= 7.11/\text{d}$$

4. Calculate the winter effluent soluble BOD_5:

$$\frac{S_e}{S_o} = \frac{1 + bt}{aKt}$$

$$S_e = \frac{1 + 0.2 \times 2}{0.5 \times 7.11 \times 2} 425$$

$$= 83.7 \text{ mg/l}$$

5. Repeat steps 2 to 4 for summer conditions:

$$T_w = \frac{12 \times 85 + 1.6 \times 2 \times 70}{12 + 1.6 \times 2}$$

$$= 81.8°\text{F or } 27.7°\text{C}$$

$$K_{27.7} = 6.3 \times 1.035^{27.7-20}$$

$$= 8.21/\text{d}$$

$$S_e = \frac{1 + 0.2 \times 2}{0.5 \times 8.21 \times 2} 425$$

$$= 72.5 \text{ mg/l}$$

6. Calculate average volatile suspended solids concentration under summer conditions:

$$X_v = \frac{aS_r}{1 + bt}$$

$$= \frac{0.5(425 - 72.5)}{1 + 0.2 \times 2}$$

$$= 126 \text{ mg/l}$$

7. Calculate oxygen requirement:

$$R_r = a'S_r + b'X_v$$

$$= 0.52 \times (425 - 72.5) \times 8.5 \times 8.34 + 0.28 \times 126 \times 8.5 \times 2 \times 8.34$$

$$= 17{,}996 \text{ lb/d (8170 kg/d)}$$

8. Compute horsepower requirement:

$$N = N_o \frac{\beta C_s - C_L}{C_{s(20)}} a\theta^{T-20}$$

$$= 3.2 \frac{0.90 \times 7.96 - 1.0}{9.2} 0.85 \times 1.024^{27.7-20}$$

$$= 2.19 \text{ lb/(hp} \cdot \text{h) (1.33 kg/kW} \cdot \text{h)}$$

$$\text{hp} = \frac{R_r}{N}$$

$$= \frac{17{,}996}{2.19 \times 24} = 342 \text{ hp (257 kW)}$$

9. Check power level:

$$\text{PL} = \frac{\text{hp}}{V}$$

$$= \frac{342}{17} = 20.1 \text{ hp/million gal (0.40 kW/m}^3\text{)}$$

For a conservative design, the minimum power level should be 14 hp/million gal (0.28 kW/m^3). If the power level is significantly less than 14 hp/million gal, then 14 hp/million gal should be used.

10. Repeat steps 1 through 9 using different detention times. The results can be tabulated as follows:

t, d	Winter or summer	T_w °F	T_w °C	K, d^{-1}	S_e, mg/l	X_v, mg/l	R_r, lb/d	hp required	PL, hp/million gal
1.0	w	79.0	26.1	7.77	131				
	s	83.2	28.4	8.41	121	127	13,727	261	30.7
2.0	w	74.3	23.5	7.11	83.7				
	s	81.8	27.7	8.21	72.5	126	17,996	342	20.1
3.0	w	70.4	21.3	6.59	68.8				
	s	80.7	27.1	8.04	56.4	115	20,436	389	15.3
4.0	w	67.3	19.6	6.21	61.6				
	s	79.8	26.6	7.91	48.4	105	22,219	476(423)†	14(12.4)†
5.0	w	64.6	18.1	5.90	57.6				
	s	79.0	26.1	7.77	43.8	95	23,480	595(449)	14(10.6)
6.0	w	62.3	16.8	5.64	55.3				
	s	78.3	25.7	7.66	40.7	87	24,528	714(469)	14(9.2)

† Minimum power level applied.

(b) Design of basin based on minimum detention time

It is assumed that the minimum power level in a facultative lagoon should be 4 hp/million gal (0.079 kW/m^3) to maintain 50 mg/l of volatile suspended solids in suspension. The design of this basin is based on selecting an effluent BOD concentration from the first basin at given detention times to be the influent to the second basin.

1. Calculate the temperature in this basin and adjust the BOD reaction rate by assuming a detention time. For example, the effluent BOD concentrations at a 2-day detention time for the first basin are 83.7 and 72.5 mg/l for winter and summer, respectively. Basin temperature:

$$T_w = \frac{DT_i + ftT_a}{D + ft}$$

Corrected influent concentration:

$$S_o' = FS_o$$

Corrected BOD reaction rate:

$$K_{20} = \frac{6.3}{425} S_o'$$

Assume $t = 5$ day.

Winter:

$$T_w = \frac{12 \times 74.3 + 1.6 \times 5 \times 34}{12 + 1.6 \times 5}$$

$$= 58.2°\text{F or } 14.6°\text{C}$$

$$K_{20} = \frac{6.3}{425} 1.0 \times 83.7$$

$$= 1.24/\text{d}$$

$$K_{14.6} = 1.24 \times 1.035^{14.6-20}$$

$$= 1.03/\text{d}$$

Summer:

$$T_w = \frac{12 \times 81.8 + 1.6 \times 5 \times 70}{12 + 1.6 \times 5}$$

$$= 77.1°\text{F or } 25.1°\text{C}$$

$$K_{20} = \frac{6.3}{425} 1.4 \times 72.5$$

$$= 1.50/\text{d}$$

$$K_{25.1} = 1.50 \times 1.035^{25.1-20}$$

$$= 1.79/\text{d}$$

2. Compute the detention time required to reduce the soluble BOD to the prescribed level:

$$S'_o = FS_o$$

$$t = \frac{S'_o(S'_o - S_e)}{KX_v S_e}$$

Winter:

$$S'_o = 83.7 \text{ mg/l}$$

$$t = \frac{83.7(83.7 - 30)}{1.03 \times 50 \times 30}$$

$$= 2.91 \text{ d}$$

Summer:

$$S'_o = 1.4 \times 72.5$$

$$= 102 \text{ mg/l}$$

$$t = \frac{102(102 - 20)}{1.79 \times 50 \times 20}$$

$$= 4.67 \text{ d}$$

Now summer conditions are controlling.

3. Repeat steps 1 and 2 until the computed detention time in step 2 is close enough to the assumed value in step 1. Final results are as follows:

$$t = 4.67 \text{ day}, V = 39.7 \text{ million gal } (150{,}300 \text{ m}^3)$$

Winter:

$$T_w = 58.8°\text{F or } 14.9°\text{C}$$

$$K_{14.9} = 1.04/\text{d}$$

$$S_e = 21.5 \text{ mg/l}$$

Summer:

$$T_w = 77.3°\text{F or } 25.2°\text{C}$$

$$K_{25.2} = 1.79/\text{d}$$

$$S_e = 20.0 \text{ mg/l}$$

4. Calculate oxygen requirement:

$$R_r = F'S_r$$

$$= 1.5 \times (1.4 \times 72.5 - 20.0) \times 8.5 \times 8.34$$

$$= 8666 \text{ lb/d } (3934 \text{ kg/d})$$

5. Calculate horsepower requirement:

$$N = N_o \frac{\beta C_s - C_L}{C_s(20)} a\theta^{T-20}$$

$$= 3.2 \frac{0.90 \times 8.36 - 1.0}{9.2} 0.85 \times 1.024^{25.2-20}$$

$$= 2.18 \text{ lb/(hp}\cdot\text{h) } (1.33 \text{ kg/kW}\cdot\text{h})$$

$$\text{hp} = \frac{R_r}{N}$$

$$= \frac{8666}{2.18 \times 24}$$

$$= 166 \text{ hp } (125 \text{ kW})$$

6. Check power level:

$$\text{PL} = \frac{\text{hp}}{V}$$

$$= \frac{166}{39.7}$$

$$= 4.2 \text{ hp/million gal } (0.083 \text{ kW/m}^3)$$

7. Repeat steps 1 through 6 using different detention times in the first basin with the adequate detention time in the second basin to meet the desired effluent quality. The results are tabulated below:

S_o, mg/l	T_w °F	T_w °C	K, d^{-1}	t, d	S_e, mg/l	X_v, mg/l	R_r, lb/d	hp required	PL, hp/million gal
131	55.0	12.8	1.51	8.57	22.1	50			
121	76.2	24.6	2.94	8.57	20.0	50	10,740	291(205)†	4.0(2.8)†
83.7	58.8	14.9	1.04	4.67	21.5	50			
72.5	77.3	25.2	1.79	4.67	20.0	50	8,666	166	4.2
68.8	59.2	15.1	0.862	3.33	22.3	50			
56.4	77.4	25.2	1.40	3.33	20.0	50	6,274	120	4.2
61.6	58.4	14.7	0.761	2.72	23.0	50			
48.4	77.2	25.1	1.19	2.72	20.0	50	5,083	97	4.2
57.6	57.3	14.1	0.697	2.34	23.8	50			
43.8	76.9	24.9	1.08	2.34	20.0	50	4,392	84	4.2
55.3	56.1	13.4	0.653	2.12	24.6	50			
40.7	76.5	24.7	0.993	2.12	20.0	50	3,934	75	4.2

† Minimum power level applied.

(c) *Optimum lagoon system*

This two-stage lagoon system can be optimized to minimize either the total detention time or the total horsepower to be installed.

Figure 7-8 shows the summarized design results of the two-stage lagoon system. The minimum total required detention time is 6.33 d, with a corresponding total installed horsepower of 509 hp (382 kW). The minimum total horsepower is 508 hp (381 kW) with a total detention time of 6.67 d. The second alternative may be the optimum system since the increased basin size will be more tolerable to the influent fluctuation in both flow rate and constituent concentration. It will also be justified by an economic gain realized by operating with less power input, although the difference in the horsepower installed is minimal. In this alternative, the deten-

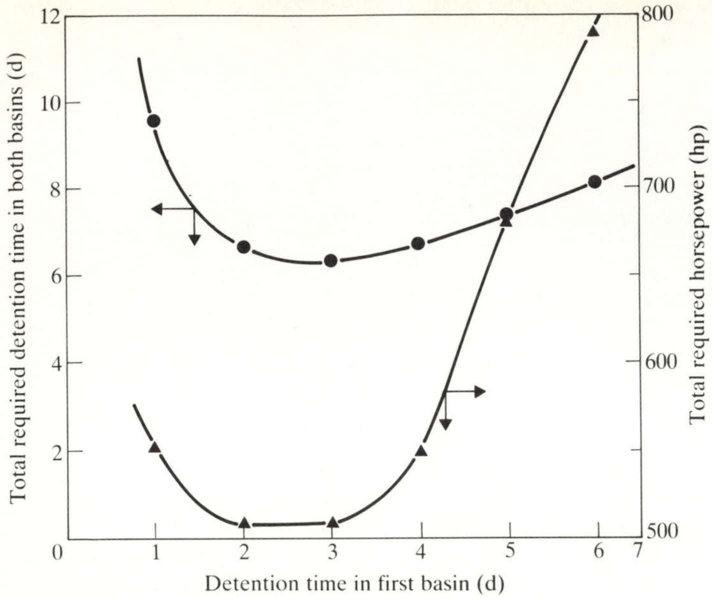

FIGURE 7-8
Effect of detention time in first basin on total required time and horsepower.

tion time in the first basin is 2 d and the detention time in the second basin 4.67 d. The results are tabulated below:

t, d			hp installed		
First basin	**Second basin**	**Total**	**First basin**	**Second basin**	**Total**
1.0	8.57	9.57	261	291	552
2.0	4.67	6.67	342	166	508
3.0	3.33	6.33	389	120	509
4.0	2.72	6.72	476	97	573
5.0	2.34	7.34	595	84	679
6.0	2.12	8.12	714	75	789

7.3 ACTIVATED SLUDGE PROCESSES

The objective of the activated sludge process is to remove soluble and insoluble organics from a wastewater stream and to convert this material into a flocculent microbial suspension that is readily settleable and will permit the use of gravitational solids liquid separation techniques. A number of different modifications or variants of the activated sludge process have been developed since the original experiments of Arden and Lockett in 1914. These variants, to a large extent, have been developed out of necessity or to suit particular circumstances that have

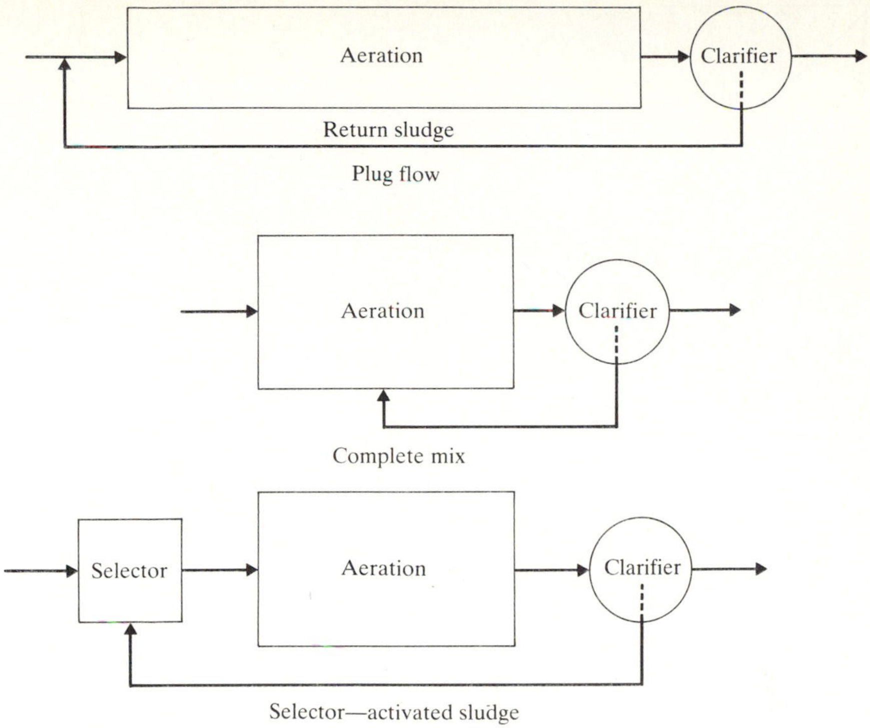

FIGURE 7-9
Types of activated sludge processes.

arisen. For the treatment of industrial wastewater, the common generic flowsheets are shown in Fig. 7-9. As discussed in Chap. 6, the nature of the wastewater will dictate the process type. Figure 7-10 illustrates the types of reaction that will occur and hence dictate the candidate process.

Plug-Flow Activated Sludge

The plug-flow activated sludge process uses long, narrow aeration basins to provide a mixing regime that approaches plug flow. Wastewater is mixed with a biological culture under aerobic conditions. The biomass is then separated from the liquid stream in a secondary clarifier. A portion of the biological sludge is wasted and the remainder returned to the head of the aeration tank with additional incoming waste. The rate and concentration of activated sludge returned to the basin determine the mixed liquor suspended solids concentration. As discussed in Chap. 6, a plug-flow regime promotes the growth of a well-flocculated, good settling sludge. If the wastewater contains toxic or inhibiting organics, they must be removed or equalized prior to entering the head end of the aeration basin. The oxygen utilization rate is high at the beginning of the aeration basin

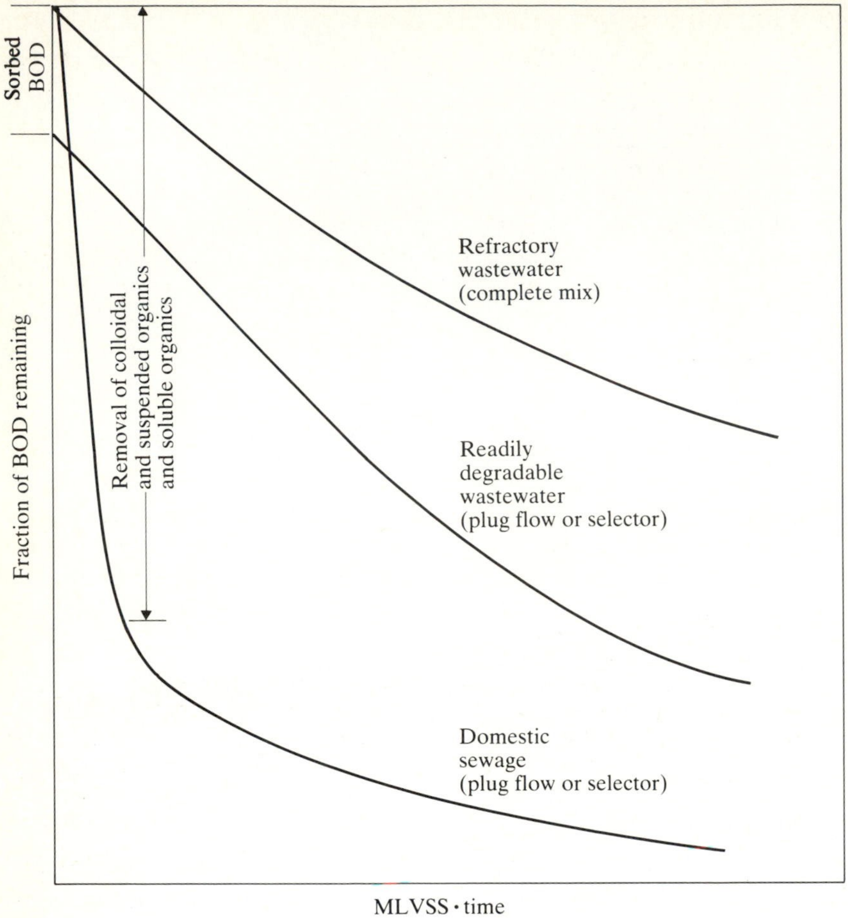

FIGURE 7-10
Removal of BOD in the activated sludge process.

and decreases with aeration time. Where complete treatment is achieved, the oxygen utilization rate approaches the endogenous level toward the end of the aeration basin.

Modification of the way in which wastewater and recycle sludge are brought into contact in a plug-flow system can have a number of benefits. The provisions of a separate zone at the inlet, with a volume of about 15 percent of the total aeration volume, together with a low-energy subsurface mechanical stirrer, can achieve controlled anoxic conditions such that nitrate associated with the recycled sludge fed to the zone can partially satisfy the BOD fed to the zone. In cases where nitrification occurs, recycle of nitrified mixed liquor from the end of the aeration basin to the anoxic zone at the head end can achieve significant denitrification.

Complete-Mix Activated Sludge

To obtain complete mixing in the aeration tank requires the proper choice of tank geometry, feeding arrangement, and aeration equipment. Through the use of complete mixing, with either diffused or mechanical aeration, it is possible to establish a constant oxygen demand as well as a uniform mixed liquor solids concentration throughout the basin volume. Hydraulic and organic load transients are dampened in these systems, giving a process that is very resistant to upset from shock loadings. Influent wastewater and recycled sludge are introduced to the aeration basin at different points. An activated sludge plant is shown in Fig. 7-11. Readily degradable wastewaters such as food-processing wastes will tend toward filamentous bulking in a complete-mix system, as has previously been discussed.

Such conditions may be minimized by the inclusion of a precontacting zone to effect a high level of substrate availability to the recycled mixed liquor. The precontacting zone should have a retention time in the order of 15 minutes to maximize biosorption. Design parameters for this contacting zone appear to be waste-specific and require bench-scale trials for their assessment. By contrast, complex chemical wastewaters do not support filamentous growth and complete-mix processes work very effectively. Wastewaters with an alkaline pH are effec-

FIGURE 7-11
Activated sludge plant.

tively treated since the CO_2 generated neutralizes the caustic alkalinity. In like manner, low-pH wastewaters containing organic acids can be effectively treated without external neutralization. A process design is given in Example 7.3.

Example 7.3. A complete-mix activated sludge plant is to be designed for the following conditions:

$$Q = 3.5 \text{ million gal/d } (1.32 \times 10^4 \text{ m}^3/\text{d})$$
$$S_o = 650 \text{ mg/l}$$
$$S \text{ (soluble)} = 20 \text{ mg/l}$$
$$X_v = 3000 \text{ mg/l}$$
$$a = 0.50$$
$$a' = 0.52$$
$$b = 0.1/\text{d at } 20°\text{C}$$
$$\theta = 1.065$$
$$K = 6.0/\text{d at } 20°\text{C}$$
$$b' = 0.14/\text{d}$$

Compute the aeration volume and the F/M.

Solution

$$t = \frac{S_o(S_o - S)}{KSX_v}$$

$$= \frac{650(630)}{6(20)(3000)} = 1.14 \text{ days}$$

$$\frac{F}{M} = \frac{S_o}{X_v t} = \frac{650}{3000 \times 1.14} = 0.19/\text{d}$$

The degradable fraction is determined by

$$X_d = \frac{0.8}{1 + 0.2b\theta_c}$$

Assuming $\theta_c = 26$ d:

$$X_d = \frac{0.8}{1 + 0.2 \times 0.1 \times 26} = 0.53$$

The aeration volume is 3.5 million gal/d × 1.14 d = 4.0 million gal (15,000 m^3). The sludge yield can be computed:

$$\Delta X_v = aS_r - bX_d X_v t$$

$$= 0.5\ (630) \text{ mg/l} - 0.1/\text{d} \times 0.53 \times 3000 \text{ mg/l} \times 1.14/\text{d}$$

$$= 134 \text{ mg/l}$$

$$\Delta X_v = 134 \text{ mg/l} \times 3.5 \text{ million gal/d} \times 8.34 \frac{\text{lb/million gal}}{\text{mg/l}}$$

$$= 3911 \text{ lb/d } (1776 \text{ kg/d})$$

Check the sludge age:

$$\theta_c = \frac{VX_v}{\Delta X_v} = \frac{3000 \times 4.0 \times 8.34}{3911} = 25.6 \text{ d} \simeq 26 \text{ d}$$

Compute the oxygen requirements:

$$\begin{aligned} O_2/\text{d} &= a'S_r Q + b'X_d X_v V \\ &= 0.52 \times 630 \times 3.5 \times 8.34 + 0.14 \times 0.53 \times 3000 \times 4.0 \times 8.34 \\ &= 16{,}989 \text{ lb/d} = 708 \text{ lb/h } (321 \text{ kg/h}) \end{aligned}$$

Compute the effluent quality at 10°C:

$$\begin{aligned} K_{10^\circ} &= 6.0 \times 1.065^{(10-20)} \\ &= 3.19/\text{d} \end{aligned}$$

$$\begin{aligned} S &= \frac{S_o^2}{KX_v t + S_o} \\ &= \frac{650^2}{3.19 \times 3000 \times 1.14 + 650} \\ &= 37 \text{ mg/l} \end{aligned}$$

Extended Aeration

In this process, sludge wasting is minimized. This results in low growth rates, low sludge yields, and relatively high oxygen requirements by comparison with the conventional activated sludge processes. The trade-off is a high-quality effluent and less sludge production. Extended aeration is a reaction defined rather than a hydraulically defined mode and can be nominally plug flow or complete mix. Design parameters include a food/microorganisms ratio F/M of 0.05 to 0.15, a sludge age of 15 to 35 days, and mixed liquor suspended solids concentrations of 3000 to 5000 mg/l. The extended aeration process can be sensitive to sudden increases in flow due to resultant high MLSS loading on the final clarifier, but is relatively insensitive to shock loads in concentration due to the buffering effect of the large biomass volume. While the extended aeration process can be used in a number of configurations, a significant number are installed as loop-reactor systems where aerators of a specific type provide oxygen and establish a unidirectional mixing to the basin contents. The use of loop-reactor systems and modifications thereof in wastewater treatment has been significant over recent years.

Oxidation Ditch Systems

A number of loop-reactor or ditch system variants are now available. In any ditch system it is necessary to adequately match basin geometry and aerator performance in order to yield an adequate channel velocity for mixed liquor solids transport. The key design factors in these systems relate to the type of

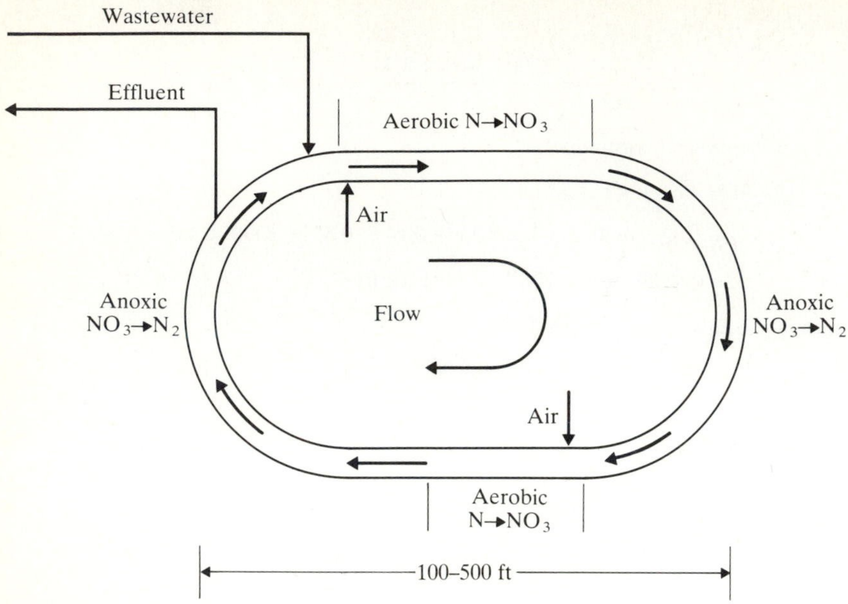

Note: ft = 0.3048 m

FIGURE 7-12
Oxidation ditch with nitrification and denitrification.

aeration that is to be provided. It is normal to design for a 1 ft/s (0.3 m/s) mid-channel velocity in order to prevent solids deposition. The ditch system is particularly amenable to those cases where both BOD and nitrogen removal is desired. Both reactions can be achieved in the same basin by alternating aerobic and anoxic zones, as shown in Fig. 7-12.

A typical oxidation ditch aeration basin consists of a single channel or multiple interconnected channels as shown in Fig. 7-13.

Intermittently Aerated and Decanted Systems

In the intermittent treatment approach, a single vessel is used to accommodate all of the unit processes and operations normally associated with conventional activated sludge treatment, i.e., primary settlement, biological oxidation, secondary settlement, and sludge digestion, as well as nitrification and the ability for substantial denitrification.[10] In using a single vessel, these processes and operations are simply timed sequences.

Figure 7-14 is a diagrammatic representation of the operating sequences for a continuous-flow intermittently aerated and decanted activated sludge system. Each sequence (t_0–t_1, t_1–t_2, t_2–t_3) of the cycle (t_0–t_3) is initiated by a time-base controller. The treatment cycle begins following the end of decantation from the

FIGURE 7-13
Oxidation ditch.

previous cycle. Aeration begins at time t_0 and continues until time t_1, during which time influent wastewater increases the volume of mixed liquor for aeration. At time t_1, aeration stops and is followed by a nonaeration sequence in which the mixed liquor undergoes settlement and anoxic processes can occur. Following the settlement/anoxic sequence (t_1–t_2), treated effluent is discharged during the

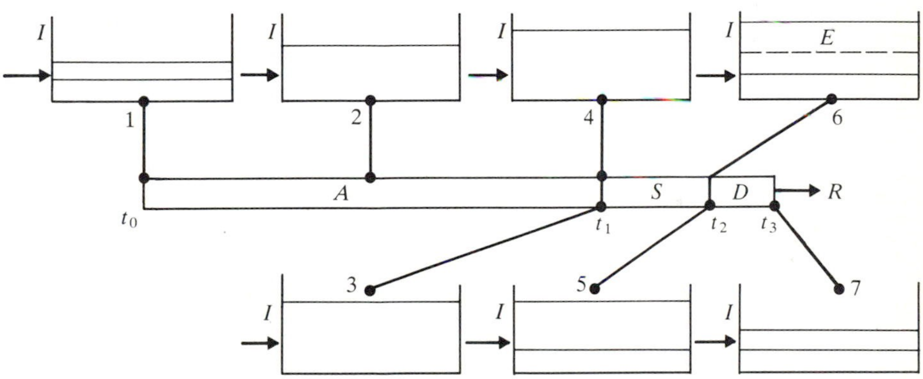

t = time
I = influent
E = effluent
A = aeration
S = settlement
D = decant
R = repeat

FIGURE 7-14
Diagrammatic representation of continuous-flow sequentially aerated activated sludge.

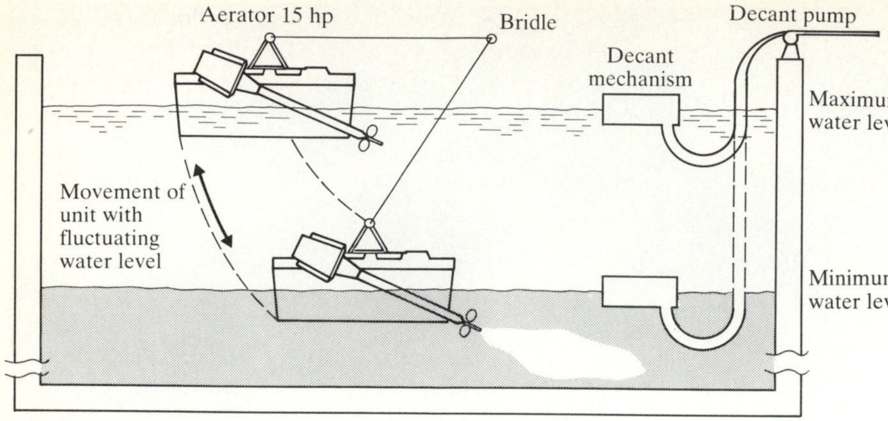

Note: hp = 0.7456 kW

FIGURE 7-15
Batch activated sludge system.

period (t_2–t_3); at the completion of t_3, the same sequence of events is repeated. Operational sequences are developed to optimize specific functions within the main process. For example, a denitrification cycle will require sufficient aeration to provide for total carbonaceous and nitrogenous oxidation within the time period (t_0–t_1), and the period (t_1–t_3) must be adequate to effect the reduction of nitrate.

An important feature of these plants is their ability to accept prolonged high-flow conditions without loss of mixed liquor solids. The hydraulic capacity of conventional continuous systems is limited by the operational capacity of the secondary settlement unit.

The decanting device in these systems is located at the vessel end opposite the inlet. The movable weir of the unit is positioned out of the mixed liquor during aeration and settlement. During the decant sequence, a hydraulic ram is activated which drives the weir trough into the surface layer of the vessel through to the design bottom water level. In this way, a surface layer of treated effluent is skimmed continuously during the decant sequence from the vessel and discharged out of the vessel by gravity via the carrier system of the decanter.

Plants can be designed on an average F/M ratio of 0.05 to 0.20 lb BOD/(lb MLSS · d) [0.05 to 0.20 kg BOD/(kg MLSS · d)], depending on the quality of effluent that is specified, at a bottom water level suspended solids concentration of up to 5000 mg/l. In calculating the volume occupied by the sludge mass, an upper sludge volume index of 150 ml/g is used. To ensure solids are not withdrawn during decant, a buffer volume is provided, the depth of which is generally in excess of 1.5 ft (0.5 m) between bottom water level and top sludge level after settlement. Batch activated sludge is similar to the intermittent system except that it is usually employed for high-strength industrial wastewaters. Wastewater is added over a short time period to maximize biosorption and flocculent sludge growth. Aeration is then continued for up to 20 h. The mixed liquor is then settled and the treated effluent decanted. A typical batch activated sludge system is shown in Fig. 7-15. A design example is shown in Example 7.4.

Example 7.4. Design a batch activated sludge plant for the following wastewater:

$$\begin{aligned} Q &= 50{,}000 \text{ gal/d } (190 \text{ m}^3/\text{d}) \\ \text{BOD } (S_o) &= 500 \text{ mg/l} \\ \text{TKN} &= 2 \text{ mg/l} \\ a &= 0.6 \\ a' &= 0.55 \\ b &= 0.1/\text{d} \\ F/M &= 0.1/\text{d} \\ S &= 10 \text{ mg/l (soluble)} \end{aligned}$$

The plant will be operated with 20 h aeration, 2 h sedimentation, and 2 h decant.

Solution. The BOD removal will be

$$S_r Q = (500 - 10) \text{ mg/l } (8.34) \frac{\text{lb/million gal}}{\text{mg/l}} (0.05) \text{ million gal} = 204 \text{ lb/d } (93 \text{ kg/d})$$

At an F/M of 0.1/d the required MLVSS is

$$X_v V = \frac{QS_o}{F/M} = \frac{0.05 \times 500 \times 8.36}{0.1}$$

$$= 2085 \text{ lb VSS } (946 \text{ kg})$$

and at 85 percent volatile sludge the MLSS is

$$\frac{2085}{0.85} = 2453 \text{ lb SS } (1113 \text{ kg})$$

Assuming the sludge has an SVI of 100 ml/g, the sludge volume will be

$$\frac{100 \text{ ml}}{\text{g SS}} \times \frac{454 \text{ g}}{\text{lb}} \times 3.53 \times 10^{-5} \frac{\text{ft}^3}{\text{ml}} = 1.6 \text{ ft}^3/\text{lb}$$

and the volume required for the settled sludge is

$$2453 \text{ lb} \times 1.6 \text{ ft}^3/\text{lb} = 3925 \text{ ft}^3 \text{ or } 29{,}360 \text{ gal } (111 \text{ m}^3)$$

If sludge is to be wasted twice per month, storage must be provided for accumulated sludge.

At an estimated degradable fraction of the VSS of 0.4, the daily accumulation of VSS is

$$\Delta X_v = aS_r Q - bX_d X_v V$$

$$= 0.6\ (204) - 0.1 \times 0.4 \times 2085$$

$$= 39 \text{ lb VSS/day } (18 \text{ kg/d})$$

Storage for 15 days will be

$$\frac{39 \text{ lb VSS/d}}{0.85 \text{ VSS/SS}} \times 15 \text{ d} \times 1.6 \text{ ft}^3/\text{lb SS} = 1100 \text{ ft}^3 \text{ or } 8240 \text{ gal } (31 \text{ m}^3)$$

The total volume of basin (excluding freeboard) will be 50,000 + 29,360 + 8240 = 87,600 gal (332 m^3). This will be a basin 35.25 ft (10.7 m) in diameter and 12 ft (3.7 m) deep. If 3 ft (0.9 m) is provided for freeboard, the operational basin dimensions will be 35.25 ft (10.7 m) in diameter by 15 ft (4.6 m) depth.

The oxygen requirements can be calculated:

$$\begin{aligned} O_2/\text{d} &= a'S_r Q + 1.4b \times X_d X_v V \\ &= 0.55\ (204) + 1.4 \times 0.1 \times 0.4 \times 2085 \\ &= 229 \text{ lb/d or } 9.54 \text{ lb/h } (4.3 \text{ kg/h}) \end{aligned}$$

The required hp at 1.5 lb $O_2/(\text{hp} \cdot \text{h})$ is

$$\frac{9.54}{1.5} = 6.4 \text{ hp (use 7.5 hp)}$$

This is equivalent to 150 hp/million gal (2.97 kW/m^3) of basin volume, which should provide adequate mixing. The nutrient requirement will be

$$\begin{aligned} N &= 0.123\frac{X_d}{0.8}\Delta X_v + 0.07\left(\frac{0.8 - X_d}{0.8}\right)\Delta X_v \\ &= 0.123 \times \frac{0.4 \times 39}{0.8} + 0.07 \times \frac{0.8 - 0.4}{0.8} \times 39 \\ &= 3.8 \text{ lb/d as N } (1.7 \text{ kg/d}) \end{aligned}$$

$$\begin{aligned} P &= 0.026 \times \frac{0.4 \times 39}{0.8} + 0.01 \times \frac{0.8 - 0.4}{0.8} \times 39 \\ &= 0.7 \text{ lb/d as P } (0.32 \text{ kg/d}) \end{aligned}$$

Oxygen Activated Sludge

The high-purity oxygen system is a series of well-mixed reactors employing concurrent gas-liquid contact in a covered aeration tank, as shown in Fig. 7-16. The process has been used for the treatment of pulp and paper-mill and organic chemical wastewaters. Feed wastewater, recycle sludge, and oxygen gas are introduced into the first stage. Gas-liquid contacting can employ submerged turbines or surface aeration.

Oxygen gas is automatically fed to either system on a pressure demand basis with the entire unit operating, in effect, as a respirometer; a restricted exhaust line from the final stage vents the essentially odorless gas to the atmosphere. Normally the system will operate most economically with a vent gas composition of about 50 percent oxygen. Based on economic considerations, about 90 percent of oxygen utilization with on-site oxygen generation is desired. Oxygen may be generated by a traditional cryogenic air-separation process for

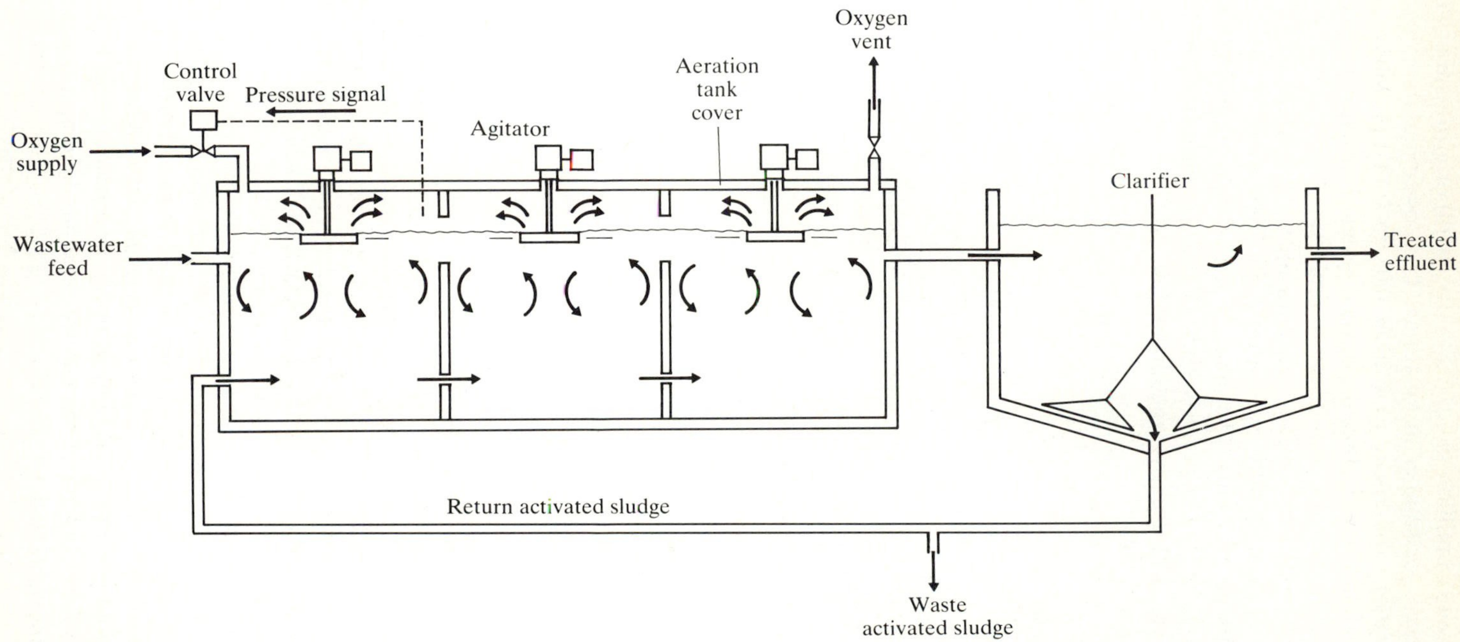

FIGURE 7-16
Schematic diagram of three-stage oxygen system.

large installations (75 million gal/d) (2.8×10^5 m^3/d) or a pressure-swing adsorption (PSA) process for smaller installations. The power requirements for surface or turbine aeration equipment vary from 0.08 to 0.14 hp/thousand gal (0.028 kW/m^3). At peak load conditions, the oxygen system is usually designed to maintain 6.0 mg/l dissolved oxygen in the mixed liquor.

Since high dissolved oxygen concentrations are maintained in the mixed liquor, the system can usually operate at high F/M levels (0.6 to 1.0) without filamentous bulking problems. The maintenance of an aerobic floc with the high zone settling velocities also permits high MLSS concentrations in the aeration tank. Solids levels will usually range from 4000 to 9000 mg/l depending on the BOD of the wastewater.

Final Clarification

The final clarifier is a key element in the activated sludge process and failure to consider major design concepts will result in process upset. Activated sludge exhibits zone settling, which has been discussed on page 65. Batch settling of activated sludge can be characterized by the relationship:

$$V_Z = V_o e^{-KC} \tag{7.13}$$

in which V_Z = zone settling velocity
C = sludge concentration at V_Z
V_o and K = empirical constants

The settling properties of activated sludge are commonly related to the sludge volume index (SVI). Daigger and Roper[11] showed that settling rates of municipal activated sludges could be related to SVI by modifying Eq. (7.13):

$$V_Z = 7.80e^{-(0.148 + 0.0021\,\mathrm{SVI}\,C)} \tag{7.13a}$$

In Eq. (7.13*a*), V_Z, C, and SVI are in meters per hour, grams per liter, and milliliters per gram, respectively. The clarifier functions primarily as a thickener in which the flux to the clarifier is related to underflow or recycle sludge concentration (see page 320). Figure 7-17 illustrates the performance characteristics of a typical final clarifier. As seen in Fig. 7-17, the performance characteristics depend on the sludge quality denoted by the SVI. A high SVI will result in lower return sludge concentrations and a lower permissible flux to the clarifier.

In order to determine the maximum allowable solids loading to a clarifier (limiting solids flux G), the subsidence flux $V_Z C$ is plotted as a function of C using the V_Z versus C relationship [Eq. (7.13)] for the sludge in question, as shown in Fig. 7-17. A line drawn tangent to the resulting flux curve will intercept on the x axis at the desired underflow concentration C_u, as shown in Fig. 7-17. The intercept on the y axis is the maximum allowable solids loading to the clarifier for that C_u based on $X_a(Q + R)$.

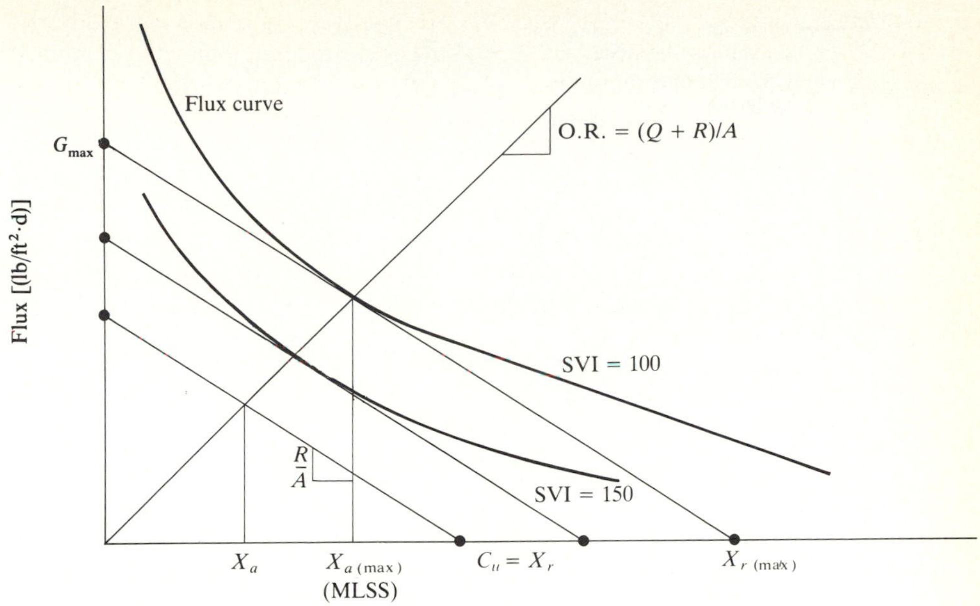

Note: lb/(ft^2·d) = 4.88 kg/(m^2·d)

FIGURE 7-17
Final clarifier relationships.

The overflow rate (O.R.) is the slope of the line $(Q + R)/A$ and is drawn through the axis to intersect with the flux line. A tangent drawn to this point of intersection will yield the maximum underflow concentration on the x axis, $X_{v(\max)}$, and the maximum permissible flux on the y axis, $G_{\max}$. A line from the tangent point to the x axis will yield the maximum MLSS to satisfy the clarification criteria. Any recycle rate below the flux curve will yield a safe operation, while values above the flux curve will result in clarifier failure. As previously noted, if the SVI increases due to sludge bulking (shown in Fig. 7-17 as an SVI of 100 and 150), the permissible flux decreases, which may result in failure to meet permit requirements. Repeating the calculation procedure for various underflow concentrations results in a series of clarifier underflow, i.e., recycle concentrations with their corresponding maximum solids flux rates in which each represents a limiting design or operating condition. Figure 7-18 represents these calculations for a range of SVI sludges. Acceptable operating points on Fig. 7-18 are those which lie below the line for the particular SVI. The dashed lines or operating lines are derived from clarifier mass balances assuming effluent suspended solids and sludge wastage are small relative to the clarifier solids feed. The state point on a given operating line is defined by the solids flux G and the corresponding clarifier underflow or recycle solids C_u. The clarifier will not be solids-limiting providing the state point lies below the limiting solids flux line for the particular SVI. This is denoted as $X_{a(\max)}$ on Fig. 7-17.

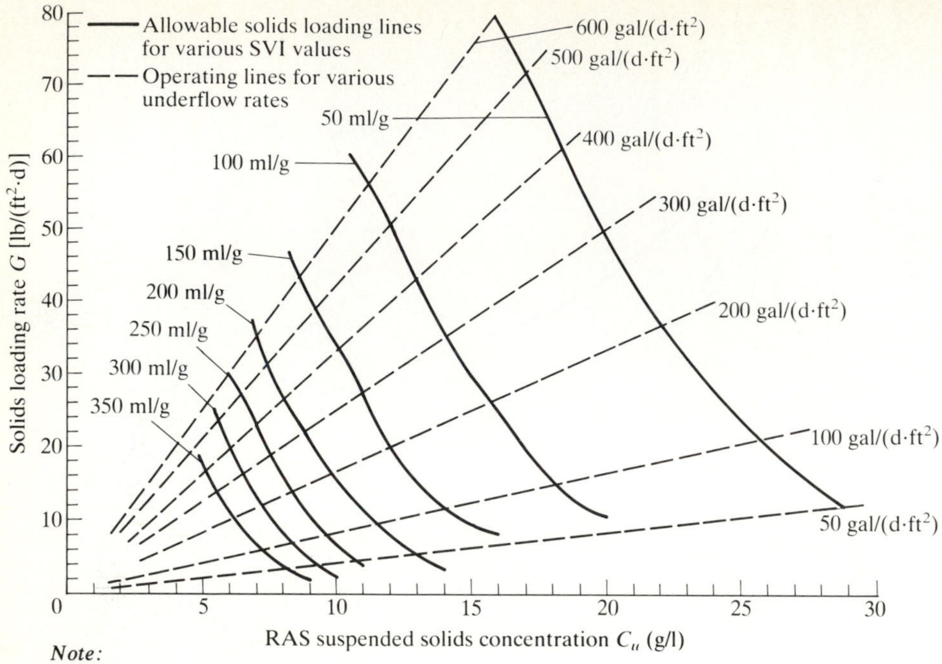

Note:
$lb/(ft^2 \cdot d) = 4.89\ kg/(m^2 \cdot d)$
$gal/(d \cdot ft^2) = 4.07 \times 10^{-2}\ m^3/(m^2 \cdot d)$

FIGURE 7-18
Clarifier design and operation diagram. (*After Daigger and Roper.*[11])

The application of Fig. 7-18 to design and operation of final clarifiers from Daigger and Roper[11] is shown in Example 7.5.

Example 7.5. A final clarifier is operating under the following conditions:

$$\text{Surface area } A = 2000 \text{ ft}^2 \ (186 \text{ m}^2)$$

$$\text{Influent flow } Q = 1.2 \text{ million gal/d } (4540 \text{ m}^3/\text{d})$$

$$\text{Sludge recycle } R = 0.6 \text{ million gal/d } (2270 \text{ m}^3/\text{d})$$

$$\text{Mixed liquor suspended solids } X_a = 3000 \text{ mg/l}$$

(*a*) Compute: The overflow rate (Q/A)
The underflow rate (R/A)
The solids flux (G)
The recycle suspended solids (X_r)
The maximum SVI for clarifier operation

The overflow rate is

$$\frac{1.2 \times 10^6}{2000} = 600 \text{ gal/(d} \cdot \text{ft}^2)[24.4 \text{ m}^3/(\text{m}^2 \cdot \text{d})]$$

The underflow rate is

$$\frac{0.6 \times 10^6}{2000} = 300 \text{ gal/(d} \cdot \text{ft}^2\text{)[12.2 m}^3\text{/(m}^2 \cdot \text{d)]}$$

The solids flux is

$$G = \frac{X_a(Q + R)}{A} = \frac{3000(1.2 + 0.6)8.34}{2000} = 22.5 \text{ lb/(ft}^2 \cdot \text{d)[110 kg/(m}^2 \cdot \text{d)]}$$

The recycle suspended solids is

$$\frac{R}{Q} = \frac{X_a}{X_r - X_a}$$

or

$$X_r = \frac{X_a(1 + R/Q)}{R/Q} = \frac{3000(1 + 0.6/1.2)}{0.6/1.2} = 9000 \text{ mg/l}$$

From Fig. 7-18 with a recycle suspended solids of 9000 mg/l and a solids flux of 22.5 lb/(ft$^2 \cdot$ d) [110 kg/(m$^2 \cdot$ d)] the maximum SVI is 200 ml/g.

(*b*) If the SVI of the sludge is 100 ml/g what could the flux and the MLSS be increased to?

From Fig. 7-18 at an SVI of 100 ml/g the flux can be increased to 35.2 lb/(ft$^2 \cdot$ d)[17.2 kg/(m$^2 \cdot$ d)]. Under these conditions, the MLSS could be increased to

$$X_a = \frac{GA}{Q + R} = \frac{35.2 \times 2000}{(1.2 \times 0.6)8.34} = 4700 \text{ mg/l}$$

(*c*) At an MLSS of 3000 mg/l compute the maximum clarifier overflow rate:

$$\frac{Q}{A} = \frac{G}{X_a} - \frac{R}{A} = \frac{35.2 \times 10^6}{3000 \times 8.34} - \frac{0.6 \times 10^6}{2000} = 1110 \text{ gal/(d} \cdot \text{ft}^2\text{)[45.1 m}^3\text{/(m}^2 \cdot \text{d)]}$$

Treatment of Industrial Wastewaters in Municipal Activated Sludge Plants

Municipal wastewater is unique in that a major portion of the organics are present in suspended or colloidal form. Typically, the BOD in municipal sewage will be 50 percent suspended, 10 percent colloidal, and 40 percent soluble. By contrast, most industrial wastewaters are almost 100 percent soluble. In an activated sludge plant treating municipal wastewater, the suspended organics are rapidly enmeshed in the flocs, the colloids are adsorbed on the flocs, and a portion of the soluble organics are absorbed. These reactions occur in the first few minutes of aeration contact. By contrast, for readily degradable wastewaters, i.e., food processing, a portion of the BOD is rapidly sorbed and the remainder removed as a function of time and biological solids concentration. Very little sorption occurs in refractory wastewaters. These phenomena are shown in Fig. 7-10. The kinetics of the activated sludge process will therefore vary depending on the percentage and type of industrial wastewater discharged to the municipal plant and must be considered in the design calculations.

The percentage of biological solids in the aeration basin will also vary with the amount and nature of the industrial wastewater. For example, municipal wastewater without primary clarification will yield a sludge that is 47 percent biomass at a 3-d sludge age. Primary clarification will increase the biomass percentage to 53 percent. Increasing the sludge age will also increase the biomass percentage as volatile suspended solids undergo degradation and synthesis. Soluble industrial wastewater will increase the biomass percentage in the activated sludge.

As a result of these considerations, there are a number of phenomena that must be considered when industrial wastewaters are discharged to municipal plants:

1. Effect on effluent quality. Soluble industrial wastewaters will affect the reaction rate K, as shown in Table 6.3. Refractory wastewaters such as tannery and chemical will reduce K while readily degradable wastewaters such as food processing and brewery will increase K.
2. Effect on sludge quality. Readily degradable wastewaters will stimulate filamentous bulking, depending on basin configuration, while refractory wastewaters will suppress filamentous bulking.
3. Effect of temperature. An increased industrial wastewater input, i.e., soluble organics, will increase the temperature coefficient θ, thereby decreasing efficiency at reduced operating temperatures.
4. Sludge handling. An increase in soluble organics will increase the percentage of biological sludge in the waste sludge mixture. This generally will decrease dewaterability, decrease cake solids, and increase conditioning chemical requirements. An exception is pulp and paper-mill wastewaters in which pulp and fiber serves as a sludge conditioner and will enhance dewatering rates.

It should be noted that most industrial wastewaters are nutrient deficient, i.e., lack nitrogen and phosphorus. Municipal wastewater, with a surplus of these nutrients, will provide the required nutrient balance.

Pretreatment of an industrial wastewater in a municipal plant is shown in Example 7.6.

Example 7.6. A 5 million gal/d (18,930 m^3/d) municipal activated sludge plant is presently operating at an F/M of 0.3 d^{-1}. A chemical complex has requested to discharge 1.2 million gal/d (4540 m^3/d) of wastewater with a BOD of 1500 mg/l to the plant. What pretreatment is required for the plant to meet their permit?

Solution

Municipal sewage

Flow = 5.0 million gal/d

S_o, BOD = 200 mg/l

Soluble BOD = 50 mg/l
$F/M = 0.3$
$X_v = 2500$ mg/l
S (soluble) = 10 mg/l
$K = 8$/d at 20°C

Industrial waste

Flow = 1.2 million gal/d
S_o, BOD = 1500 mg/l
$K = 2.2$/d at 20°C

The estimated maximum MLVSS is 3500 mg/l.

Present detention time is

$$\frac{F}{M} = \frac{S_o}{X_v t}$$

$$t = \frac{200}{2500 \times 0.3}$$

$$= 0.27 \text{ d}$$

Average reaction rate k will be

$$\frac{5.0\ (8) + 1.2\ (2.2)}{6.2} = 6.88/\text{d}$$

The new detention time is $0.27 \times 5/6.2 = 0.22$ d. The influent BOD to the plant to meet the permit can be calculated:

$$\frac{S_o - S_e}{X_v t} = k\,\frac{S_e}{S_o}$$

$$S_o^2 - SS_o - SKX_v t = 0$$

$$S_o = \frac{S + \sqrt{S^2 + 4SKX_v t}}{2} = \frac{10 + \sqrt{100 + 4 \times 10 \times 6.88 \times 3500}}{2}$$

$$= 235 \text{ mg/l of soluble BOD}$$

The concentration of BOD in the pretreated industrial wastewater can then be calculated by a material balance:

$$Q_s(S_{o,\,s}) + Q_I(S_{o,\,I}) = (Q_s + Q_I)S_{o,\,S+I}$$

$$5.0\ (50) + 1.2\ (S_{o,\,I}) = 6.2\ (235)$$

$$S_{o,\,I} = 1000 \text{ mg/l}$$

Pretreatment is required to reduce 33 percent of the BOD in the industrial wastewater.

Temperature effects
Determine the change in MLVSS that will be required when the temperature coefficient θ increases from 1.015 to 1.04 due to an increase in soluble industrial wastewater BOD. The temperature change is from 20 to 10°C.

$$\frac{K_{20}}{K_{10}} = (1.015)^{10} = 1.16 \text{ sewage}$$

$$\frac{K_{20}}{K_{10}} = (1.04)^{10} = 1.48 \text{ sewage – industrial waste mixture}$$

The increase in MLVSS can be calculated:

$$\frac{1.48}{1.16} \times 2500 \frac{\text{mg/l}}{1} = 3190 \text{ mg/l}$$

Effluent Suspended Solids Control

In municipal wastewater treatment, the activated sludge process can be expected to produce an effluent with a suspended solids content of less than 20 mg/l. In the

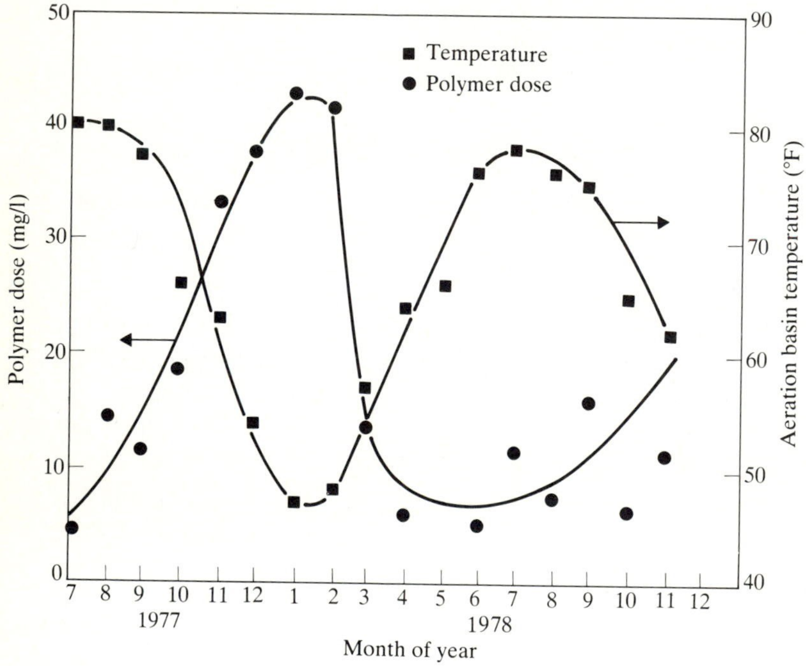

Note: °C = $\frac{5}{9}$ (°F − 32)

FIGURE 7-19
Polymer addition for effluent suspended solids control.

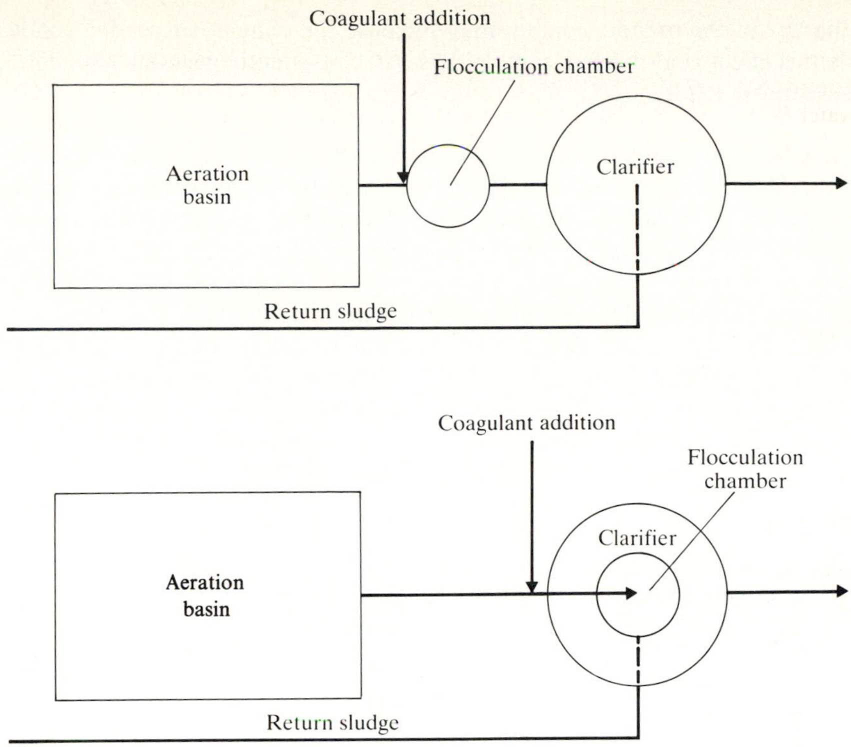

FIGURE 7-20
Coagulant addition for suspended solids control.

case of industrial wastewaters, however, high effluent suspended solids of a dispersed character can result from one of several causes:

1. High TDS (salt content) may cause an increase in nonsettleable suspended solids. While the specific cause of floc dispersion has not been defined, increasing the TDS has generally resulted in an increase in nonsettleable, dispersed solids. High TDS will also increase the specific gravity of the liquid, thereby reducing the settling rate of the biological sludge. The salt content appears to have little effect on the kinetics of the process under acclimated conditions.
2. Dispersed suspended solids increase as the aeration basin temperature decreases. For example, an aerated sludge plant in West Virginia treating an organic chemicals wastewater had an effluent suspended solids of 42 mg/l during summer operation and 104 mg/l during winter operation. The change in effluent suspended solids as evidenced by coagulant dosage at a Tennessee organic chemicals plant is shown in Fig. 7-19.
3. Dispersed suspended solids increase with a decrease in surface tension. At one deinking mill the effluent suspended solids was directly related to the surfactant usage in the mill.

4. The nature of the organic content may increase the effluent suspended solids. While this effect is ill-defined, some plants will consistently generate high effluent suspended solids.

The effluent suspended solids can be reduced by the addition of a coagulant prior to the final clarifier. It is important that there is sufficient time for flocculation to occur. This can be achieved by a flocculation chamber between the aeration basin and the clarifier or by a flocculation well within the clarifier, as shown in Fig. 7-20.

Cationic polyelectrolytes or alum or iron salts may be used as a coagulant. Choice of coagulant depends on a test program to select the most economic solution. Data reported by Paduska[12] using a cationic polyelectrolyte are shown in Fig. 7-19. It is important when using a cationic polymer to avoid overdosing which will cause a charge reversal and a redispersion of the solids. The temperature effect as previously noted is also shown in Fig. 7-19.

TABLE 7.3
Comparison of biodegradability of toxic organic pollutants as predicted by EPA study to the five-plant study results

Volatile compound	EPA: degradation, %†	Five-plant study: removal, %‡
Acrylonitrile	100	99
Benzene	100	100
Bromomethane	48	100
Bromodichloromethane	67	89
Carbon tetrachloride	100	100
Chlorobenzene	100	56§
Chloroethane	NA	91
Chloroform	100	99
Dibromochloromethane	55	100
1,1-Dichloroethane	100	79§
1,2-Dichloroethane	100	99
1,1-Dichloroethene	100	100
t-1,2-Dichloroethene	100	98
1,2-Dichloropropane	92	97
1,3-Dichloropropane	100	98
Ethylbenzene	100	99
Methylene chloride	100	75
1,1,2,2-Tetrachloroethane	36	93
Tetrachloroethene	100	27§
1,1,1-Trichloroethane	100	38§
1,1,2-Trichloroethane	59	72§
Trichloroethene	100	40§
Toluene	100	100
Vinyl chloride	NA	100

TABLE 7.3 *(continued)*

Base/neutral compound	**EPA: degradation, %†**	**Five-plant study: removal, %‡**
Acenaphthene	100	95
Acenaphthylene	98	93
Anthracene	92	97
Benzo (a) anthracene	35	56§
Benzo (b) fluoranthene	100	—§
Benzo (a) pyrene	NA	—§
Bis (2-ethylhexyl)phthalate	95	64
Butylbenzylphthalate	100	87§
Dibenzo (a, h) anthracene	NA	—§
Di-*n*-butylphthalate	100	99
1,3-Dichlorobenzene	35	100§
1,2-Dichlorobenzene	29	94
Diethylphthalate	100	99
Dimethylphthalate	100	99
Dioctylphthalate	94	90
Fluoranthene	100	83§
Fluorene	77	93
Isophorone	100	100
Naphthalene	100	100
Nitrobenzene	100	98
Pyrene	100	84§
1,2,4-Trichlorobenzene	24	82
Acid compound	**EPA: degradation, %†**	**Five-plant study: removal, %‡**
2-Chlorophenol	100	41
2,4-Dichlorophenol	100	91
2,4-Dimethylphenol	100	100
2,4-Dinitrophenol	100	84
2-Nitrophenol	100	77
Pentachlorophenol	100	36
Phenol	100	98
2,4,6-Trichlorophenol	100	45

† Includes volatilization losses at 25°C and initial concentration of 5 mg/l.
‡ Based on arithmetic means.
§ Influent concentration less than 40 ppb.
NA = not available.

Priority Pollutant Removal

Increasing emphasis is being placed on the removal of priority pollutants (toxic organics) present in low concentrations in wastewaters. Most of these pollutants are removed in the activated sludge process by biooxidation, air stripping, or adsorption to the flocs. It has been shown that in most cases adsorption plays a minor role in the removal process. Depending on the organic, stripping may be the primary removal mechanism. Table 7.3 summarizes data reported to date on removal of these organics.

7.4 TRICKLING FILTRATION

A trickling filter is a packed bed of media covered with slime growth over which wastewater is passed. As the waste passes through the filter, organic matter present in the waste is removed by the biological film.

Plastic packings are employed in depths up to 40 ft (12.2 m), with hydraulic loadings as high as 4.0 gal/(min · ft²) [0.16 m³/(min · m²)]. Depending on the hydraulic loading and depth of the filter, BOD removal efficiencies as high as 90 percent have been attained on some wastewaters. In one industrial plant, a minimum hydraulic loading of 0.5 gal/(min · ft²) [0.02 m³/(min · m²)] was required in order to avoid the generation of filter flies (psychoda). Figure 7-21 shows an installation of a plastic-packed filter.

Theory

As wastewater passes through the filter, nutrients and oxygen diffuse into the slimes, where assimilation occurs, and byproducts and CO_2 diffuse out of the

FIGURE 7-21
Plastic-packed trickling filter.

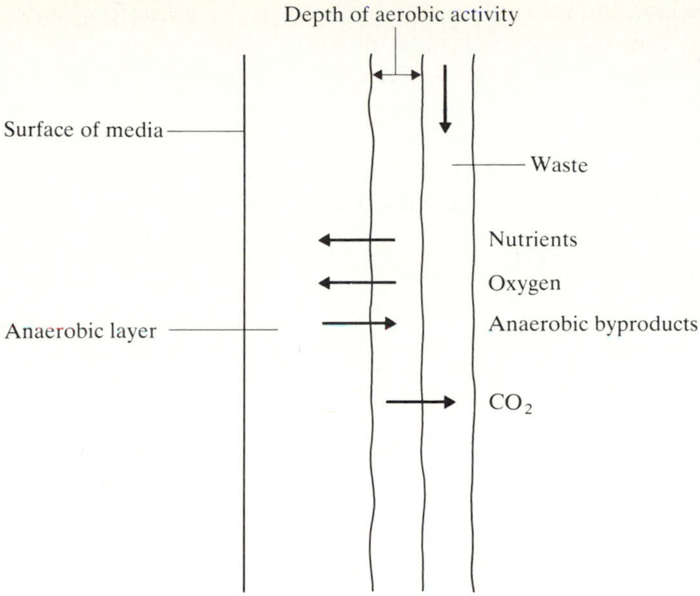

FIGURE 7-22
Mode of operation of a trickling filter.

slime into the flowing liquid. As oxygen diffuses into the biological film, it is consumed by microbial respiration, so that a defined depth of aerobic activity is developed. Slime below this depth is anaerobic, as shown in Fig. 7-22.

In a manner analogous to activated sludge, BOD removal through a trickling filter is related to the available biological slime surface and to the time of contact of wastewater with that surface.

The mean time of contact of liquid with the filter surface is related to the filter depth, the hydraulic loading, and the nature of the filter packing:[13]

$$t = \frac{CD}{Q^n} \tag{7.14}$$

where t = mean detention time
D = filter depth
Q = hydraulic loading

C and n are constants related to the specific surface and the configuration of the packing.

There are a number of commercial packings available. These include vertical-flow, random-packed, and cross-flow media. Investigations of several types of packings showed that

$$C = C'A_v^m \tag{7.15}$$

in which A_v is the specific surface expressed in square feet per cubic foot (*note*: $ft^2/ft^3 = 3.28\ m^2/m^3$). For spheres, rock, and Polygrid plastic media without slime, C' has a value of 0.7 and m a value of 0.75.[14] This relationship will vary for others that have a different general configuration.

The exponent n in Eq. (7.14) has been observed to decrease with decreasing specific surface A_v. Equations (7.14) and (7.15) can be combined to yield a general expression for the mean retention time through any type of filter packing:

$$\frac{t}{D} = \frac{C'A_v^m}{Q^n} \tag{7.16}$$

The mean retention time increases considerably in the presence of filter slime, in some cases being as much as four times that prevailing with the non-slimed surface.[14]

In order to avoid filter plugging, a maximum specific surface of 30 ft^2/ft^3 (98 m^2/m^3) is recommended for the treatment of carbonaceous wastewaters. Specific surfaces in excess of 100 ft^2/ft^3 (328 m^2/m^3) can be used for nitrification because of the low yield of biological cellular material.

Recent data have shown improved performance with cross-flow media with the same specific surface as other media due to the increased time of contact of the wastewater passing through the filter.

In Chap. 6 it was shown that soluble BOD removal in the plug-flow activated sludge process could be expected to follow the relationship

$$\frac{S}{S_o} = e^{-K_b X_v t/S_o} \tag{6.4}$$

These equations may be analogously applied to trickling filters under the following conditions:

1. The specific surface must remain constant. This is true for any specific filter medium but varies from medium to medium.
2. The medium must have a uniform, thin slime cover. This does not always occur, particularly in the case of rock filters.

The second condition also requires uniform distribution of the hydraulic loading to the filter. This condition is violated when heavy slime buildup short-circuits the filter.

Employing the retention time as defined by Eq. (7.14) and considering the available bacterial surface to be proportional to the specific surface A_v,

$$\frac{S}{S_o} = e^{-KA_v D/Q^n S_o} \tag{7.17}$$

It has been shown that in many cases BOD removal can be increased by recirculation of filter effluent around the filter. The recirculated flow serves as a diluent to the influent waste. When recirculation is used, the applied BOD to the filter S_o becomes

$$S_o = \frac{S_a + NS}{1 + N} \tag{7.18}$$

where S_o is the BOD of waste applied to the filter after mixing with the recirculated flow, S_a is the BOD of the influent, S is the BOD of filter effluent, and N is the recirculation ratio, R/Q.

Combining Eqs. (7.17) and (7.18) yields the BOD removal relationship:

$$\frac{S}{S_a} = \frac{e^{-KA_vD/Q^nS_o}}{(1 + N) - Ne^{-KA_vD/Q^nS_o}} \tag{7.19}$$

When the BOD of the waste exhibits a decreasing removal rate with decreasing concentration, the BOD in the recirculated flow will be removed at a lower rate than that prevailing in the influent. In this case, one must apply a coefficient of retardation to the recirculated flow.

Oxygen Transfer and Utilization

Oxygen is transferred from air passing through the filter to the films of liquid passing over the slimes. Since the rate of oxygen transfer is related to fluid mixing and turbulence, it can be expected that hydraulic loading and media configu-

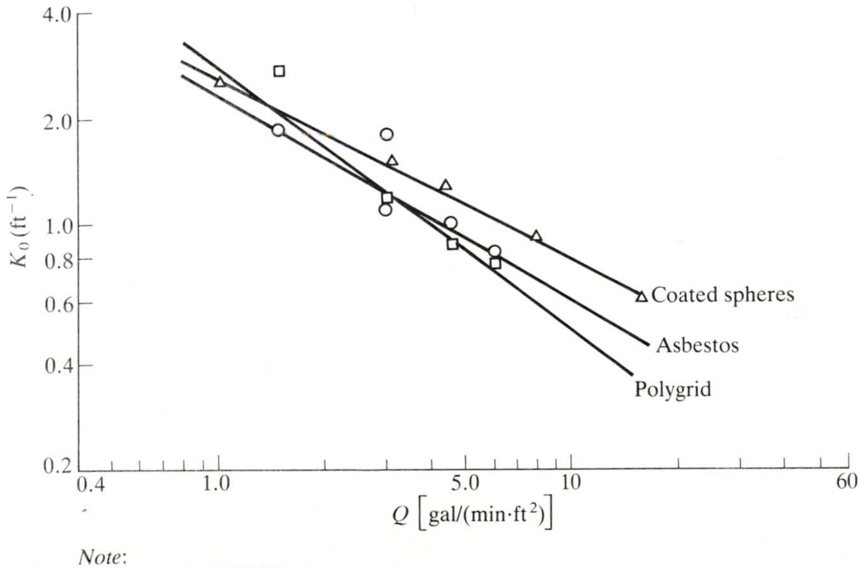

Note:
ft = 0.3048 m
gal/(min · ft^2) = 4.075×10^{-2} m^3/(min · m^2)

FIGURE 7-23
Relationship between the oxygen-transfer-rate coefficient and hydraulic loading.

ration will influence the transfer rate. Experimentation on various types of filter media has shown that the following relationship applies:[15]

$$\frac{-dC}{dD} = K_o(C_s - C_L) \tag{7.20}$$

where D is filter depth, C_L is concentration of dissolved oxygen in the liquid passing over the filter, and K_o is the transfer-rate coefficient. This integrates to

$$\frac{(C_s - C_L)_1}{(C_s - C_L)_2} = e^{-K_o(D_2 - D_1)} \tag{7.21}$$

The transfer-rate coefficient K_o is related to the hydraulic loading to the filter. The oxygen-transfer relationship is shown in Fig. 7-23.

The oxygen transfer can be expressed in mass of oxygen per hour per unit volume of filter medium by the following relationship:

$$N = 5.0 \times 10^{-4} K_o(C_s - C_L)Q \tag{7.22}$$

where N = lb $O_2/(\text{ft}^3 \cdot \text{h})$
C_s = oxygen saturation, mg/l
C_L = dissolved oxygen concentration, mg/l
Q = hydraulic loading, gal/(min · ft^2)
K_o = transfer-rate coefficient, ft^{-1}

or

$$N = 0.06\, K_o\, (C_s - C_L)Q$$

where N = kg $O_2/(\text{m}^3 \cdot \text{h})$
$Q = \text{m}^3/(\text{m}^2 \cdot \text{min})$
$K_o = \text{m}^{-1}$

The ability of a filter to remove BOD is limited by the aerobic activity of the filter film. This in turn is limited by the amount of oxygen transferred to the slime from the flowing liquid.

The activity of the filter films is measured by the surface oxygen-utilization rate. British studies[16] of the treatment of domestic sewage showed an oxygen-utilization rate of 0.028 mg $O_2/(\text{cm}^2 \cdot \text{h})$. Studies of the treatment of dilute black liquor[14] at a filter loading of 400 lb BOD/(thousand $\text{ft}^3 \cdot \text{d}$) [6.41 kg/($\text{m}^3 \cdot \text{d}$)] yielded a utilization rate of 0.0434 mg $O_2/(\text{cm}^2 \cdot \text{h})$.

The total quantity of aerobic film can be estimated from the relationship:

$$h = \sqrt{\frac{2D_L C_L}{K_r \rho}} \tag{7.23}$$

where h = depth of film to which oxygen penetrates
D_L = diffusivity of oxygen through the filter film
C_L = concentration of dissolved oxygen in the liquid passing the film surface
ρ = density of the filter film
K_r = unit oxygen-utilization rate of the filter film

It is possible to estimate the maximum BOD which can be assimilated by the filter film from Eqs. (7.22) and (7.23).

When treating high-concentration industrial wastewaters, a maximum influent concentration must be maintained in order to avoid anaerobic conditions prevailing in the filter slimes and resultant escape of anaerobic products and odors to the atmosphere. Depending on the degradability of the wastewater this may vary from 600 to 1200 mg/l. Higher influent BOD concentrations require recirculation for dilution of the influent strength.

Effect of Temperature

The performance of trickling filters will be affected by changes in the temperature of the filter films and the liquid passing over the films. It is usually assumed that these two temperatures will be essentially the same when only the aerobic portion of the film is considered. A decrease in temperature results in a decrease in respiration rate, a decrease in oxygen-transfer rate, and an increase in oxygen saturation. The combined effect of these factors results in an increase of aerobic film at a lower activity level, yielding a somewhat reduced efficiency at lower temperatures. The relationship of efficiency and temperature can be expressed as

$$E_T = E_{20} \times 1.035^{(T-20)} \tag{7.24}$$

where E = filter efficiency
T = temperature, °C

Trickling Filter Applications

In most cases, the reaction rate K for soluble industrial wastewaters is relatively low and hence filters are not economically attractive for high treatment efficiency (85 percent BOD reduction) of such wastewaters. Plastic-packed filters, however, have been employed as a pretreatment for high-strength wastewaters in which BOD removals in the order of 50 percent have been achieved at hydraulic and organic loadings of greater than 4 gal/(min · ft^2) [0.16 m^3/(min · m^2)] and 500 lb BOD/(thousand ft^3 · d) [8.0 kg/(m^3 · d)]. Performance characteristics for several industrial wastewaters are shown in Table 7.4. Example 7.7 shows a trickling filter design.

Example 7.7. One million gallons per day (3785 m^3/d) of an industrial wastewater with a BOD of 900 mg/l is to be pretreated to an effluent BOD of 300 mg/l. A plastic tower with a specific surface of 30 ft^2/ft^3 (98 m^2/m^3) and a depth of 20 ft (6.1 m) is to be used. Determine the hydraulic loading, the recirculation ratio, and the filter area required. The maximum BOD applied to the filter is 600 mg/l. The reaction rate coefficient for this wastewater is 1.0. The media has a coefficient n of 0.5.

TABLE 7.4
High-rate trickling filtration performance

Waste	Hydraulic loading, MGAD	Depth, ft	Raw BOD	Recycle ratio	BOD removal (clarified), %	Temperature	BOD loading, lb/thousand ft^3
Sewage	126	21.6	145	3	88		54
	252		131	3	82		110
	252		175	1	70		250
	63		173	0	78		95
	126		152	0	76		67
	189	10.8	166	0	45		549
	135		165	0	57		390
	95		185	0	51		304
Citrus	72	21.6	542	3	69		199
	189		464	2	42		612
Citrus and sewage	189	21.6	328	2	53		384
Kraft mill	365	18.0	250	0	10	34	
	185	18.0	250	0	24	36	
	200	21.6	250	0	23	40	
	90	21.6	250	0	31	33	
Black liquor	47	18	400	0	73	24	200
	95	18	400	0	58	29	380
	189	18	400	0	58	35	780

Note:
MGAD = 0.9354 m^3/(m$^2 \cdot$ d)
ft = 0.3048 m
lb/thousand ft^3 = 16 × 10^{-3} kg/m^3

Solution. Calculate the required recycle ratio:

$$S_o = \frac{S_a + NS_e}{1 + N}$$

$$600 = \frac{900 + 300N}{1 + N}$$

$$N = 1.0$$

Calculate the hydraulic loading:

$$\frac{S}{S_o} = e^{-KA_vD/Q^nS_o}$$

$$\frac{300}{600} = e^{-(1 \times 30 \times 20)/(Q_o^{0.5} \times 600)} = e^{-x}$$

$$x = 0.69 = \frac{1}{Q_o^{0.5}}$$

$$Q_o = 2 \text{ gal/(min} \cdot \text{ft}^2\text{) } [0.082 \text{ m}^3\text{/(min} \cdot \text{m}^2\text{)}]$$

Since $N = 1.0$, $R = Q$, and the wastewater flow is 1.0 gal/(min · ft²) [0.041 m³/(min · m²)],

$$A = \frac{10^6 \text{ gal/d} \times 6.94 \times 10^{-4} \text{ d/min}}{1.0 \text{ gal/(min} \cdot \text{ft}^2)} = 694 \text{ ft}^2 \ (64.5 \text{ m}^2)$$

The filter volume is $V = AD = 694 \times 20 = 13{,}880 \text{ ft}^3$ (393 m³)

Check for oxygen requirements
The BOD removed through the filter is

$$2 \text{ million gal/d} \times (600 - 300)(\text{mg/l}) \times 8.34 \frac{\text{lb/million gal}}{\text{mg/l}} = 5000 \text{ lb/d (2270 kg/d)}$$

Assuming 0.8 lb O_2/lb BOD removed the oxygen requirement is 4000 lb/d (1816 kg/d).

From Fig. 7-23 at a hydraulic loading of 2 gal/(min · ft²) the oxygen transfer rate coefficient K_o is 2.0/ft.

The oxygen transfer can be calculated from Eq. (7.22):

$$N = 5.0 \times 10^{-4} \times 2.0/\text{ft} \times (9 - 1) \text{ mg/l} \times 2 \text{ gal/(min} \cdot \text{ft}^2)$$
$$= 160 \times 10^{-4} \text{ lb } O_2/(\text{ft}^3 \cdot \text{h}) \ [0.26 \text{ kg/(m}^3 \cdot \text{h})]$$

Assuming an α value of 0.85,

$$N = 160 \times 10^{-4} \times 0.85 = 136 \times 10^{-4} \text{ lb } O_2/(\text{ft}^3 \cdot \text{h}) \ [0.22 \text{ kg/(m}^3 \cdot \text{h})]$$
$$\text{lb } O_2/\text{d} = 136 \times 10^{-4} \text{ lb/(ft}^3 \cdot \text{h}) \times 24 \text{ h/d} \times 13{,}880 \text{ ft}^3 = 4341 \text{ lb/d (2055 kg/d)}$$

7.5 ROTATING BIOLOGICAL CONTACTORS

The rotating biological contactor consists of large-diameter plastic media mounted on a horizontal shaft in a tank as shown in Fig. 7-24. The contactor is slowly rotated with approximately 40 percent of the surface area submerged. A 1 to 4 mm layer of slime biomass is developed on the media. (This would be equivalent to 2500 to 10,000 mg/l in a mixed system.) As the contactor rotates, it carries a film of wastewater through the air, resulting in oxygen and nutrient transfer. Additional removal occurs as the contactor rotates through the liquid in the tank. Shearing forces cause excess biomass to be stripped from the media in a manner similar to a trickling filter. This biomass is removed in a clarifier. The attached biomass is shaggy with small filaments resulting in a high surface area for organic removal to occur. Present media consists of high-density polyethylene with a specific surface of 37 ft²/ft³ (121 m²/m³). Single units are up to 12 ft (3.7 m) in diameter and 25 ft (7.6 m) long, containing up to 100,000 ft² (9290 m²) of surface in one section.

The primary variables affecting treatment performance are:

1. Rotational speed
2. Wastewater retention time
3. Staging

FIGURE 7-24
Rotating biological contactor. (*Courtesy of Envirex Inc.*)

4. Temperature
5. Disk submergence

In the treatment of low-strength wastewaters (BOD to 300 mg/l) performance increased with rotational speed to 60 ft/min (18 m/min) with no improvement noted at higher speeds. Increasing rotational speed increases contact, aeration, and mixing, and would therefore improve efficiency for high BOD wastewaters. However, increasing rotational speed rapidly increases power consumption, so that an economic evaluation should be made between increased power and increased area.

In the treatment of domestic wastewater performance increases with liquid volume to surface areas up to 0.12 gal/ft^2 (0.0049 m^3/m^2). No improvement was noted above this value.

In many cases, significant improvement was observed by increasing from two to four stages with no significant improvement with greater than four stages. Several factors could account for these phenomena. The reaction kinetics would favor plug-flow or multistage operation. With a variety of wastewater constituents, acclimated biomass for specific constituents may develop in different stages. Nitrification will be favored in the later stages where low BOD levels permit a higher growth of nitrifying organisms on the media. In the treatment of industrial wastewaters with high BOD levels or low reactivity more than four stages may be desirable. For high-strength wastewaters, an enlarged first stage may be

TABLE 7.5
RBC performance on goldmine tailings wastewater

	SCN^-, mg/l	CN_t^-, mg/l	CN_c^-, mg/l	Cu, mg/l	NH_3-N, mg/l
Influent	51.9	6.94	4.60	0.99	5.5
Effluent	<1.0	0.35	0.05	0.03	0.5

CN_c^- = cyanide amenable to chlorination
CN_t^- = total cyanide

Results across four stages:

		Stage			
	Influent	1	2	3	4
SCN^- mg/l	51.9	20.3	3.0	<1.0	<1.0
NH_3-N mg/l	2.2	8.4	3.4	0.9	0.3

employed to maintain aerobic conditions. An intermediate clarifier may be employed where high solids are generated to avoid anaerobic conditions in the contactor basins. Performance on minewater tailings are shown in Table 7.5.

Design for industrial wastewaters will usually require a pilot plant study. A kinetic model similar to that developed for activated sludge may be employed:

$$\frac{Q}{A}(S_o - S) = kS \tag{7.25}$$

in which Q is the flow rate, A the surface area, S_o is the influent substrate concentration, S is the effluent substrate concentration, and k is the reaction rate, or for wastewaters with a highly variable influent strength:

$$\frac{Q}{A}(S_o - S) = K\frac{S}{S_o} \tag{7.26}$$

Equation (7.25) for several industrial wastewaters is shown in Fig. 7-25. For high BOD wastewaters, performance can be improved by using enriched oxygen surrounding the media to enhance oxygen transfer and BOD removal.

Several factors become apparent from Fig. 7-25. The maximum BOD removal rate $Q/A(S_o - S)$ for a given operating condition (rotational speed, gas oxygen content, etc.) will relate to both the concentration of influent BOD and the biodegradability of the wastewater. The performance of multiple contactors in series can be defined by the relationship:

$$\frac{S}{S_o} = \left(\frac{1}{1 + kA/Q}\right)^n \tag{7.27}$$

in which n is the number of stages. Since at some loading oxygen will be limiting, it is important that the loading for each stage is checked for a multistage system.

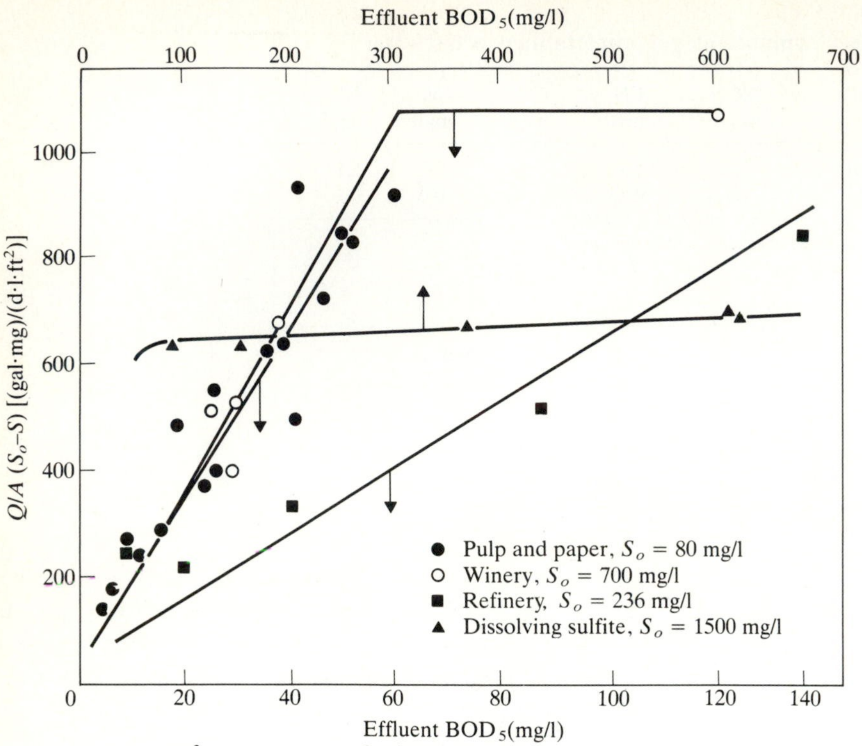

Note: gal/(ft^2·d) = 4.08 × 10^{-2} m^3/ (m^2·d)

FIGURE 7-25
BOD removal characteristics of an RBC treating industrial wastewaters.

As previously noted, the economics of the alternatives should be evaluated for each application. A process design is shown in Example 7.8.

Example 7.8. An RBC is to be designed for an effluent soluble BOD of 20 mg/l for a wastewater with an initial soluble BOD of 300 mg/l. The flow is 0.5 million gal/d (1900 m^3/d). Compute the total number of stages and the required area. The performance relationship is shown in Fig. 7-26.

Solution. The maximum loading to an RBC occurs at that point where oxygen becomes limiting, as shown in Fig. 7-26. The maximum hydraulic loading can be computed:

$$\frac{Q}{A} = \frac{[(Q/A)S_r]_{\max}}{S_o - [(Q/A)S_r]_{\max}/k}$$

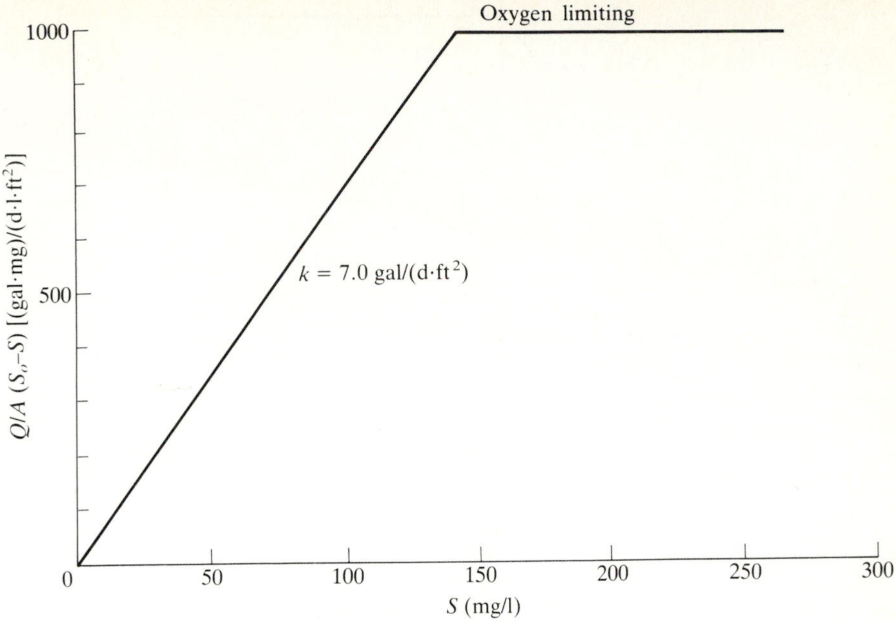

Note: gal/(d·ft²) = 4.075 × 10^{-2} m³/(d·m²)

FIGURE 7-26
BOD removal relationship.

The performance of multiple contactors in series can be defined by the relationship:

$$\frac{S_e}{S_o} = \left(\frac{1}{1 + kA/Q}\right)^n$$

$$\frac{Q}{A} = \frac{k}{(S_o/S_e)^{1/n} - 1}$$

The maximum hydraulic loading can be computed:

$$\left(\frac{Q}{A} S_r\right)_{\max} = 1000 \frac{\text{gal/(d}\cdot\text{ft}^2)}{\text{mg/l}}$$

$$\frac{Q}{A} = \frac{1000}{300 - 1000/7} = 6.37 \text{ gal/(d}\cdot\text{ft}^2) \text{ [}0.26 \text{ m}^3/(\text{d}\cdot\text{m}^2)]$$

The maximum hydraulic loading can also be obtained graphically as the slope of a line drawn from the influent concentration (300 mg/l) to the limiting loading point, as shown in Fig. 7-27.

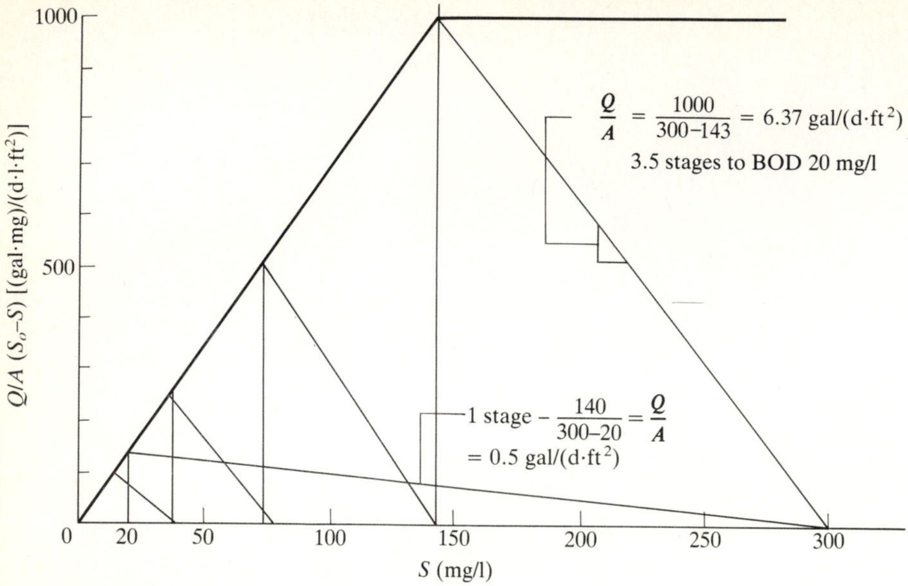

Note: gal/(d·ft²) = 4.075 × 10^{-2} m³/(d·m²)

FIGURE 7-27
Graphical design solution.

The required area for various numbers of stages can be computed as tabulated below:

Number of stages, n	Q/A, gal/(d·ft²)	$(Q/A)/n$, gal/(d·ft²·stage)	$A \times 10^{-3}$, ft²/stage
1	0.5	0.5	1000
2	2.44	1.22	410
3	4.83	1.61	310
4	7.29	1.82	275
5	10.0	2.0	250

$$\frac{Q}{A} = \frac{7.0}{(S_o/S_e)^{1/n} - 1}$$

The hydraulic loading (Q/A) can be plotted versus the number of stages, as shown in Fig. 7-28. In order not to exceed the maximum loading to the first stage, the maximum number of stages would be 3.5. In this case, the plant would be designed for three stages.

The calculations can also be made graphically, as shown on Fig. 7-27.

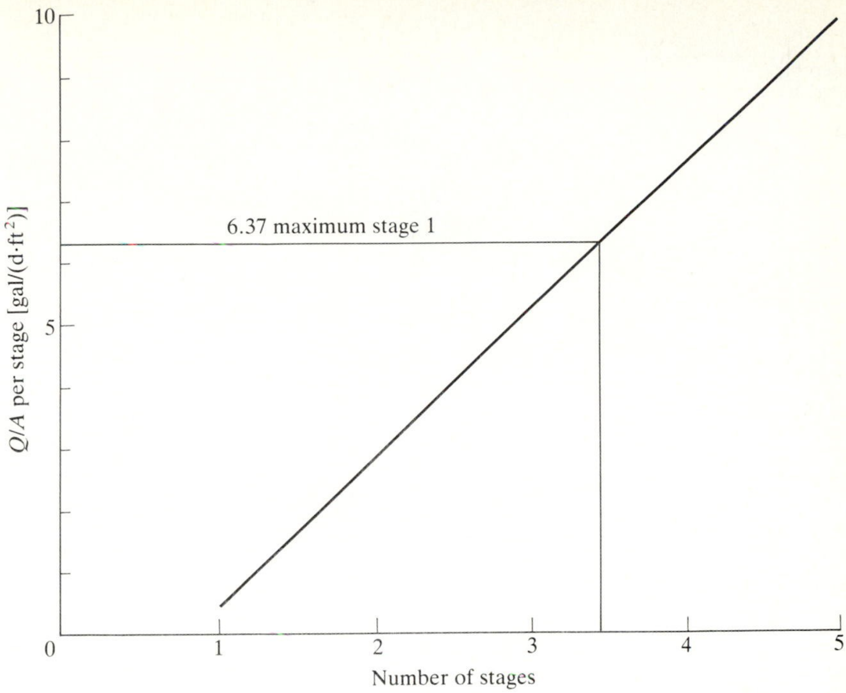

Note: gal/(d·ft²) = 4.075×10^{-2} m³/(d·m²)

FIGURE 7-28
Calculation of number of stages.

7.6 ANAEROBIC DECOMPOSITION

Anaerobic decomposition involves the breakdown of organic wastes to gas (methane and carbon dioxide) in the absence of oxygen. Although the process kinetics and material balances are similar to the aerobic systems, certain basic differences require special consideration.

The conversion of organic acids to methane gas yields little energy; hence the rate of growth is slow and the yield of organisms by synthesis is low. The kinetic rate of removal and the sludge yield are both considerably less than in the activated sludge process. The quantity of organic matter converted to gas will vary from 80 to 90 percent. Since there is less cell synthesis in the anaerobic process, the nutrient requirements are correspondingly less than in the aerobic system. High process efficiency requires elevated temperatures and the use of heated reaction tanks. The methane gas produced by the reaction can be used to provide this heat. Wastes of low COD or BOD concentration will not provide sufficient methane for heating, and a supplementary source of heat is necessary.

The anaerobic process is operated in one of several ways:

1. The anaerobic contact process[17,18] (shown in Fig. 7-29) provides for separation and recirculation of seed organisms, thereby allowing process operation

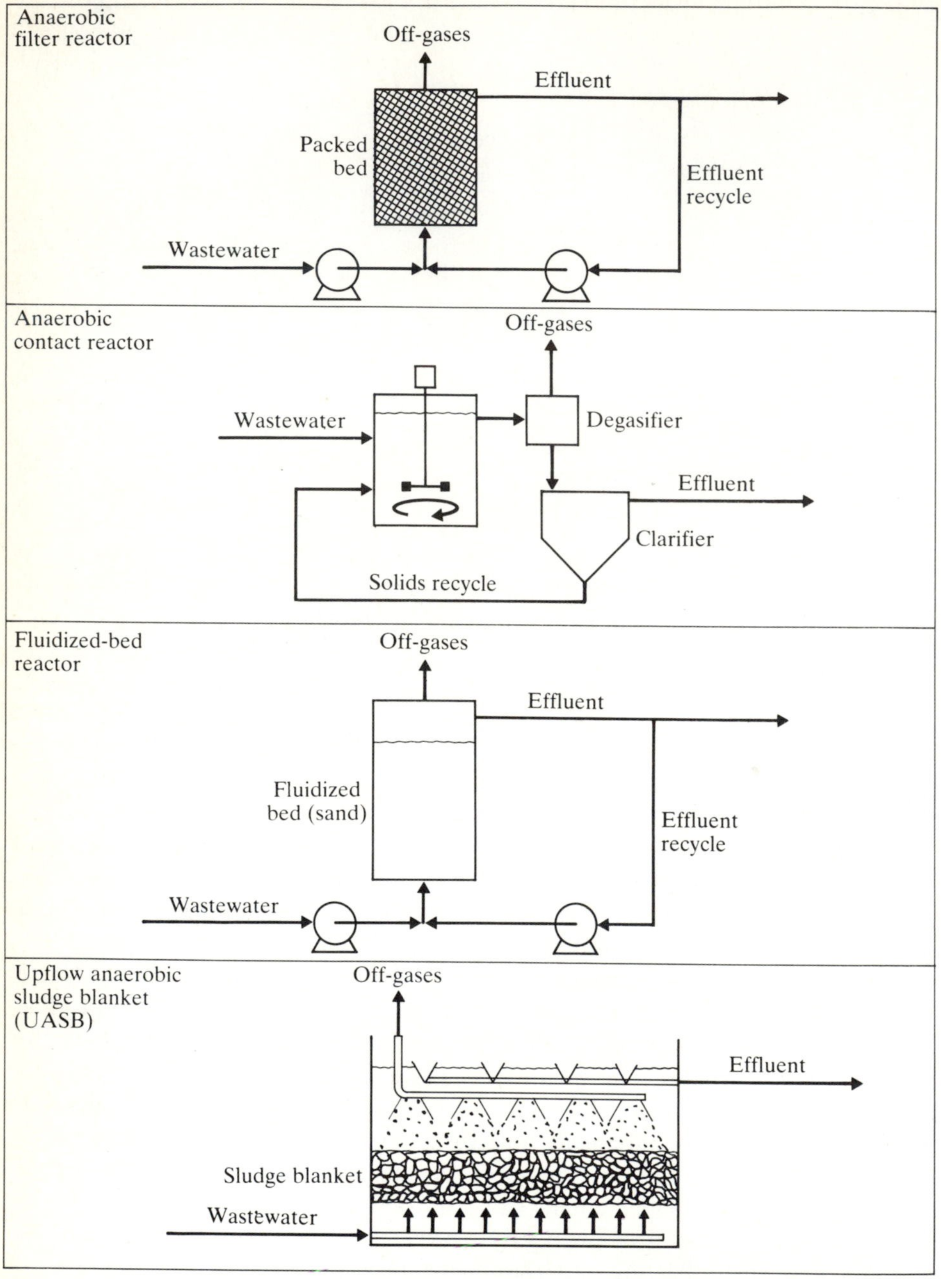

FIGURE 7-29
Anaerobic wastewater treatment processes.

at retention periods of 6 to 12 h. A degasifier is usually needed to minimize floating solids in the separation step. For high-degree treatment, the solids retention time has been estimated at 10 d at 90°F (32°C); the estimate doubles for each 20°F (11°C) reduction in operating temperature. A full-scale anaerobic contact process (30 to 35°C) treating a meat-packing wastewater at a loading of 2.5 kg COD/($m^3 \cdot d$), an HRT of 13.3 h, and a 90 percent removal of COD was reported by Steffen and Bedker.[17] The SRT was about 13.3 d.

2. The anaerobic filter establishes growth of the anaerobic organisms on a packing. The filter may be operated upflow, as shown in Fig. 7-29, or downflow. The packed filter media, while retaining biological solids, also provides a mechanism for separating the solids and the gas produced in the digestion process. Jennet and Dennis,[19] treating a pharmaceutical wastewater, were able to achieve a 97 percent removal of COD at a loading of 3.5 kg COD/($m^3 \cdot d$) (37°C). Sachs *et al.*[20] achieved an 80 percent COD reduction at a loading of 0.56 kg COD/($m^3 \cdot d$) (35°C) and a 36 h HRT for a synthetic organic chemicals wastewater. Obayashi and Roshanravan[21] obtained a 70 percent COD reduction at a loading of 2 kg COD/($m^3 \cdot d$) on a rendering wastewater.

3. In the upflow anaerobic sludge blanket (UASB) (Fig. 7-29) wastewater is directed to the bottom of the reactor where it must be distributed uniformly. The wastewater flows upward through a blanket of biologically formed granules which consume the waste as it passes through the blanket. Methane and carbon dioxide gas bubbles rise and are captured in the gas dome. Liquid passes into the settling portion of the reactor where solids-liquid separation takes place. The solids return to the blanket area while the liquid exits over the wiers. Formation of granules and their maintenance is extremely important in the operation of the process. Pol *et al.*[22] reported that 150 mg/l Ca^{2+} promotes granulation and Fe^{2+} at 5 to 10 mg/l helped reduce filamentous growth. To keep the blanket in suspension, an upflow velocity of 2 to 3 ft/h (0.6 to 0.9 m/h) has been used. Waste stabilization occurs as the waste passes through the sludge bed, with solids concentrations in the sludge bed having been reported to reach as high as 100 to 150 g/l.[22] Loadings up to 600 lb COD/(hundred $ft^3 \cdot d$) [96 kg/($m^3 \cdot d$)] have been used successfully in certain wastewaters. In pilot studies organic loadings of 15 to 40 kg COD/($m^3 \cdot d$) at liquid residence times of 3 to 8 h successfully treated high-strength wastewaters. A full-scale plant treating beet sugar wastewater achieved 80 percent removal with loadings of 10 kg COD/($m^3 \cdot d$) and an HRT of 4 h.

4. In the fluidized bed reactor (FBR) (Fig. 7-29) wastewater is pumped upward through a sand bed on which microbial growth has been developed. Biomass concentrations exceeding 30,000 mg/l have been reported. Recycle of effluent is returned to mix with the feed in quantities dictated by the strength of the wastewater and the fluidization velocity. Organic removal efficiencies of 80 percent were achieved at loadings of 4 kg COD/($m^3 \cdot d$) on dilute wastewaters.[23]

FIGURE 7-30
The ADI-BVF anaerobic process.

5. The ADI-BVF process is a low-rate anaerobic reactor with intermittent mixing and sludge recycle. The reactor has two zones: a reaction zone at the inlet end and a clarification zone at the outlet end. The reactor can be an above-ground tank or a lined earthen basin with a floating insulated membrane cover for gas recovery and temperature and odor control. Because of the large reactor volume, equalization requirements are minimized. A typical unit is shown in Fig. 7-30 and performance data in Table 7.6.

Mechanism of Anaerobic Fermentation

In anaerobic fermentation roughly four groups of microorganisms sequentially degrade organic matter. Hydrolytic microorganisms degrade polymer-type material such as polysaccharides and proteins to monomers. This reduction results in no reduction of COD. The monomers are then converted into fatty acids (VFA) with a small amount of H_2. The principal acids are acetic, propionic, and butyric with small quantities of valeric. In the acidification stage there is minimal reduction of COD. Should a large amount of H_2 occur, some COD reduction will result. This seldom exceeds 10 percent. All acids higher than acetic

TABLE 7.6
Performance of anaerobic processes

Wastewater	Process	Loading, $kg/(m^3 \cdot d)$	HRT, h	Temperature, °C	Removal, %	Reference
Meat packing	Anaerobic	3.2 (BOD)	12	30	95	18
Meat packing	contact	2.5 (BOD)	13.3	35	95	17
Keiring		0.085 (BOD)	62.4	30	59	24
Slaughter house		3.5 (BOD)	12.7	35	95.7	24
Citrus		3.4 (BOD)	32	34	87	25
synthetic	Upflow	1.0 (COD)	—	25	90	26
Pharmaceutical	filter	3.5 (COD)	48	35	98	19
Pharmaceutical		0.56 (COD)	36	35	80	20
Guar gum		7.4 (COD)	24	37	60	27
Rendering		2.0 (COD)	36	35	70	21
Landfill leachate		7.0 (COD)	—	25	89	28
Paper-mill foul condensate		10–15 (COD)	24	35	77	29
Synthetic	Expanded	0.8–4.0 (COD)	0.33–6	10–3	80	23
Paper-mill foul condensate	bed	35–48 (COD)	8.4	35	88	29
Skimmed milk	USAB	71 (COD)	5.3	30	90	30
Sauerkraut		8–9 (COD)	—	—	90	30
Potato		25–45 (COD)	4	35	93	30
Sugar		22.5 (COD)	6	30	94	30
Champagne		15 (COD)	6.8	30	91	30
Sugar beet		10 (COD)	4	35	80	31
Brewery		95 (COD)	—	—	83	32
Potato		10 (COD)	—	—	90	32
Paper-mill foul condensate		4–5 (COD)	70	35	87	29
Potato	ADI-BFV	0.2 (COD)	360	25	90	—
Corn starch		0.45 (COD)	168	35	85	—
Dairy		0.32 (COD)	240	30	85	—
Confectionary		0.51 (COD)	336	37	85	—

acid are converted into acetate and H_2 by acetogenic microorganisms. The conversion of propionic acid is

$$C_3H_6O_2 + 2H_2O \longrightarrow C_2H_4O_2 + CO_2 + 3H_2$$

In this reaction, COD reduction does occur in the form of H_2. This reaction will occur only if the concentration of H_2 is very low. The breakdown of organic acids to CH_4 and CO_2 is shown in Fig. 7-31. The acetic acid and H_2 are converted to CH_4 by methanogenic organisms:

Acetic acid:

$$C_2H_4O_2 \longrightarrow CO_2 + CH_4$$

$$CH_3COO^- + H_2O \longrightarrow CH_4 + HCO_3^-$$

Hydrogen:

$$HCO_3^- + 4H_2 \longrightarrow CH_4 + OH^- + 2H_2O$$

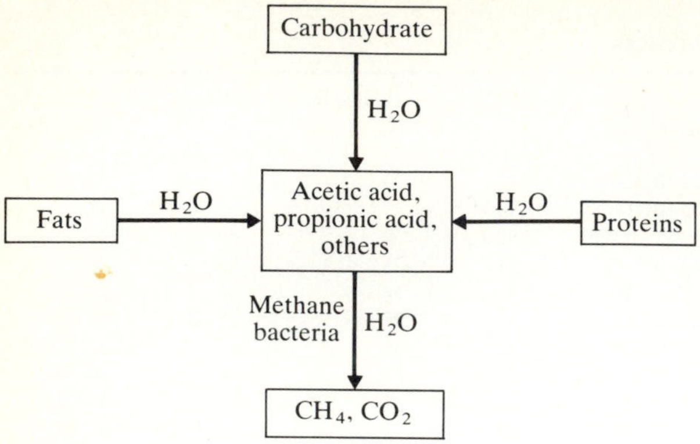

FIGURE 7-31
Anaerobic degradation of organics.

The specific activity of typical anaerobic processes treating soluble industrial wastewaters is approximately 1 kg COD utilized/(kg biomass · d). There are two classes of methanogenes that convert acetate to methane, namely *Methanothrix* and *Methanosarcina.*[33] *Methanothrix* has a low specific activity so that they will predominate in systems with a low steady-state acetate concentration. In highly loaded systems, *Methanosarcina* will predominate with a higher specific activity (3 to 5 times as high as *Methanothrix*) if trace nutrients are available. These trace nutrients are iron, cobalt, nickel, molybdenum, selenium, calcium, magnesium, and microgram per liter levels of vitamin B_{12}.

At standard temperature and pressure, 1 lb (0.454 kg) of COD or ultimate BOD removed in the process will yield 5.62 ft^3 (0.16 m^3) of methane.

The quantity of cells produced during methane fermentation will depend on the strength of the waste, the character of the waste, and the retention of the cells in the system. As in an aerobic system a portion of the cells produced will be destroyed by endogenous metabolism. The data of McCarty and Vath[34] are shown in Fig. 7-32. A relationship similar to Eq. (6.10) can be used to estimate cell yield. The relationships obtained by McCarty and Vath are:

Amino and fatty acids: $A = 0.054F - 0.038M$
Carbohydrates: $A = 0.46F - 0.088M$
Nutrient broth: $A = 0.076F - 0.014M$

where A = biological solids accumulated
M = mixed liquor volatile suspended solids
F = COD utilized

McCarty has estimated the cell composition as $C_5H_9NO_3$ in which the nitrogen requirement is 11 percent of the net cell weight. The phosphorus requirement has been estimated as 2 percent of the biological cell weight. The COD of the cells is 1.21 mg/mg volatile suspended solids.

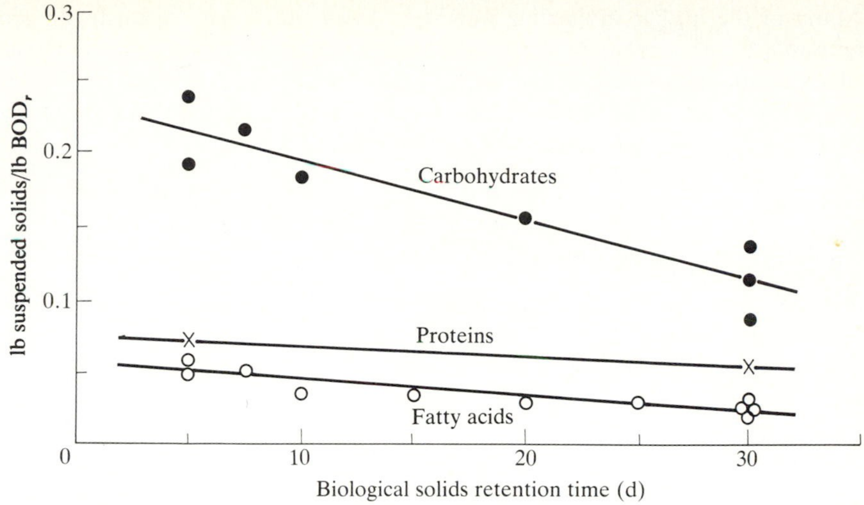

FIGURE 7-32
Biological solids production resulting from methane fermentation. (*After McCarty and Vath.*[34])

Biodegradation of Organic Compounds under Anaerobic Conditions

Anaerobic processes can break down a variety of aromatic compounds. It is known that anaerobic breakdown of the benzene nucleus can occur by two different pathways, namely, photometabolism and methanogenic fermentation. It has been shown that benzoate, phenylacetate, pheny/propionate, and annamate were completely degraded to CO_2 and CH_4. Lower fatty acids were detected as reaction intermediates. While long acclimation periods were required to initiate gas production, the time required could be reduced by adapting the bacteria to an acetic acid substrate before adapting them to the aromatic.

Chmielowski *et al.*[35] showed that phenol, *p*-cresol, and resorcinol yielded complete conversion to CH_4 and CO_2. Kirsch *et al.*[36] showed that after acclimation pentachlorophenol underwent anaerobic degradation.

Food-processing and brewery wastes are readily broken down anaerobically with BOD removals in the range of 85 to 95 percent. In the anaerobic treatment of distillery waste, special precautions must be taken to reduce the inhibitory tendency of sulfates contained in the waste by either dilution or adjusting loading rates. While animal wastes are readily treated anaerobically, ammonia toxicity may be a problem when dealing with fresh waste that contains a large amount of urine.[37] Vath[38] showed that linear anionic and nonionic ethoxylated surfactants underwent degradation as observed by a loss of surfactant properties. Numerous investigators have shown that a variety of pesticides including lindane and isomers of benzene hexachloride will degrade under anaerobic conditions.

Many of the higher molecular weight hydrocarbons that make up oil are decomposable by anaerobic bacteria. Shelton and Hunter[39] demonstrated that anaerobic decomposition of oil occurred naturally in bottom deposits.

An excellent review of the anaerobic treatment of industrial wastewaters by anaerobic processes has been presented by Obayashi and Gorgan.[40]

Factors Affecting Process Operation

The anaerobic process will effectively function over two temperature ranges: the mesophilic range of 85 to 100°F (29 to 38°C) and the thermophilic range of 120 to 135°F (49 to 57°C). Although the rates of reaction are much greater in the thermophilic range, the maintenance of higher temperatures is usually not economically justifiable.

The methane organisms function over a pH range of 6.6 to 7.6 with an optimum near pH 7.0. When the rate of acid formation exceeds the rate of breakdown to methane, a process unbalance results in which the pH decreases, gas production falls off, and the CO_2 content of the gas increases.

pH control is therefore essential to ensure a high rate of methane production. Lime is commonly used to raise the pH of an anaerobic system when there is a process imbalance. Caution must be taken, since excess application of lime will result in precipitation of calcium carbonate. Sodium bicarbonate can alternatively be used for pH adjustment. It is desirable to have a bicarbonate alkalinity in the range of 2500 to 5000 mg/l in order to provide a buffer capacity to handle volatile acids increase with a minimal decrease in pH. If sufficient amounts of alkalinity are not present in the feed, then the alkalinity can be controlled by reducing the feed rate or adding alkalinity to the wastewater.

Inorganic salts in low concentration may provide stimulation; in high concentration, they may be toxic.[41] In some cases, adaptation will increase the tolerance level of the organisms. The stimulatory and inhibitory concentrations of some common cations are summarized in Table 7.7.

TABLE 7.7
Cation toxicity to anaerobic decomposition
Maximum ion concentration, mol/l

	Other ions < 10 mg/l		Antagonist present	
Ion	**Slug**	**Acclimated**	**Slug**	**Acclimated**
Na^+	0.2	0.3	0.25–0.3	>0.35
NH_4^+	0.1		0.15–0.18	
K^+	0.09	0.15	0.15–0.2	>0.35
Ca^{2+}	0.07	>0.2	0.13	>0.2
Mg^{2+}	0.05	0.075	0.1	>0.14

The presence of antagonistic ions may sharply reduce the inhibitory effect of specific cations. Kugelman and McCarty[42] have shown that 300 mg/l potassium will reduce by 80 percent the inhibitory effect of 7000 mg/l sodium. The inhibitory effect will be completely eliminated by the addition of 150 mg/l calcium, but the calcium without the potassium will exert no beneficial effect. Ammonia is toxic at concentrations in excess of 3000 mg/l and inhibitory at levels greater than 1500 mg/l. The toxicity is related to the pH, which indicates how much ammonia is present in gaseous form. Gaseous ammonia is more inhibitory than the ammonium ion.

The maximum nontoxic concentration of soluble sulfide in an anaerobic system is 200 mg/l. Since sulfide is present as H_2S (gaseous), HS^-, and precipitated sulfide, the total sulfide concentration may be considerably higher. The loss of sulfide in the gas would also permit higher concentrations of either sulfate or sulfide in the waste feed to the system.

The heavy metals, such as copper, zinc, and nickel, are toxic at low concentrations. In the presence of sulfides, however, some of these metals are precipitated. The toxicity to organisms is related to their concentration in the digester itself.

When toxic materials are present, they must be removed before treatment or diluted below the toxic levels. The proper approach will depend principally on the overall waste composition.

In order to avoid a decrease in pH, sufficient alkalinity must be present to compensate for the high CO_2 content. At a 30 percent CO_2 content in the digester gas, 1500 mg/l of alkalinity is necessary. A laboratory apparatus for developing anaerobic digestion data is shown in Figs. 7-33 and 7-34.

A batch fermentation will yield information on the anaerobic degradability of a wastewater. Anaerobic sludge is mixed with the wastewater and the contents gently stirred. The gas exiting the reactor is captured and the CH_4 content measured. Changes in COD and VFA are recorded. Sampling intervals are usually selected after generation of 500 ml of methane gas. Typical data are shown in Fig. 7-35.

Methane production can be estimated by the relationship

$$G = 5.62(S_r - 1.42\Delta X_v) \tag{7.28}$$

where G = CH_4 produced per day, ft^3/d
S_r = BOD removed, lb/d
X_v = VSS produced, lb/d

or

$$G = 0.351(S_r - 1.42\Delta X_v) \tag{7.28a}$$

where $G = m^3/d$; S_r and ΔX_v = kg/d

One cubic foot ($0.0283\ m^3$) of methane has a net heating value of 960 Btu (1.01×10^6 J). The performance of anaerobic systems treating industrial wastes is summarized in Table 7.8. The design of an anaerobic contact process is shown in Example 7.9.

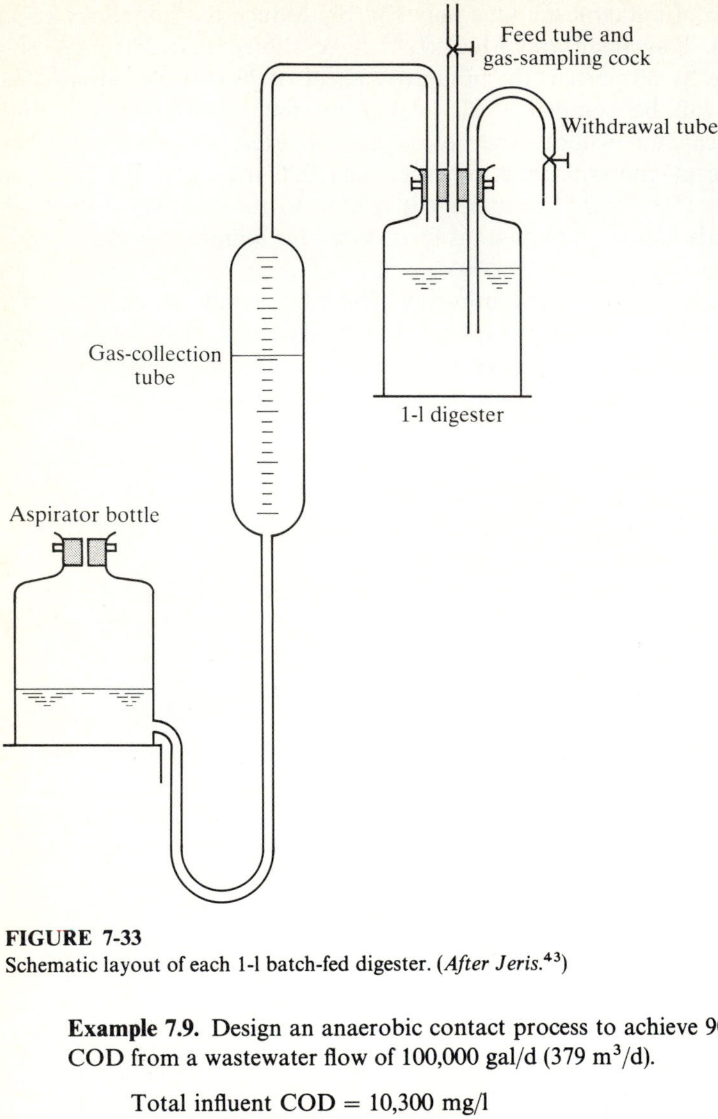

FIGURE 7-33
Schematic layout of each 1-l batch-fed digester. (*After Jeris.*[43])

Example 7.9. Design an anaerobic contact process to achieve 90 percent removal of COD from a wastewater flow of 100,000 gal/d (379 m^3/d).

Total influent COD = 10,300 mg/l
Nonremovable COD = 2200 mg/l
Removable COD (COD_R) = 8100 mg/l
COD to be removed = 90 percent

Process parameters:

SRT = 20 days minimum
Temperature = 35°C
$a = 0.136$ mg VSS/mg COD_R
$b = 0.021$ mg VSS/(mg VSS · day)
$k = 0.0004$ l/(mg · d)
$X_v = 5000$ mg/l

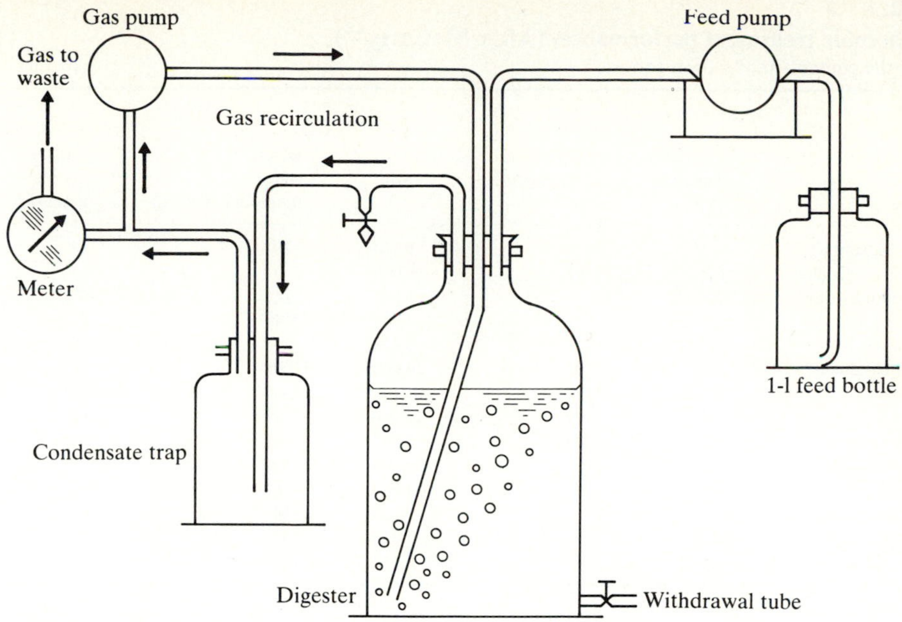

FIGURE 7-34
Schematic layout of continuous-fed digester. (*After Jeris.*[43])

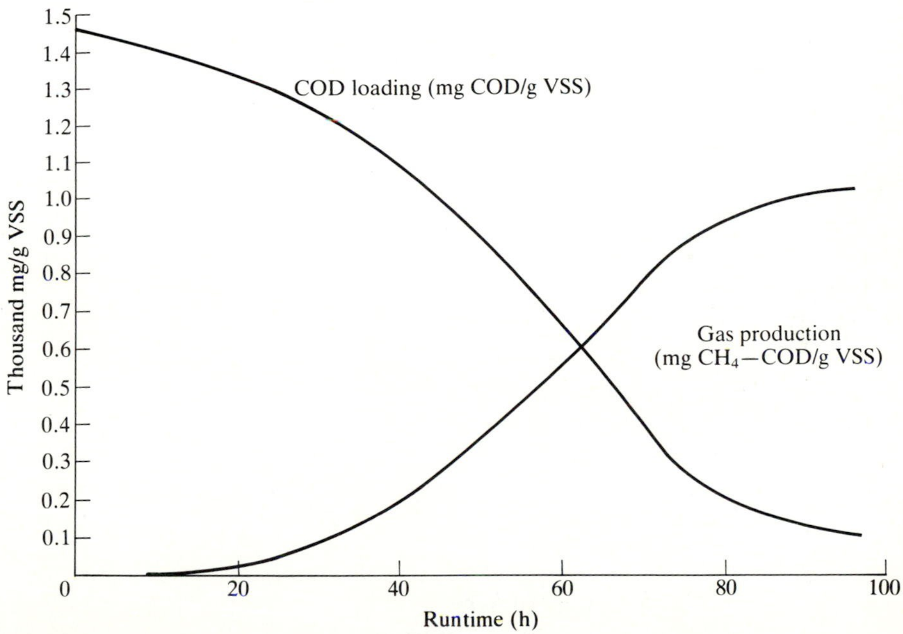

FIGURE 7-35
Batch anaerobic digestion of an organic wastewater.

TABLE 7.8
Anaerobic treatment performance. (After McCarty[44])
For the contact-stabilization process

Waste	Hydraulic detention time, d	Digestion temperature, °F	BOD$_5$ Raw waste, mg/l	BOD$_5$ added, lb/(thousand ft^3·d)	Removed, %
Maize starch	3.3	73	6,280	110	88
Whisky distillery	6.2	92	25,000	250	95
Cotton kiering	1.3	86	1,600	74	67
Citrus	1.3	92	4,600	214	87
Brewery	2.3		3,900	127	96
Starch-gluten	3.8	95	14,000†	100†	80†
Wine	2.0	92	23,400†	730†	85†
Yeast	2.0	92	11,900†	372†	65†
Molasses	3.8	92	32,800†	546†	69†
Meat packing	1.3	92	2,000	110	95
Meat packing	0.5	92	1,380	156	91
Meat packing	0.5	95	1,430	164	95
Meat packing	0.5	85	1,310	152	94
Meat packing	0.5	75	1,110	131	91

† Volatile suspended solids, rather than BOD$_5$.

For the conventional process

Waste	Hydraulic detention time, d	Digestion temperature, °F	Volatile solids added, lb/(thousand ft^3·d)	Volatile solids stabilized, %	BOD$_5$ added, lb/(thousand ft^3·d)	BOD stabilized, %
Pea blancher	3.5	131	700	83		
Pea blancher	6.0	99	400	85		
Winery		97	200	87		
Butanol	10.0		190		110	68
Rye fermentation	2.0	130	930†			54
Corn fermentation	4.0	130	330†			76
Whey waste	29.0	130	150†			72
Acetic acid	30.0	95	1370		975	90
Butyric acid	30.0	95	830		1000	91

† Total solids.

Note:

°C $= \frac{5}{9}(°F - 32)$

lb/(thousand ft^3·d) = 0.016 kg/(m^3·d)

Digester volume

From the kinetic relationship:

$$t = \frac{S_r}{X_v kS}$$

$$= \frac{(8100)(0.9)}{(5000)(0.0004)(810)} = 4.5 \text{ d}$$

The digester volume is therefore

$$(4.5\ \text{d})(0.1\ \text{million gal/d}) = 0.45\ \text{million gal}\ (1700\ \text{m}^3)$$

Check SRT (sludge age):

$$\text{SRT} = \frac{X_v t}{\Delta X_v} = \frac{X_v t}{aS_r - bX_v t}$$

$$= \frac{(5000)(4.5)}{(0.136)(7290) - (0.021)(5000)(4.5)} = 43.4\ \text{d}$$

This is in excess of the recommended SRT of 20 days to insure the growth of methane formers.

A vacuum degasifier or flash aerator should be provided between the digester and the clarifier to purge gas from the sludge. A flash aerator inhibits further methane production in the clarifier with resulting floating sludge.

Sludge yield
The sludge yield from the process is

$$\Delta X_v = aS_r - bX_v t$$

$$= (0.136)(7290) - (0.021)(5000)(4.5) = 519\ \text{mg/l}$$

$$\Delta X_v = 519\ \text{mg/l} \times 0.1\ \text{million gal/d} \times 8.34\ \frac{\text{lb/million gal}}{\text{mg/l}}$$

$$= 433\ \text{lb/d}\ (196\ \text{kg/d})$$

Gas production

$$G = 5.62(S_r - 1.42\Delta X_v)$$

$$= 5.62(7290)(0.1)(8.34) - (1.42)(433)$$

$$= 30{,}700\ \text{ft}^3\ \text{CH}_4\text{/d}\ (869\ \text{m}^3\text{/d})$$

Heat requirements
These can be estimated by calculating the energy required to raise the influent wastewater temperature to 95°F (35°C) and allowing 1°F (0.56°C) heat loss per day of detention time.
Average wastewater temperature = 75°F (23.9°C)

Heat-transfer efficiency = 50 percent

$$\text{Btu} = \frac{W(T_i - T_e)}{E} \times (\text{specific heat})$$

$$\text{Btu}_{\text{required}} = \frac{(0.1\ \text{million gal/d})(8.34\ \text{lb/gal})(95^\circ + 4.5^\circ - 75^\circ)}{0.5} \times \left(\frac{1\ \text{Btu}}{\text{lb}\cdot{}^\circ\text{F}}\right)$$

$$= 40{,}900{,}000\ \text{Btu/d}\ (4.3 \times 10^{10}\ \text{J/d})$$

The heat available from gas production is

$$\text{Btu}_{\text{available}} = (30{,}700 \text{ ft}^3 \text{ CH}_4/\text{d})(960 \text{ Btu/ft}^3 \text{ CH}_4)$$
$$= 29{,}500{,}000 \text{ Btu/d } (3.1 \times 10^{10} \text{ J/d})$$

External heat of 40,900,000 − 29,500,000 = 11,400,000 Btu/d (1.2×10^{10} J/d) should be supplied to maintain the reactor at 95°F (35°C).

Nutrient requirements
The nitrogen requirement is

$$\text{N} = 0.12\Delta X_v = 0.12 \times 433 \text{ lb/d}$$
$$= 52 \text{ lb/d } (23.6 \text{ kg/d})$$

The phosphorus requirement is

$$\text{P} = 0.025\Delta X_v = 0.025 \times 433 \text{ lb/d}$$
$$= 11 \text{ lb/d } (5 \text{ kg/d})$$

PROBLEMS

7.1. A wastewater is presently being treated in four lagoons in series. Compute the efficiency of the lagoon system in the summer ($T = 25$°C) and the winter ($T = 10$°C)

Data
Flow = 1.8 million gal/d (6810 m^3/d)
BOD = 1000 mg/l

First lagoon—6 million gal (22,700 m^3), 12 ft deep (3.66 m)
Second lagoon—10 million gal (37,850 m^3), 10 ft (3.05 m) deep
Third lagoon—45 million gal (170,300 m^3), 6 ft (1.83 m) deep
Fourth lagoon—45 million gal, 6 ft deep

$K_{\text{aerobic}} = 0.5$/d at 20°C
$K_{\text{anaerobic}} = 0.05$/d at 20°C
$\theta = 1.08$

7.2. A wastewater of 3 million gal/d (11,350 m^3) has a BOD of 2100 mg/l and a VSS content of 400 mg/l. It is to be pretreated in an aerobic lagoon with a retention period of 1.2 d. The K is 8.0/d. Compute:

(*a*) The effluent soluble BOD in mg/l
(*b*) The effluent VSS in mg/l
(*c*) The oxygen required in mass/day if

$a = 0.5$
$a' = 0.55$
$b = 0.15$/d

7.3. An industrial facility is to treat its waste in a series of aerated lagoons. The final effluent should have a maximum soluble BOD_5 of 20 mg/l for the summer months and 30 mg/l for the winter months.

$$\text{Flow} = 8.5 \text{ million gal/d } (32{,}170 \text{ m}^3\text{/d})$$
$$BOD_{5\,\text{inf}} = 425 \text{ mg/l}$$
$$\text{Temperature of waste} = 95°\text{F } (35°\text{C})$$
$$K = 6.3/\text{d at } 20°\text{C}$$
$$\theta = 1.065$$
$$a = 0.5$$
$$a' = 0.52$$
$$b = 0.20/\text{d}$$
$$\alpha = 0.88$$
$$C_1 = 1.0 \text{ mg/l}$$
$$T_a = 30°\text{F (winter)}$$
$$= 65°\text{F (summer)}$$

Design the lagoons and optimize the system with respect to area. Design and specify the aeration system. Present a diagram showing basin geometries and aerator locations.

7.4. A batch activated sludge plant is to be designed for the following conditions:

$$Q = 40{,}500 \text{ gal/d } (153 \text{ m}^3\text{/d})$$
$$S_o = 975 \text{ mg/l}$$
$$S_e = 10 \text{ mg/l}$$
$$\text{Period of wastewater discharge} = 11.0 \text{ h}$$

A laboratory bench scale study yielded the following operational parameters:

$$(F/M)_{\text{av}} = 0.2/\text{d}$$
$$a = 0.53$$
$$b = 0.16/\text{d}$$
$$a' = 0.43$$
$$\text{Eff SS} = 50 \text{ mg/l (85\% volatile)}$$
$$\text{Equilibrium sludge density} = 0.8 \text{ lb sludge/ft}^3 \ (12.8 \text{ kg/m}^3)$$

The sludge is to be stored for a period of 30 days when the reactor will be emptied and cleaned. A value of 45 lb BOD removed/(hp · d) [27.4 kg/(kW · d)] was also found to be in effect.

(*a*) Calculate the following:
  - (i) Volume of active sludge required to oxidize each day's accumulation of degradable solids
  - (ii) The 30 days sludge storage volume
  - (iii) The daily volume for the waste
  - (iv) The freeboard at 5 percent of the waste volume
  - (v) The total volume and basin dimensions

(*b*) Calculate the amount of oxygen per day required.

(*c*) Calculate the effluent BOD total.

(*d*) Estimate the aeration power requirements.

7.5. The municipal plant in Example 7.6 is planning to accept 0.5 million gal/d (1890 m^3/d) of a brewery wastewater with a BOD of 1200 mg/l. Determine what changes need to be made to avoid filamentous bulking. Use the F/M dissolved relationship shown in Fig. 6-14.

7.6. A municipal plant receives fruit-processing wastewaters and has a combined BOD of 600 mg/l. The return sludge has a concentration of 8000 mg/l ISS. What recycle is required for a contactor with a floc load of 150 mg BOD/g VSS?

7.7. Compute the required clarifier area for a wastewater flow of 1.0 million gal/d (3785 m^3/d), an MLSS of 3000 mg/l, an SVI of 150 ml/g, and an R/Q of 0.5. What is the maximum MLSS if the SVI increases to 250 ml/g and the recycle rate stays the same?

7.8. A trickling filter presently achieves 52 percent BOD reduction from a brewery wastewater under the following conditions:

$N = 3{:}1$
$Q = 3$ gal/(min · ft^2) [0.122 m^3/(m^2 · min)]
$d = 20$ ft (6.1 m)
$S_o = 850$ mg/l
Flow = 0.8 million gal/d (3030 m^3/d)
$n = 0.5$

In-plant changes will reduce the flow to 0.5 million gal/d and the BOD to 700 mg/l. Using the same recycle, what will the new BOD removal efficiency be?

7.9. The following data were obtained on an RBC for a bleached pulp-mill effluent. The test unit had four stages.

Average flow, million gal/(d · ft^2)	Influent, mg/l	Stages, mg/l 1	2	3	4
2.0	76	26	10	7	6
4.0	72	41	24	16	12
6.0	81	42	25	15	10
8.0	78	52	38	25	18
10.0	83	60	47	36	30

Correlate the data and design a system for 80 percent BOD removal, assuming an influent BOD of 85 mg/l.

7.10. Consider Example 7-9. At present, an air-activated sludge plant follows the anaerobic process. The activated sludge plant operates under the following conditions to produce an effluent with a COD of 30 mg/l (degradable):

$T = 20°C$
$F/M = 0.3\ d^{-1}$
$X_v = 2500$ mg/l

The load to the anaerobic plant is to be increased to 120,000 gal/d (454 m^3/d) with the same removable COD.

(*a*) Compute the new effluent from the anaerobic plant assuming X_v remains the same. What will the new gas production be?

(*b*) What modifications to the aerobic plant must be made to maintain the same effluent quality? Assume the sludge settling characteristics are the same as originally. Compute:

(i) The new sludge recycle and the new MLSS. The volatile content of the sludge is 74 percent.
(ii) The new F/M.
(iii) Calculate the increase in power required assuming 1.5 lb O_2/(hp · h) (0.91 kg/kW).

The process parameters are:

$a = 0.5$

$a' = 0.6$

$b = 0.15$ at 20°C

REFERENCES

1. Gloyna, E. F.: *Waste Stabilization Pond Concepts and Experiences*, Division of Environmental Health, World Health Organization, Geneva, 1965.
2. Oswald, W. J.: *Advances in Water Quality Improvement* (Eds. E. F. Gloyna and W. W. Eckenfelder), vol. 1, University of Texas Press, Austin, Texas, 1968.
3. Gellman, I., and H. F. Berger: *Advances in Water Quality Improvement* (Eds. E. F. Gloyna and W. W. Eckenfelder), vol. 1, University of Texas Press, Austin, Texas, 1968.
4. Hermann, E. R., and E. F. Gloyna: *Sewage Ind. Wastes*, vol. 30, p. 963, 1958.
5. Marais, G. V., and V. A. Shaw: *Trans. S. Afr. Inst. Civ. Engrs*, vol. 3, p. 205, 1961.
6. Steffen, A. J.: *J. Water Pollut. Control Fed.*, vol. 35, pt. 4, p. 440, 1963.
7. Cooper, R. C.: *Dev. Appl. Microbiol.*, vol. 4, pp. 95–103, 1963.
8. Argamon, Y., and C. Adams: *Proceedings of 8th International Conference on Water Pollution Research*, Pergamon Press, Oxford, 1976.
9. Eckenfelder, W. W., C. Adams, and S. McGee: *Advances in Water Pollution Research*, vol. 2, Pergamon Press, Oxford, 1972.
10. Goronszy, M.: *J. Water Pollut. Control Fed.*, vol. 41, pt. 2, p. 274, 1979.
11. Daigger, G. T., and R. E. Roper: *J. Water Pollut. Control Fed.*, vol. 57, p. 859, 1985.
12. Paduska, R. A.: *Proc. 24th Ind. Waste Conf.*, 1979, Purdue University.
13. Howland, W. E.: *Proc. 12th Ind. Waste Conf.*, Purdue University, vol. 94, p. 435, 1958.
14. Eckenfelder, W. W., and E. L. Barnhart: *J. Water Pollut. Control Fed.*, vol. 35, p. 535, 1963.
15. Eckenfelder, W. W.: *Proc. ASCE SA*, vol. 4, pt. 2, p. 860, 33, July 1961.
16. Department of Scientific and Industrial Research: *Water Pollution Research, 1956 (British)*, Her Majesty's Stationery Office, London, 1957.
17. Steffen, A. J., and M. Bedker: *Proc. 16th Ind. Waste Conf.*, 1961, Purdue University.
18. Schropfer, G. J., *et al.*: *Sewage Ind. Wastes*, vol. 27, p. 460, 1955.
19. Jennet, J. C., and N. D. Dennis: *J. Water Pollut. Control Fed.*, vol. 47, p. 104, 1975.
20. Sachs, E. F., *et al.*: *Proc. 33d Ind. Waste Conf.*, 1978, Purdue University.
21. Obayashi, A. W., and M. Roshanravan: Unpublished report, Illinois Institute of Technology, Chicago, Ill., 1980.
22. Pol, H. L. W., *et al.*: IAWPR Seminar on *Anaerobic Treatment of Wastewater in Fixed Film Reactors*, Copenhagen, June 1982.
23. Jewell, W. J., *et al.*: *J. Water Pollut. Control Fed.*, vol. 53, p. 482, 1981.
24. Pettet, A. E., *et al.*: Institution of Public Health Engineers, New York, N.Y., 1959.
25. McNary, R. R., *et al.*: *Proc 17th Ind. Waste Conf.*, 1962, Purdue University.
26. Young, J. C., and P. L. McCarty: Technical Report 87, Department of Civil Engineering, Stanford University, 1968.
27. Witt, E. R., *et al.*: *Proc. 34th Ind. Waste Conf.*, 1979, Purdne University.
28. Dewalle, F. B., and E. S. K. Chian: *Biotech. Bioengng*, vol. 18, p. 1275, 1976.
29. Donovan, G.: Personal communication.
30. Lettinga, G., and W. de Zeeuw: *Proc. 35th Ind. Waste Conf.*, 1980, Purdue University.
31. Pette, K. C., *et al.*: CSM Suiker, Amsterdam, Netherlands.
32. J. Oat Corp.: *Biothane Dig.*, vol. 1, pp. 1–4, 1981.
33. Speece, R. E.: *2d Natl. Conf. on Anaerobic Treatment of Industrial Wastewaters*, Chicago, Ill., 1986.
34. McCarty, P. L., and C. A. Vath: *Int. J. Air Water Pollut.*, vol. 6, p. 65, 1962.
35. Chmielowski, J., *et al.*: *Zesz. Nauk. Politech Slaska Inz. (Polish)*, vol. 8, p. 97, 1965.
36. Kirsch, E. J. *et al.*: U.S. EPA, March 1981.

37. Lapp, *et al.*: *Proc. 3d Int. Symp. on Livestock Wastes*, University of Illinois, 1975.
38. Vath, C. A.: Soap and Chem. Specif., March 1964.
39. Shelton, T. B., and J. B. Hunter: *J. Water Pollut. Control Fed.*, vol. 47, pt. 9, p. 2257, 1975.
40. Obayashi, A. W., and J. M. Gorgan: *Management of Industrial Pollutants by Anaerobic Processes*, Lewis Publishers, Chelsea, Mich., 1985.
41. McCarty, P. L., *et al.*: *J. Water Pollut. Control Fed.*, vol. 35, pt. 1, p. 501, 1963.
42. Kugelman, I. J., and P. L. McCarty: *Proc. 19th Ind. Waste Conf.*, 1964, Purdue University.
43. Jeris, J.: Personal communication.
44. McCarty, P. L.: Public Works, September to December 1964.

CHAPTER

8

ADSORPTION

Many industrial wastes contain organics which are refractory and which are difficult or impossible to remove by conventional biological-treatment processes. Examples are ABS and some of the heterocyclic organics. These materials can frequently be removed by adsorption on an active-solid surface. The most commonly used adsorbent is activated carbon.

8.1 THEORY OF ADSORPTION

A solid surface in contact with a solution tends to accumulate a surface layer of solute molecules because of the unbalance of surface forces. Chemical adsorption results in the formation of a monomolecular layer of the adsorbate on the surface through forces of residual valence of the surface molecules. Physical adsorption results from molecular condensation in the capillaries of the solid. In general, substances of the highest molecular weight are most easily adsorbed. There is a rapid formation of an equilibrium interfacial concentration, followed by slow diffusion into the carbon particles. The overall rate of adsorption is controlled by the rate of diffusion of the solute molecules within the capillary pores of the carbon particles. The rate varies reciprocally with the square of the particle diameter, increases with increasing concentration of solute, increases with increasing temperature, and decreases with increasing molecular weight of the solute. Morris and Weber[1] found the rate of adsorption to vary as the square root of the time of contact with the adsorbent (Fig. 8-1). The rate also increases with decreasing pH because of changes in surface charges of the carbon.

The adsorptive capacity of a carbon for a solute will likewise be dependent on both the carbon and the solute.

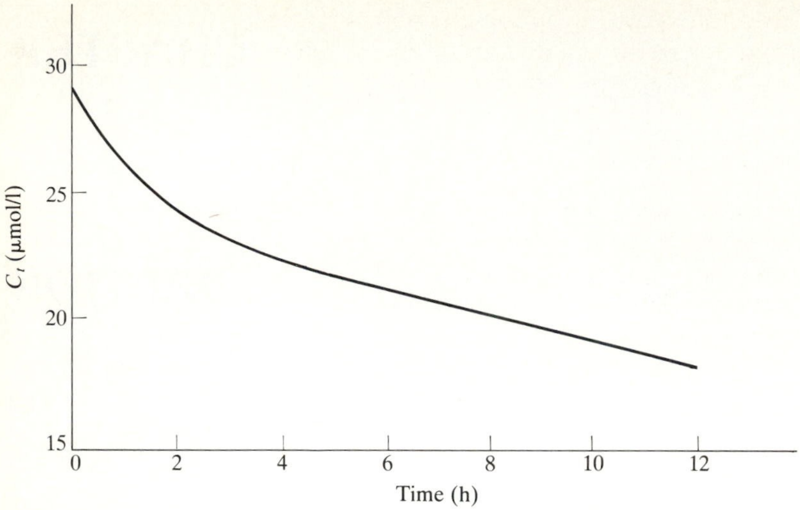

FIGURE 8-1
Rate of adsorption of 2-dodecyl benzene sulfonate by Columbia carbon (30°C, 75 mg/l, 0.273 mm diameter). (*After Morris and Weber.*[1])

TABLE 8.1
Influence of molecular structure and other factors on adsorbability

1. An increasing solubility of the solute in the liquid carrier decreases its adsorbability.
2. Branched chains are usually more adsorbable than straight chains. An increasing length of the chain decreases solubility.
3. Substituent groups affect adsorbability:

Substituent group	Nature of influence
Hydroxyl	Generally reduces adsorbability. Extent of decrease depends on structure of host molecule.
Amino	Effect similar to that of hydroxyl but somewhat greater. Many amino acids are not adsorbed to any appreciable extent.
Carbonyl	Effect varies according to host molecule. Glyoxylic acid more adsorbable than acetic but similar increase does not occur when introduced into higher fatty acids.
Double bonds	Variable effects as with carbonyl.
Halogens	Variable effects.
Sulfonic	Usually decreases adsorbability.
Nitro	Often increases adsorbability

4. Generally, strong ionized solutions are not as adsorbable as weakly ionized ones; i.e., undissociated molecules are in general preferentially adsorbed.
5. The amount of hydrolytic adsorption depends on the ability of the hydrolysis to form an adsorbable acid or base.
6. Unless the screening action of the carbon pores intervene, large molecules are more sorbable than small molecules of similar chemical nature. This is attributed to more solute carbon chemical bonds being formed, making desorption more difficult.
7. Molecules with low polarity are more sorbable than highly polar ones.

Most wastewaters are highly complex and vary widely in the adsorbability of the compounds present. Molecular structure, solubility, etc., all affect the adsorbability. These effects are shown in Table 8.1.

Formulation of Adsorption

The degree to which adsorption will occur and the resulting equilibrium relationships have been correlated according to the empirical relationship of Freundlich and the theoretically derived Langmuir relationship. For practical application, the Freundlich isotherm usually provides a satisfactory correlation. The Freundlich isotherm is expressed as

$$\frac{X}{M} = kC^{1/n} \tag{8.1}$$

where X = weight of substance adsorbed
M = weight of adsorbent
C = concentration remaining in solution

k and n are constants depending on temperature, the adsorbent, and the substance to be adsorbed. The Freundlich constants for several priority pollutants are shown in Table 8.2.[2]

The Langmuir equation is based on an equilibrium between condensation and evaporation of adsorbed molecules, considering a monomolecular adsorption layer:

$$\frac{X}{M} = \frac{abC}{1 + aC} \tag{8.2}$$

This can be reexpressed in linear form as

$$\frac{1}{X/M} = \frac{1}{b} + \frac{1}{ab}\frac{1}{C} \tag{8.2a}$$

where b = amount adsorbed to form a complete monolayer on the surface
a = constant which increases with increasing molecular size

Since most wastewaters contain more than one substance which will be adsorbed, direct application of the Langmuir equation is not possible. Morris and Weber[1] have developed relationships from the Langmuir equation for competitive adsorption of two substances:

$$\frac{X_A}{M} = \frac{a_A b_A C_A}{1 + a_A C_A + a_B C_B} \tag{8.3a}$$

$$\frac{X_B}{M} = \frac{a_B b_B C_B}{1 + a_A C_A + a_B C_B} \tag{8.3b}$$

TABLE 8.2
Summary of Freundlich parameters at neutral pH

Compound	*K*, mg/g	1/*n*
Hexachlorobutadiene	360	0.63
Anethole	300	0.42
Phenyl mercuric acetate	270	0.44
p-Nonylphenol	250	0.37
Acridine yellow	230	0.12
Benzidine dihydrochloride	220	0.37
n-Butylphthalate	220	0.45
N-Nitrosodiphenylamine	220	0.37
Dimethylphenylcarbinol	210	0.33
Bromoform	200	0.83
β-Naphthol	100	0.26
Acridine orange	180	0.29
α-Naphthol	180	0.31
α-Naphthylamine	160	0.34
Pentachlorophenol	150	0.42
p-Nitroaniline	140	0.27
1-Chloro-2-nitrobenzene	130	0.46
Benzothiazole	120	0.27
Diphenylamine	120	0.31
Guanine	120	0.40
Styrene	120	0.56
Dimethyl phthalate	97	0.41
Chlorobenzene	93	0.98
Hydroquinone	90	0.25
p-Xylene	85	0.16
Acetophenone	74	0.44
1,2,3,4-Tetrahydronaphthalene	74	0.81
Adenine	71	0.38
Nitrobenzene	68	0.43
Dibromochloromethane	63	0.93

More complex relationships could similarly be developed for multicomponent mixtures. It should be noted that although the equilibrium capacity for each individual substance adsorbed in a mixture is less than that of the substance alone, the combined adsorption is greater than that of the individuals alone. In industrial application, contact times of less than 1 h are usually used. Equilibrium is probably closely realized when high carbon dosages are employed, since the rate of adsorption increases with carbon dosage.

8.2 PROPERTIES OF ACTIVATED CARBON

Activated carbons are made from a variety of materials including wood, lignin, bituminous coal, lignite, and petroleum residues. Granular carbons produced from medium volatile bituminous coal or lignite have been most widely applied to the treatment of wastewater. Activated carbons have specific properties depending on the material source and the mode of activation. Property standards are helpful in specifying carbons for a specific application. In general, granular carbons from bituminous coal have a small pore size, a large surface area, and the highest bulk density. Lignite carbon has the largest pore size, least surface area, and the lowest bulk density. Adsorptive capacity is the effectiveness of the carbon in removing desired constituents such as COD, color, phenol, etc., from the wastewater. Several tests have been employed to characterize adsorptive capacity. The phenol number is used as an index of a carbon's ability to remove taste and odor compounds. The iodine number relates to the ability of activated carbon to adsorb low-molecular-weight substances (micropores having an effective radius of less than 2 μm) while the molasses number relates to the carbon's ability to adsorb high-molecular-weight substances (pores ranging from 1 to 50 μm). In general, high iodine numbers will be most effective on wastewaters with predominantly low-molecular-weight organics, while high molasses numbers will be most effective for wastewaters with a dominance of high-molecular-weight organics. Properties of commercial carbons are shown in Table 8.3.

Depending on the characteristics of the wastewater, one type of carbon may be superior to another (Fig. 8-2) since the capacity is greater at equilibrium effluent concentrations. For TOC, carbon-2 is better for carbon column operation since the capacity is greater in equilibrium with the influent (C_0) at column exhaustion, while carbon-1 would be better for batch treatment.

Laboratory Evaluation of Adsorption

In order to evaluate the feasibility and economics of adsorption, a laboratory adsorption study should be conducted. If granular carbon is to be evaluated, the carbon must first be ground to pass a 325-mesh screen. Grinding the carbon will not significantly affect its adsorptive capacity, but will increase the rate of adsorption. The time of contact required to approach equilibrium should first be evaluated. A carbon dosage of 500 mg/l is mixed with waste for various periods of time, and the degree of adsorption is determined at selected time intervals. A mixing time sufficient to achieve 90 percent or more of equilibrium should be used for subsequent studies. Usually, a 2-h contact is sufficient to attain greater than 90 percent of equilibrium, though in some cases a longer contact time is required. The initial testing should include a 24-h contact time. If the equilibrium value after 2 h is greater than 90 percent of the 24-h value, the 2-h test can be used.

Various dosages of carbon are then mixed with waste for the time interval selected. The carbon is then filtered off and the concentration remaining in solution is measured. These data are then plotted in accordance with the Freundlich isotherm to determine the adsorption characteristics. A crude estimate of the carbon required can be made from this plot (see Fig. 8-2) for any required removal.

TABLE 8.3
Properties of commercially available carbons[3]

	ICI America Hydrodarco 3000 (lignite)	Calgon Filtrasorb 300 (8 × 30) (bituminous)	Westvaco Nuchar WV-L (8 × 30) (bituminous)	Witco 517 (12 × 30) (bituminous)
Physical properties				
Surface area, m^2/g(BET)	600–650	950–1050	1000	1050
Apparent density, g/cm^3	0.43	0.48	0.48	0.48
Density, backwashed and drained, lb/ft^3	22	26	26	30
Real density, g/cm^3	2.0	2.1	2.1	2.1
Particle density, g/cm^3	1.4–1.5	1.3–1.4	1.4	0.92
Effective size, mm	0.8–0.9	0.8–0.9	0.85–1.05	0.89
Uniformity coefficient	1.7	1.9 or less	1.8 or less	1.44
Pore volume, cm^3/g	0.95	0.85	0.85	0.60
Mean particle diameter, mm	1.6	1.5–1.7	1.5–1.7	1.2
Specifications				
Sieve size (U.S. standard series)				
Larger than No. 8 (max. %)	8	8	8	†
Larger than No. 12 (max. %)	†	†	†	5
Smaller than No. 30 (max. %)	5	5	5	5
Smaller than No. 40 (max. %)	†	†	†	†
Iodine No.	650	900	950	1000
Abrasion No., minimum	‡	70	70	85
Ash, %	‡	8	7.5	0.5
Moisture as packed (max. %)	‡	2	2	1

† Not applicable to this size carbon.
‡ No available data from the manufacturer.
Note: $lb/ft^3 = 16\,kg/m^3$

Continuous Carbon Filters

Although batch laboratory adsorption studies provide useful information on the application of adsorption to the removal of specific waste constituents, continuous carbon filters provide the most practical application of this process in waste treatment. The reasons for this are:

1. High capacities in equilibrium with the influent concentration rather than the effluent concentration can be approached.
2. Biological activity in the presence of degradable organics will affect the apparent carbon capacity.

The carbon filter can be considered as a non-steady-state process in which, as an increasing quantity of water is passed through the bed, the adsorbent is removed in an increasing amount. Consider the system shown in Fig. 8-3. As

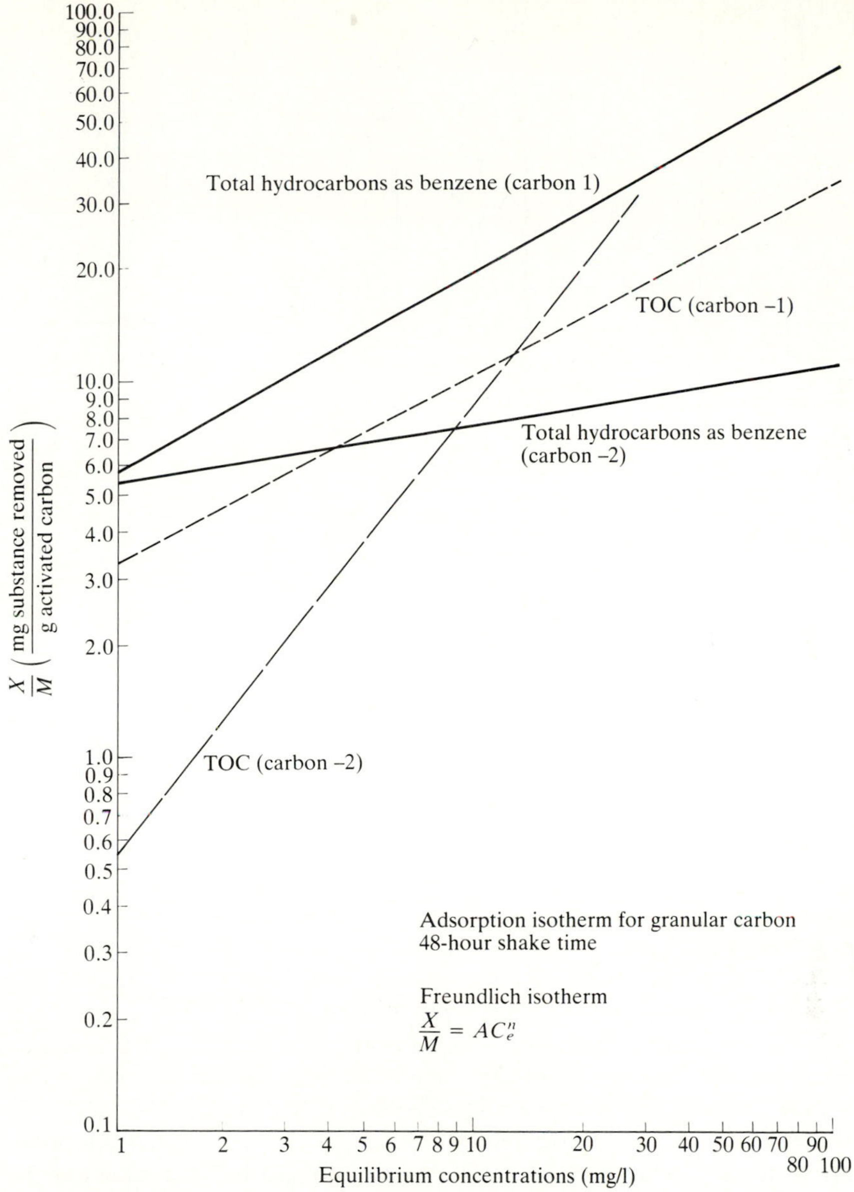

FIGURE 8-2
Freundlich isotherms for total hydrocarbons and total organic carbon.

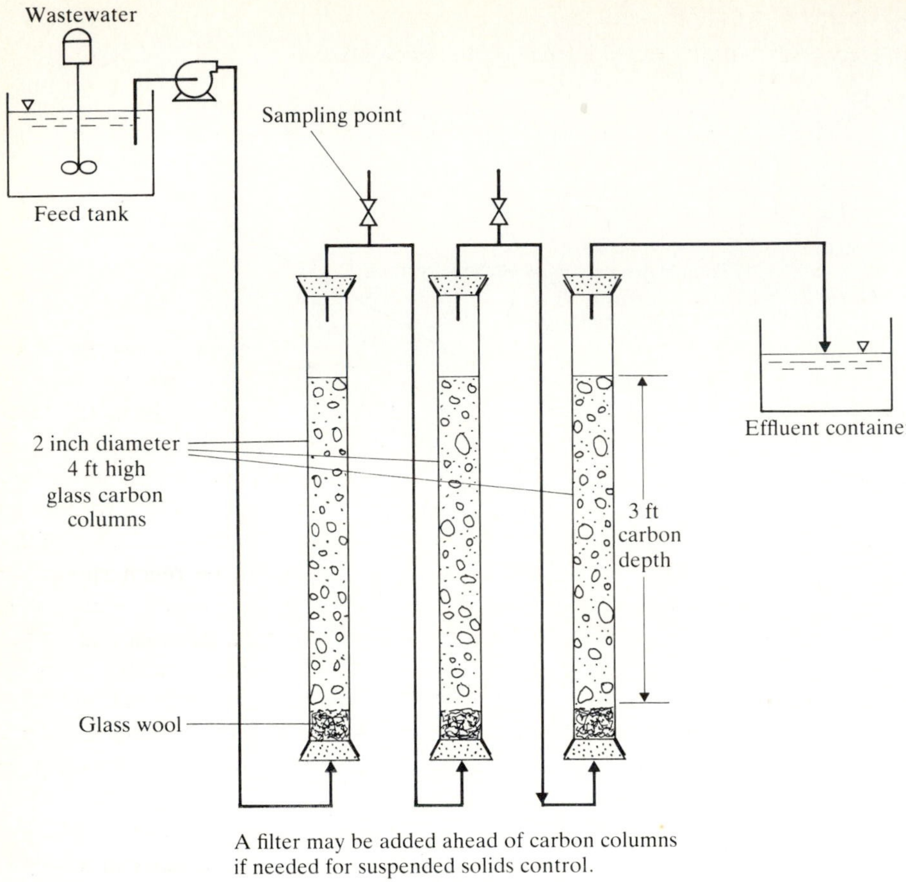

FIGURE 8-3
Granular activated carbon columns schematic (laboratory scale).

water initially passes through the uppermost layers, rapid adsorption occurs in equilibrium with the effluent concentration. As this water passes through the bed, the equilibrium shifts with the decreasing concentration of remaining solute; this results in a substantially solute-free effluent. With continuing flow of water the adsorption zone, in equilibrium with the influent concentration, moves downward in the bed. As this zone approaches the bottom of the bed, the concentration of solute in the effluent increases. The breakpoint is defined as the volume of water passed through the bed before a maximum effluent concentration is reached. As the adsorption zone falls to the bottom of the bed, the effluent concentration increases until it equals the influent concentration.

The breakthrough point:

1. Decreases with decreased bed height
2. Decreases with increasing particle size of adsorbent

3. Decreases with increased flow rate
4. Decreases with increased initial solute concentration

A granular carbon pilot plant for the development of breakthrough curves is shown in Fig. 8-3.

Carbon Regeneration

It is generally feasible to regenerate spent carbon for economic reasons. In the regeneration process, the object is to remove from the carbon pore structure the previously adsorbed materials. The modes of regeneration are thermal, steam, solvent extraction, acid or base treatment, and chemical oxidation. The latter methods (excluding thermal) are usually to be preferred when applicable, since they can be accomplished in situ. The difficulty arises that adsorption from multicomponent wastewaters usually does not lend itself to high-efficiency regeneration by these methods. Exceptions are phenol, which can be regenerated with caustic in which the phenol is converted to the more soluble phenate and a single chlorinated hydrocarbon which can be removed with steam. In most wastewater cases, however, thermal regeneration is required. Thermal regeneration is the process of drying, thermal desorption, and high-temperature heat treatment (1200 to 1800°F) (650 to 980°C) in the presence of limited quantities of water vapor, flue gas, and oxygen. Multiple-hearth furnaces or fluidized-bed furnaces can be used.

Weight losses of carbon result from attrition and carbon oxidation. Depending on the type of carbon and furnace operation, this usually amounts to 5 to 10 percent by weight of the carbon regenerated. There is also a change in carbon capacity through regeneration that may be caused by a change in pore size (usually an increase resulting in a decrease in iodine number) and a loss of pores by deposition of residual materials. In the evaluation of carbons for a wastewater-treatment application, the change in capacity through successive regeneration cycles should be evaluated. In most cases, three regeneration cycles will define the maximum capacity loss.

Adsorption System Design

In granular carbon columns, the carbon capacity at breakthrough as related to exhaustion is a function of the waste complexity, as shown in Fig. 8-4. A single organic such as dichloroethane will yield a sharp breakthrough curve such that the column is greater than 90 percent exhausted when breakthrough occurs. By contrast, a multicomponent petrochemical wastewater shows a drawn-out breakthrough curve due to varying rates of sorption and desorption, as shown in Fig.

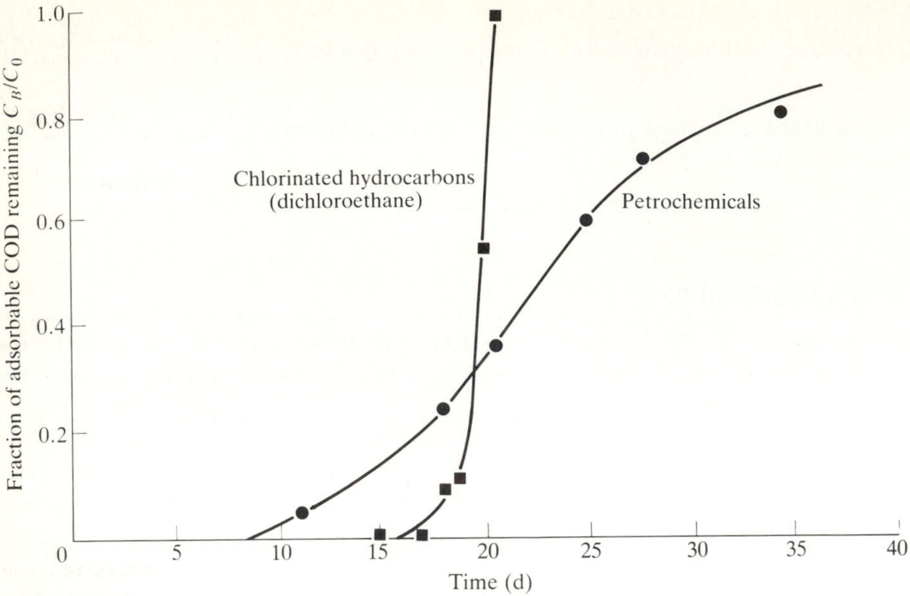

FIGURE 8-4
Continuous carbon column breakthrough curves.[4]

8-5. Depending on the nature of the wastewater one of several modes of carbon column design may be employed:

1. Downflow. These are fixed beds in series. When breakthrough occurs in the last column, the first column is in equilibrium with the influent concentration (C_0) in order to achieve a maximum carbon capacity. After carbon replacement in the first column, it becomes the last column in a series, etc. (Fig. 8-6).
2. Multiple units. These are operated in parallel with the effluent blended to achieve the final desired quality. The effluent from a column ready for regeneration or replacement, which is high in COD, is blended with the other effluents from fresh carbon columns to achieve the desired quality (Fig. 8-6). This mode of operation is most adaptable to waters in which the capacity at breakthrough/capacity at exhaustion ratio is near 1.0, as in the dichloroethane case previously mentioned.
3. Upflow. Expanded beds are used when suspended solids are present in the influent or when biological action occurs in the bed (Fig. 8-6).
4. Continuous counterflow. These are column or pulsed beds with the spent carbon from the bottom (in equilibrium with influent solute concentration) sent to regeneration. Since this design cannot be backwashed, residual biodegradable organic content in the influent should be very low to avoid plugging. Regenerated and makeup carbon is fed to the top of the reactor (Fig. 8-6). A granular carbon system is shown in Fig. 8-7.

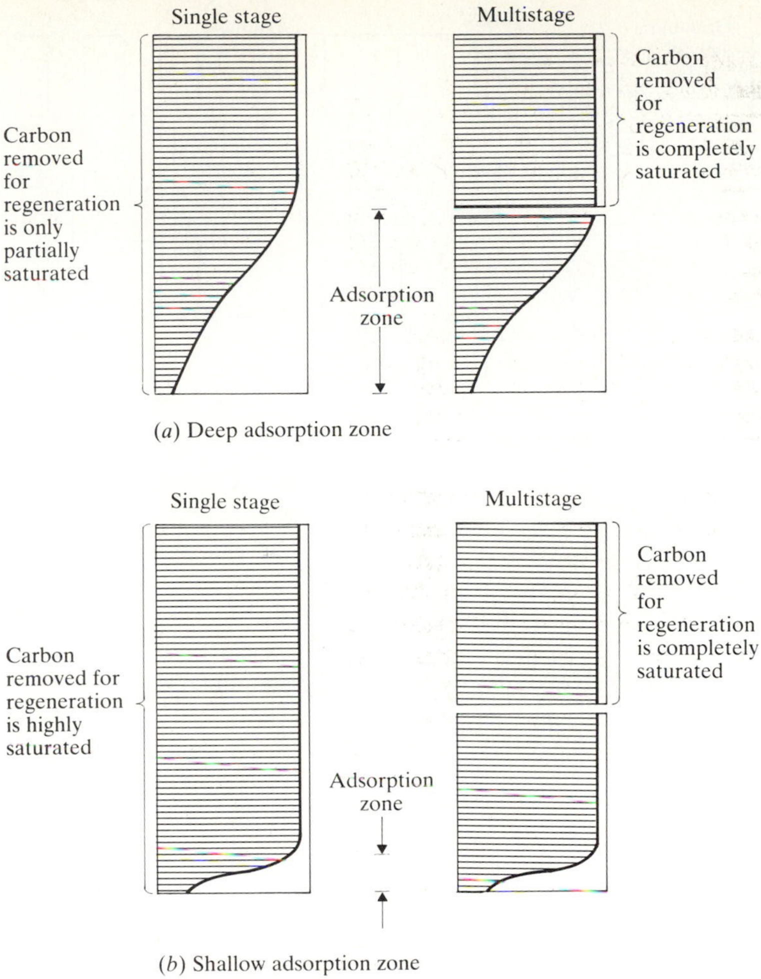

FIGURE 8-5
Adsorption zones for single- and multicomponent wastewaters.

5. The upflow-downflow concept provides a countercurrent two-bed series system. The two beds are arranged so that the gravity, open-top structures are operated in a series upflow "roughing" contactor and a downflow "polishing" contactor. Once breakthrough occurs, the pair of columns are taken off-line, the spent upflow column regenerated, and the unused capacity of the downflow column is used by reversing the flow and employing it as the upflow reactor, using the former upflow column containing regenerated carbon as the downflow polishing unit.

Bohart and Adams[5] developed a relationship, based on a surface-reaction-rate theory, which can be used to predict the performance of continuous carbon

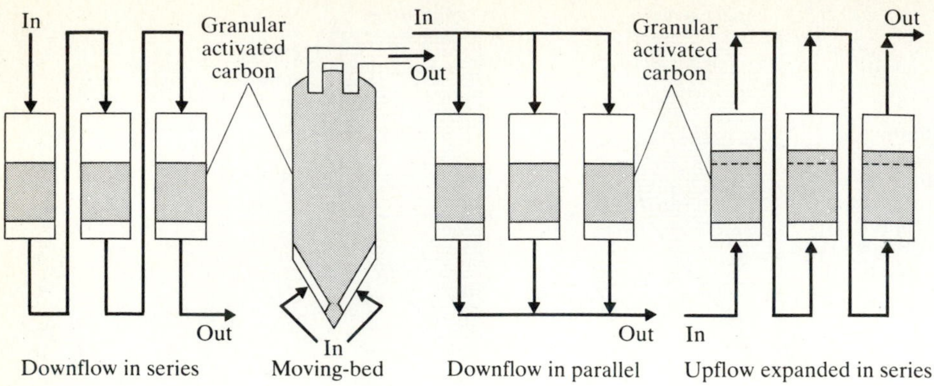

FIGURE 8-6
Types of GAC column design. (*Calgon Carbon Corporation.*)

columns:

$$\ln\left(\frac{C_0}{C_B}-1\right)=\ln\left(e^{KN_0X/v}-1\right)-KC_0t \tag{8.4}$$

Since $e^{KN_0X/v}$ is much greater than unity, Eq. (8.4) can be simplified to

$$t=\frac{N_0}{C_0v}\left[X-\frac{v}{KN_0}\ln\left(\frac{C_0}{C_B}-1\right)\right] \tag{8.5}$$

where t = service time
v = linear flow rate
X = depth of bed
K = rate constant
N_0 = adsorptive capacity
C_0 = influent concentration
C_B = allowable effluent concentration

The bed depth, which theoretically is just sufficient to prevent penetration of concentration in excess of C_B at zero time, is defined as the critical bed depth and is determined from Eq. (8.5) when $t = 0$:

$$X_0=-\frac{v}{KN_0}\ln\left(\frac{C_0}{C_B}-1\right) \tag{8.6}$$

From Eq. (8.5) it can be shown that the adsorptive capacity N_0 can be determined from the slope of a linear plot of t versus X. The rate constant K is then calculated from the intercept of this plot:

$$b=-\frac{1}{C_0K}\ln\left(\frac{C_0}{C_B}-1\right) \tag{8.7}$$

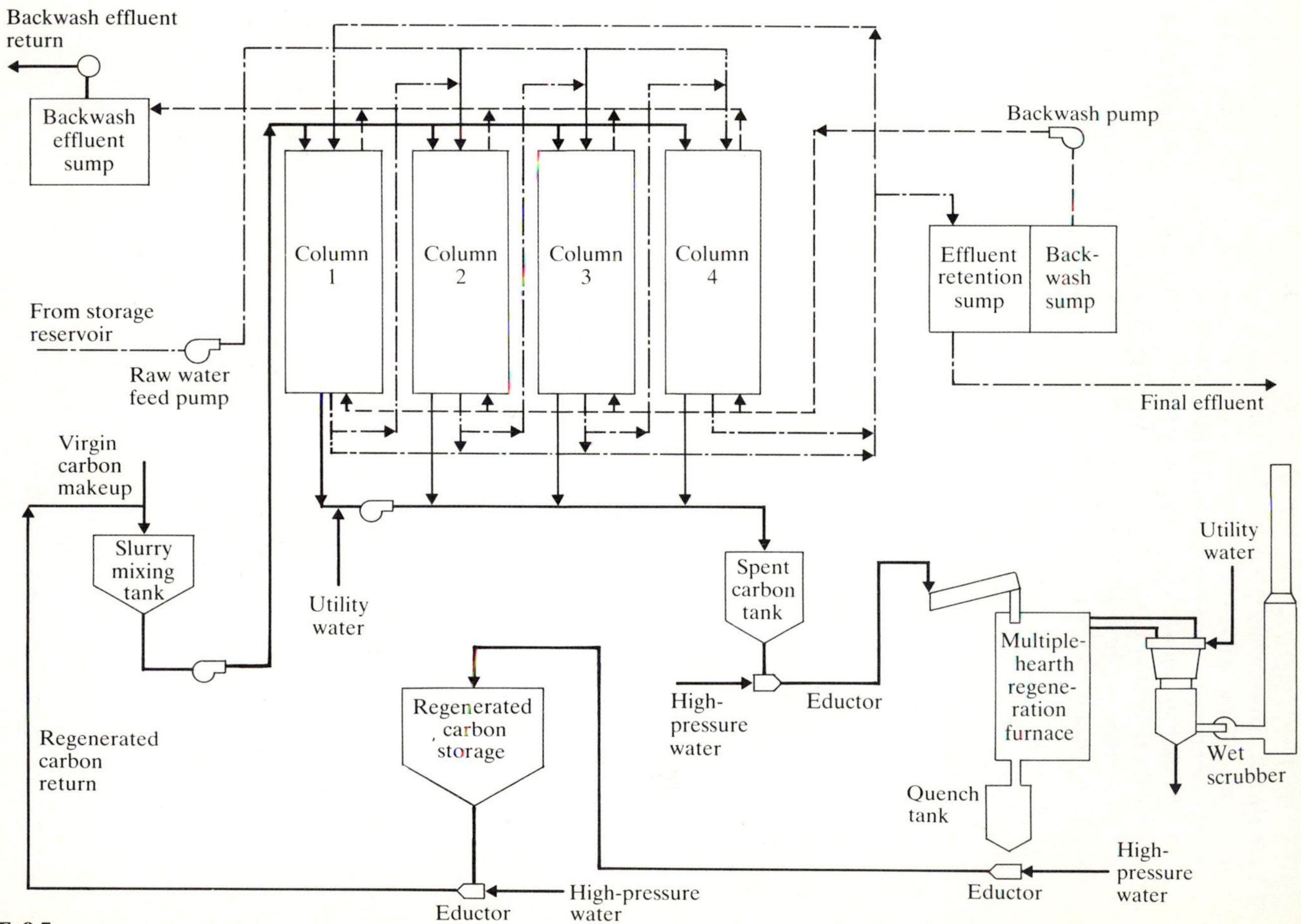

FIGURE 8-7
GAC process flowsheet.

The application of the Bohart-Adams equation to the removal of ABS[4] in continuous carbon columns is shown in Example 8.1.

Example 8.1. Experiments in a carbon column 1 in (2.54 cm) in diameter were conducted on a 10-ppm ABS solution.[4] The data obtained are given in Table 8.4.

Compute: (*a*) the coefficients in the Bohart-Adams equation for this carbon and (*b*) the carbon required per year to treat 100,000 gal (380 m^3) of water per week with an ABS content of 10 mg/l to a residual of 0.5 ppm.

The tower will be 5 ft (1.5 m) in depth and 2 ft (0.6 m) in diameter; compute the adsorption efficiency of the carbon.

Solution

(*a*)
$$t = \frac{N_0}{C_0 v}\left[X - \frac{v}{KN_0}\ln\left(\frac{C_0}{C_B} - 1\right)\right]$$

where $N_0 = M$ ABS/L^3 C
$X = L$
$C_0 = M/L^3$
$v = L/T$
$K = L^3/ML$
$t = T$

Area of column A_c (1-in diameter) = 0.00545 ft^2 (5.06 cm^2)

Calculation of N_0 and K

$$C_0 = 10 \text{ mg/l} \times 62.4 \times 10^{-6} \text{ lb/ft}^3 = 0.000624 \text{ lb/ft}^3$$

TABLE 8.4
Column capacity data for the adsorption of ABS from a 10-ppm aqueous solution (capacity to 0.5 ppm breakpoint)

Flow rate, gal/(min · ft^2)	Bed depth, ft	Throughput volume, gal
2.5	2.5	363
	5.0	1216
	7.5	2148
5.0	2.5	141
	5.0	730
	10.0	2190
10.0	5.0	332
	10.0	1380
	15.0	2760

Note:
gal/(min · ft^2) = 4.07 × 10^{-2} m^3/(min · m^2)
ft = 0.3048 m
gal = 3.785 × 10^{-3} m^3

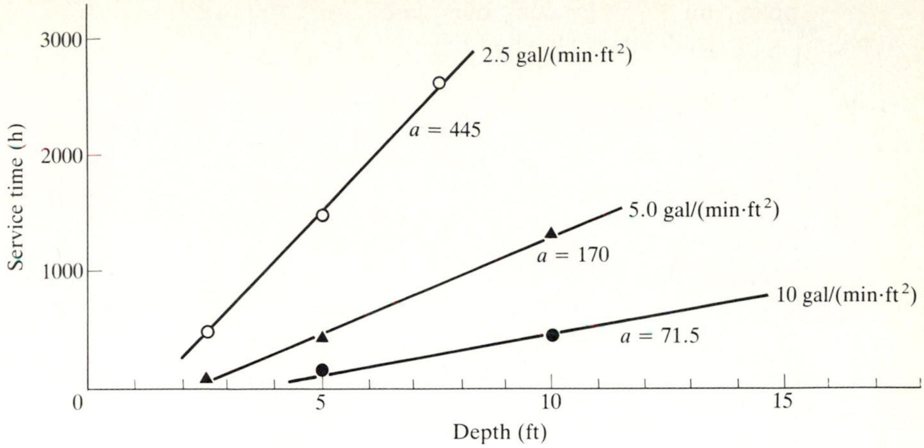

Note:
gal/(min·ft^2) = 4.07×10^{-2} m^3/(min · m^2)
ft = 0.3048 m

FIGURE 8-8

From the equation for t above, N_0 is computed from the slope of the t versus X plot (see Fig. 8-8 and Table 8.5). To prepare Table 8.5 the following considerations apply:

v(ft/h) = q[gal/(min · ft^2)] × 8.02(ft/h)/[gal/(min · ft^2)]

V(ft^3) = throughput (gal) × 0.134 (ft^3/gal)

$$t = \frac{V}{vA_c}$$

TABLE 8.5

Flow		Depth,	Volume treated	
q, gal/(min · ft^2)	v, ft/h	ft	V, ft^3	t, h
2.5	20	2.5	49	440
		5.0	162	1480
		7.5	268	2620
5.0	40	2.5	19	87
		5.0	98	440
		10.0	290	1340
10	80	5.0	44	102
		10.0	184	420
		15.0	370	835

Note: gal/(min · ft^2) = 8.02 ft/h = 4.07×10^{-2} m^3/(min · m^2)

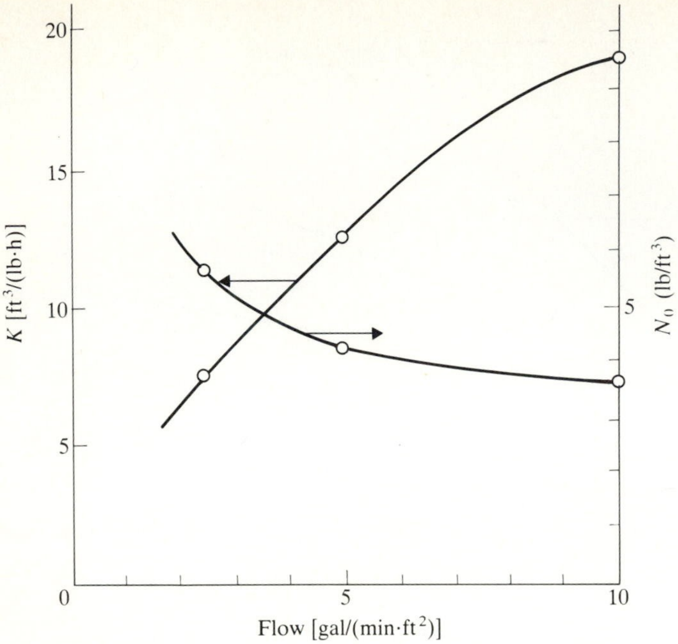

Note:
ft^3/(lb·h) = 6.24×10^{-2} m^3/kg·h)
gal/(min·ft^2) = 4.07×10^{-2} m^3/(min·m^2)
lb/ft^3 = 16.0 kg/m^3

FIGURE 8-9

At 2.5 gal/(min · ft^2):

$$N_0 = C_0 v_a$$
$$= 0.000624 \times 20 \times 445 = 5.55 \text{ lb/ft}^3 \text{ (89 kg/m}^3\text{)}$$

At 5.0 gal/(min · ft^2):

$$N_0 = 0.000624 \times 40 \times 170 = 4.24 \text{ lb/ft}^3 \text{ (67.9 kg/m}^3\text{)}$$

At 10 gal/(min · ft^2):

$$N_0 = 0.000624 \times 80 \times 71.5 = 3.57 \text{ lb/ft}^3 \text{ (57.2 kg/m}^3\text{)}$$

$$b = -\frac{1}{C_0 K} \ln\left(\frac{C_0}{C_B} - 1\right) \qquad \text{and} \qquad K = -\frac{1}{C_0 b} \ln\left(\frac{C_0}{C_B} - 1\right)$$

From Fig. 8-8 the values of b are taken and K can be calculated, giving the values presented in Table 8.6.

TABLE 8.6

Flow, gal/(min · ft²)	*b*, h	*K*, ft³/(lb · h)
2.5	−630	7.5
5.0	−370	12.7
10.0	−250	18.8

Note:
ft³/(lb · h) = 6.24×10^{-2} m³/(kg · h)
gal/(min · ft²) = 4.07×10^{-2} m³/(min · m²)
lb/ft³ = 16.0 kg/m³

Calculation of critical layer [Eq. (8.6)]

$$X_0 = \frac{v}{KN_0} \ln\left(\frac{C_0}{C_B} - 1\right)$$

$$X_0 = \frac{20}{7.5 \times 5.69 \times 2.95} = 1.41 \text{ ft (0.43 m)} \quad \text{for} \quad 2.5 \text{ gal/(min} \cdot \text{ft}^2)$$

In this manner $X_0 = 2.19$ ft (0.67 m) for 5.0 gal/(min · ft²)

$X_0 = 3.51$ ft (1.07 m) for 10.0 gal/(min · ft²)

(*b*) Using a tower of 24 in diameter (0.61 m), 5 ft deep (1.53 m), 3.15 ft² (0.29 m²), 15.75 ft³ (0.45 m³),

$$\frac{100{,}000 \text{ gal/week}}{5 \text{ d/week} \times 1440 \text{ min/d}} = 13.9 \text{ gal/min } (0.05 \text{ m}^3/\text{min})$$

$$\text{or } 4.4 \text{ gal/(min} \cdot \text{ft}^2)[0.18 \text{ m}^3/(\text{min} \cdot \text{m}^2)]$$

At 4.4 gal/(min · ft²), $v = 35$ ft/h, $K = 11.5$, and $N_0 = 4.2$,

$$t = \frac{N_0}{C_0 v}\left[X - \frac{v}{KN_0} \ln\left(\frac{C_0}{C_B} - 1\right)\right]$$

$$= \frac{4.2}{0.000624 \times 35}\left(5 - \frac{35}{11.5 \text{ ft}^3/(\text{lb} \cdot \text{h}) \times 4.2} 2.95\right)$$

$$= 551 \text{ h}$$

At a flow rate of 834 gal/h the total volume treated before breakthrough is (551)(834) = 460,000 gal. For a yearly volume of 5.2 million gallons, 11 carbon charges are required.

Bed efficiency

$$\text{Total ABS adsorbed} = 460{,}000 \times 8.34 \times 9.5 = 36.4 \text{ lb}$$

$$\text{Total capacity} = 4.2 \text{ lb ABS/ft}^3 \text{ C} \times 15.7 \text{ ft}^3 \text{ C} = 66 \text{ lb}$$

$$\% \text{ efficiency} = \frac{36.4}{66} \times 100 = 55\%$$

Efficiency computed from X_0:
Using Eq. (8.6),

$$X_0 = \frac{35}{11.5 \times 4.2} \ln \left(\frac{10}{0.5} - 1 \right) = 2.13 \text{ ft}$$

$$\% \text{ efficiency} = \frac{X - X_0}{X} \times 100 = \frac{5 - 2.13}{5} \times 100 = 57\%$$

Hutchins[6] presented a modification of a Bohart-Adams equation which requires only three column tests to collect the necessary data. This is called the bed depth–service time (BDST) approach. The Bohart-Adams equation can be expressed as

$$t = aX + b \tag{8.8}$$

where a = slope = $N_0/C_0 v$
b = intercept = $-1/KC_0 \times \ln (C_0/C_B - 1)$

If a value of a is determined for one flow rate, values for other flow rates can be computed by multiplying the original slope by the ratio of the original and new flow rates. The b value change is insignificant with respect to changing flow rates. Adjustment for changing initial concentration can be made as follows:

$$a_2 = a_1 \frac{Q_1}{Q_2}$$

$$b_2 = b_1 \frac{C_1}{C_2} \frac{\ln (C_2/C_F - 1)}{\ln (C_1/C_B - 1)}$$

in which C_F and C_B are the effluent concentrations at C_2 and C_1, respectively. In order to develop a BDST correlation, a number of pilot columns of equal depth are operated in series and breakthrough curves plotted for each, as shown in Fig. 8-10. These data are then used to plot a BDST correlation by recording the operating time required to reach a certain removal at each depth. A BDST plot of the data from Fig. 8-10 is shown in Fig. 8-11.

The slope of the BDST line is equal to the reciprocal velocity of the adsorption zone and the X intercept is the critical depth defined as the minimum bed depth required to obtain the desired effluent quality at time zero.

If the adsorption zone is arbitrarily defined as the carbon layer through which the liquid concentration varies from 90 to 10 percent of the feed concentration then this zone is defined by the horizontal distance between these two lines in the BDST plot.

In order to design an adsorption system with maximum carbon utilization, the carbon removed should be near saturation in equilibrium with the influent concentration. In multistage columns, the first run of three columns will be made with all fresh carbon. When the third column breaks through, the first column should be exhausted and a fourth column with fresh carbon is placed at the end of the train. This process is repeated each time the last column breaks through.

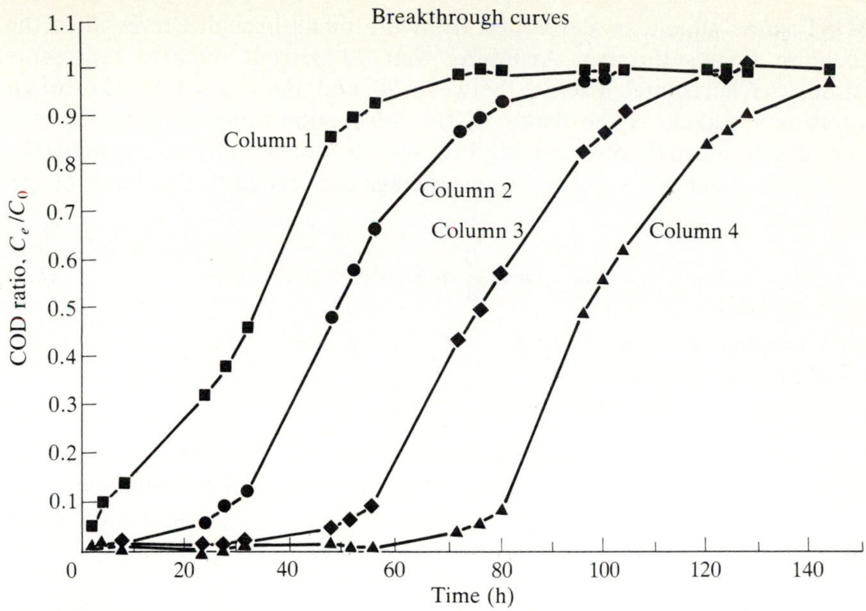

FIGURE 8-10
Column breakthrough curves.

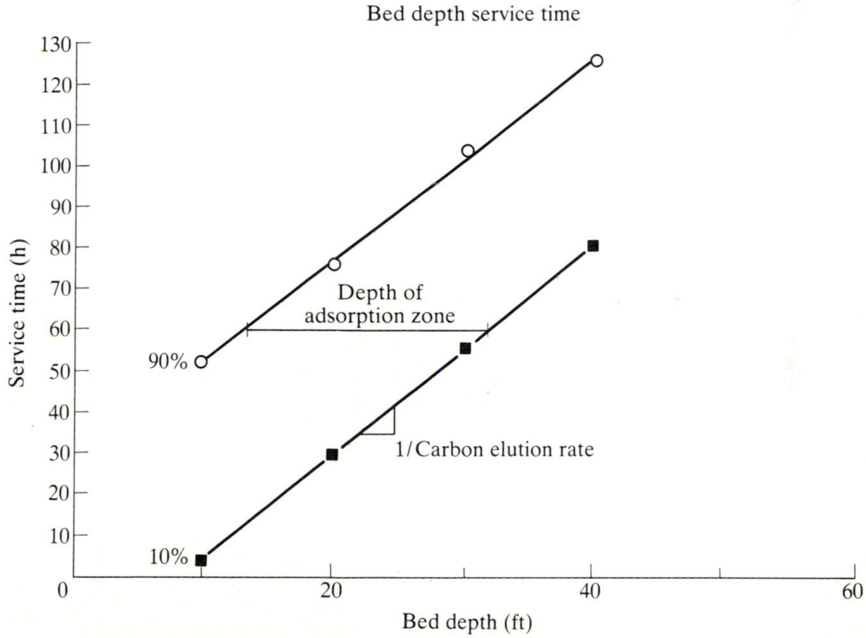

Note: ft = 0.3048 m

FIGURE 8-11
BDST design curves.

The BDST curve should be developed from the breakthrough curves after the third column breaks through. Assuming that 90 percent removal represents exhaustion, the horizontal distance between 90 and the desired breakthrough concentration is taken as the depth of the adsorption zone. This is also the minimum depth in a pulsed or moving-bed system. For a multistage system, the number of stages and the bed depth in each stage are related to the depth of the adsorption zone:

$$n = \frac{D}{d} + 1 \tag{8.9}$$

where n = number of stages in series
D = depth of adsorption zone
d = depth of single stage

Selection of d should be based on practical considerations and should be an integer fraction of D. Selecting a small d will result in small-size equipment with lower carbon inventory, but a high number of stages and consequently more costly equipment. Example 8.2 illustrates a BDST design.

Example 8.2. A petrochemical washwater with a flow of 85,000 gal/d (322 m^3/d) has to be treated to an effluent standard of 50 mg/l. A four-column pilot plant was operated with a carbon that had a density of 30 lb/ft^3 (481 kg/m^3). The columns were 10 ft (3 m) long and loaded at a hydraulic rate of 5 gal/(min · ft^2) [0.20 m^3/(min · m^2)]. The pilot plant was operated in series: the effluent from column 1 passed to the top of column 2 and sequentially to columns 3 and 4.

Calculate the depth of the adsorption zone, the required number of columns, the time required to exhaust a column, the column diameter, the daily carbon use, and the carbon adsorption loading.

The data should be plotted as sequential breakthrough curves (Fig. 8-10) of the bed effluent concentration as a ratio of the influent concentration of the system (630 mg/l).

The breakthrough curves for the four columns are symmetrical. The adsorption zone is defined as the carbon layer through which the concentration varies from 10 percent to 90 percent of the feed concentration. The breakthrough times are plotted as the service times and are a function of total carbon bed depth (Fig. 8-11).

In this case, the lines are almost parallel and the depth of the adsorption zone is between 18 and 19 ft. The total bed depth will be the adsorption zone plus one additional column:

$$\text{Number of columns} = \frac{19}{10} + 1 = 2.9, \text{ round up to } 3$$

The bed depth service time (Fig. 8-11) data fits well in the Bohart-Adams equation:

$$t = \frac{N_0}{C_0 v}(X) - \frac{1}{C_0 K} \ln_e \left(\frac{C_0}{C_e} - 1 \right)$$

At 10 percent,

$$t = 2.57(X) - 21.5$$

with correlation coefficient (r) = 0.9999; at 90 percent,

$$t = 2.50(X) - 27.0$$

with correlation coefficient (r) = 0.9981.

$$\text{Adsorption velocity} = \frac{1}{\text{slope}} = \frac{1}{2.57\ \text{h/ft}} = 0.39\ \text{ft/h or } 9.34\ \text{ft/d } (2.85\ \text{m/d})$$

The rest of the design and operation can be calculated from the above determinations:

$$\text{Time to exhaust a column} = \frac{10\ \text{ft}}{9.34\ \text{ft/d}} \times 24\ \text{h/d} = 26\ \text{h}$$

$$\text{Area required} = \frac{Q}{A}$$

$$= \frac{85{,}000\ \text{gal/d}}{5\ \text{gal/(min}\cdot\text{ft}^2)} \times \frac{\text{d}}{1440\ \text{min}}$$

$$= 11.8\ \text{ft}^2\ (1.1\ \text{m}^2)$$

$$\text{Diameter} = 3.88\ \text{ft } (1.2\ \text{m})$$

$$\text{Carbon use} = 11.8\ \text{ft}^2 \times 9.34\ \text{ft/d} \times 39\ \text{lb/ft}^3$$

$$= 3300\ \text{lb/d } (1500\ \text{kg/d})$$

$$\text{Carbon adsorption loading} = \frac{(630 - 50)\ \text{mg/l} \times Q}{\text{carbon use}}$$

$$= \frac{580\ \text{mg/l} \times 85{,}000\ \text{gal/d}}{3300\ \text{lb/d}} \times \frac{8.34\ \text{lb/gal}}{10^6\ \text{mg/l}}$$

$$= 0.125\ \text{lb COD/lb carbon}$$

Performance of Activated Carbon Systems

Activated carbon columns are employed for the treatment of toxic or nonbiodegradable wastewaters and for tertiary treatment following biological oxidation.

When degradable organics (BOD) are present in the wastewater, biological action provides biological regeneration of the carbon, thus increasing the apparent capacity of the carbon. Biological activity may be an asset or a liability. When the applied BOD is in excess of 50 mg/l, anaerobic activity in the columns may cause serious odor problems while aerobic activity may cause plugging due to the biomass generation by aerobic activity.

Most heavy metals are removed through carbon columns. In order to avoid reduced capacity after regeneration, the carbon should be acid-washed prior to reuse. The effectiveness of activated carbon in the treatment of various industrial wastewaters is shown in Table 8.7.

TABLE 8.7
Results from adsorption isotherms on various industrial wastewaters

Type of industry	Initial TOC (or phenol), mg/l	Initial color OD	Average reduction, %	Carbon exhaustion rate, lb/thousand gal
Food and kindred products	25–5300	—	90	0.8–345
Tobacco manufacturers	1030	—	97	58
Textile mill products	9–4670	—	93	1–246
	—	0.1–5.4	97	0.1–83
Apparels and allied products	390–875	—	75	12–43
Paper and allied products	100–3500	—	90	3.2–156
	—	1.4	94	3.7
Printing, publishing, and allied industries	34–170	—	98	4.3–4.6
Chemicals and allied products	19–75,500	—	85	0.7–2905
	(0.1–5325)	—	99	1.7–185
	—	0.7–275	98	1.2–1328
Petroleum refining and related industries	36–4400	—	92	1.1–141
	(7–270)	—	99	6–24
Rubber and miscellaneous plastic products	120–8375	—	95	5.2–164
Leather and leather products	115–9000	—	95	3–315
Stone, clay, and glass products	12–8300	—	87	2.8–300
Primary metal industries	11–23,000	—	90	0.5–1857
Fabricated metal products	73,000	—	25	606

Note: lb/thousand gal = 0.120 kg/m^3

8.3 THE PACT® PROCESS

Recently, powdered activated carbon (PAC) has been added to the activated sludge process for enhanced performance (the PACT® process). The flowsheet for this process is shown in Fig. 8-12. The addition of PAC has several process advantages, namely, decreasing variability in effluent quality and removal by adsorption of nondegradable organics—principally, color, reduction of inhibition in industrial wastewater treatment, and removal of refractory priority pollutants. PAC offers the advantage of being able to be integrated into existing biological-treatment facilities at minimum capital cost. Since the addition of PAC enhances

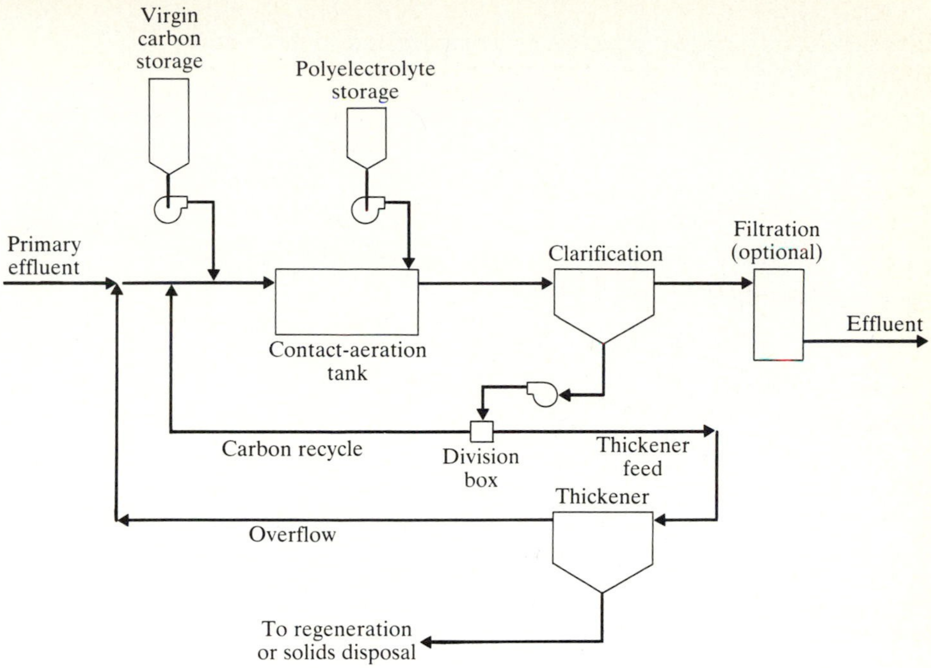

FIGURE 8-12
PACT® wastewater-treatment system general process diagram.

sludge settleability, conventional secondary clarifiers will usually be adequate, even with high carbon dosages. In some industrial waste applications, nitrification is inhibited by the presence of toxic organics. The application of PAC has been shown to reduce or eliminate this inhibition. Batch isotherm screening tests are used on the biological effluent in order to select the optimal carbon. Bench-scale continuous reactors as shown in Fig. 6-22 can be used to develop process design criteria. Several reactors are run in parallel, a control with no PAC and several with varying dosages of PAC.

The PAC dosage and the PAC mixed liquor solids concentration are related to the sludge age:

$$X_p = \frac{X_i \theta_c}{t} \tag{8.10}$$

in which X_p = equilibrium PAC MLSS content
X_i = PAC dosage
t = hydraulic retention time

The sludge age affects the PAC efficiency with higher sludge ages enhancing the organic removal per unit of carbon; affects the molecular configuration of the adsorbate based on varying biological uptake patterns and end products; and establishes the equilibrium biological solids level in the aeration basin. There is

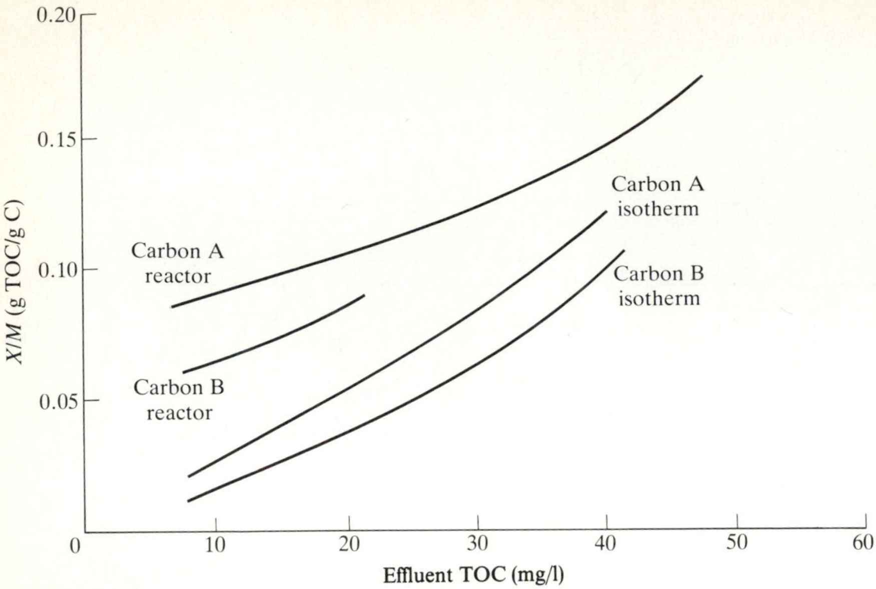

FIGURE 8-13
Performance relationship of PACT® reactors with isotherm data.

some evidence that the attached biomass degrades some of the low-molecular-weight compounds that are adsorbed, as demonstrated by superior TOC removal rates for PAC when added to an aeration basin as opposed to isotherm predictions of adsorption capacity. Figure 8-13 was developed by evaluating the difference in TOC removal for biological units operated in parallel with and without PAC. As can be seen, the performance of the carbon in the bioreactor was significantly greater than predicted by the isotherm only. The mechanisms felt to be responsible for this phenomena include:

1. Additional biodegradation of organics due to decreased biological toxicity or inhibition via activated carbon.

TABLE 8.8
Effect of PAC on nitrification of coke plant wastewaters

PAC feed, mg/l	SRT, d	TOC, mg/l	TKN, mg/l	NH_3–N, mg/l	NO_2–N, mg/l	NO_3–N, mg/l
0	40	31	72	68	4.0	0
33	30	20	6.3	1	4.0	9.0
50	40	26	6.4	1	1.0	13.0

Influent conditions: TOC = 535 mg/l, TKN = 155 mg/l, NH_3–N = 80 mg/l.

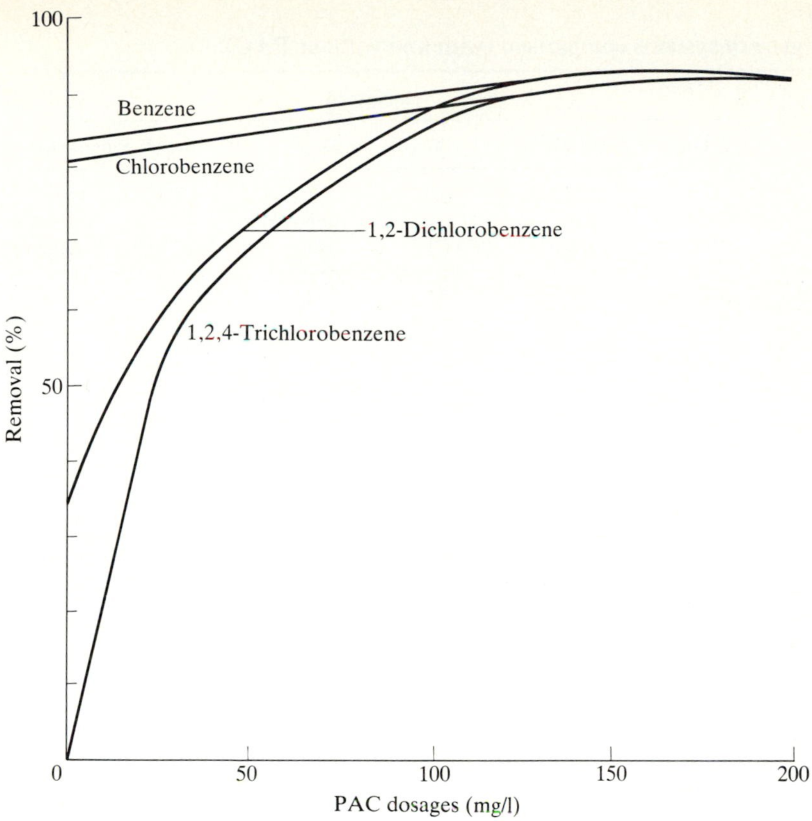

FIGURE 8-14
Effect of PAC dosage on chlorinated benzenes.[7]

TABLE 8.9

Larger doses of powdered activated carbon result in greater removal of organic carbon, color, and heavy metals

	Wastewater composition, mg/l							Bioassay†
	BOD	TOC	TSS	Color	Cu	Cr	Ni	LC_{50}
Influent	320	245	70	5365	0.41	0.09	0.52	
Biotreatment	3	81	50	3830	0.36	0.06	0.35	11
+50 mg/l PAC	4	68	41	2900	0.30	0.05	0.31	25
+100 mg/l PAC	3	53	36	1650	0.18	0.04	0.27	33
+250 mg/l PAC	2	29	34	323	0.07	0.02	0.24	>75
+500 mg/l PAC	2	17	40	125	0.04	<0.02	0.23	>87

† Percentage of wastewater in which 50 percent of aquatic organisms survive for 48 h.

TABLE 8.10
Suspended growth systems comparison (with and without PAC)

	Industrial					
	Organic chemicals[5]		Textile finishing[7]		Berndt-Polkowski[8]	
Operating conditions	**PACT®†**	**Activated sludge**	**PACT®**	**Activated sludge**	**PACT®†**	**Air activated sludge**
Aeration *d*, d	6	6	2.4	2.4	3.2–4.5	‡
SRT, d	25	45†				
Temperature, °C	25	25			13	16
Performance results					20	20
Influent characteristics, mg/l						
BOD_5	4,035	4,035	660	Unknown		
TOC	2,965	2,965			134	128
COD	10,230	10,230	1362	1590	364	320
TKN	120	120	Unknown	106	39.4	32.0
NH_3–N	76	76	74	31	19.5	19.8
Chlorinated hydrocarbons	5–67	5				
Phenol	8.1	8.1				
Effluent characteristics, mg/l						
BOD_5	11	17	5	26	1.2	24
TOC	25	65			49.8	63
COD	102	296	116	270	3.7	6.4
TKN	4	—	6	29	0.2	3.6
NH_3–N	0.8	—	3.6	15		
Chlorinated hydrocarbons	0.1	0.9				
Phenol	0.01	0.22	0.44	1.6		
Color, APHA	94	820	240	600		
Detergents			0.8	11.4		

† Includes wet oxidation carbon regeneration.

‡ Contact stabilization facility; aeration d_t = 4.2 h, stabilization d = 6.2 h.

2. Degradation of normally nondegradable substances due to increased exposure time to the biomass through adsorption on the carbon. The carbon with adsorbed material remains in the system for one sludge age, typically 10 to 30 d, while without carbon the substances would remain in the system for only one hydraulic retention time, typically 6 to 36 h.
3. Substitution/adsorption phenomena, replacement of low-molecular-weight compounds with high-molecular-weight compounds, resulting in improved adsorption efficiency and lower toxicity.

Elimination of nitrification inhibition by the addition of PAC for a coke plant wastewater is shown in Table 8.8.

When there is a small or intermittent application of PAC, the carbon is disposed of with the excess sludge. Continuous application at larger plants, however, requires regeneration of the carbon. This can be accomplished by the use of wet air oxidation (WAO).

In the WAO process, the biological carbon sludge mixture is treated in a reactor at 450°C and 750 lb/in^2 (51 atm) for 1 h in the presence of oxygen. The biological sludge is oxidized and solubilized under these conditions and the carbon regenerated. The exothermic reaction will provide energy for the reaction providing that the influent solids content exceeds 10 percent. The decant liquor from the reactor will contain 5000 mg/l BOD which is recycled back to the aeration basins. In some cases, there is an ash buildup which must be removed from the system. Depending on the characteristics of the wastewater and the type of carbon used, there may be significant losses in carbon capacity through regeneration. This phenemona should be evaluated by a pilot plant study for any specific application.

Carbon dosages may vary from 20 to 200 mg/l depending on the results desired. Since the carbon is abrasive, equipment selection should consider this fact. The effect of PAC dosage on the removal of chlorinated benzenes is shown in Fig. 8-14. Performance of the PACT® process for an organic chemicals wastewater is shown in Table 8.9. Activated sludge process performance with and without PAC is summarized in Table 8.10.

PROBLEMS

8.1. Given the following data for the adsorption of napthlene on Filtersorb 400 carbon:

Carbon dose, mg/l	Effluent C_f mg/l
0	9.94
11.2	5.3
22.3	3.0
56.1	0.71
168.3	0.17
224.4	0.06

(*a*) Develop the adsorption isotherm and the Freundlich parameters.
(*b*) What is the adsorption capacity at an initial concentration of 10 mg/l?
(*c*) What is the carbon dose required to reduce the concentration from 1 to 0.01 mg/l?

8.2. The breakthrough curves for a pilot plant operation are shown in Fig. P8-2. The pilot columns were 3.5 ft (1.07 m) long. Design a carbon column system to a breakthrough TOC concentration of 20 mg/l. The influent wastewater has a flow of 3100 gal/min (11.7 m^3/min) and an influent TOC of 60 mg/l. Use a hydraulic loading of 4 gal/(min · ft^2) [0.163 m^3/(min · m^2)] and a carbon density of 28 lb/ft^3 (449 kg/m^3).

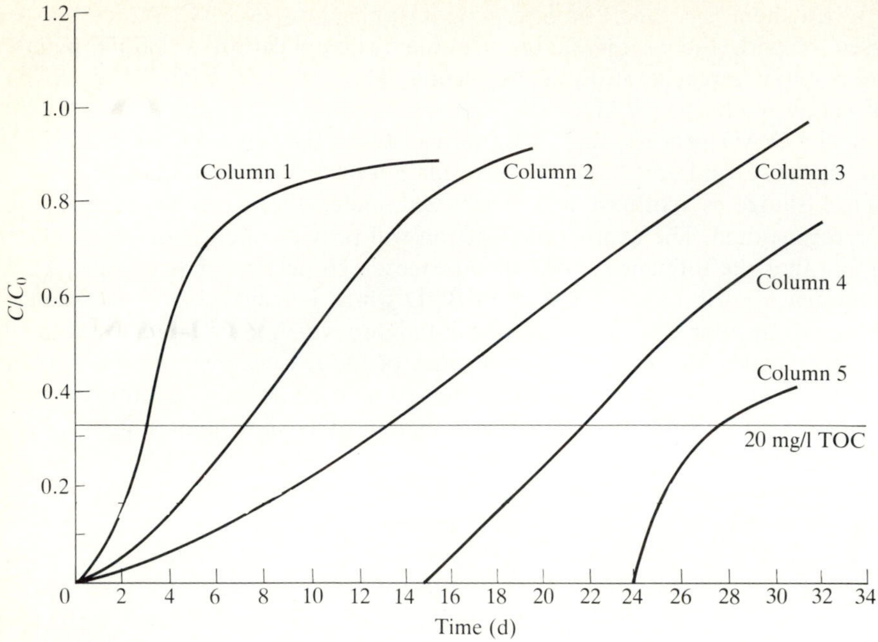

FIGURE P8-2
Breakthrough curves for GAC columns.

REFERENCES

1. Morris, J. C., and W. J. Weber: "Adsorption of Biochemically Resistant Materials from Solution," Environmental Health Series AWTR-9, May 1964.
2. EPA: "Carbon Adsorption Isotherms for Toxic Organics," EPA-600/8-80-023, April 1980.
3. U.S. EPA: *Process Design Manual for Carbon Adsorption*, Technology Transfer, 1973.
4. Argamon, Y., and W. W. Eckenfelder: *Water 1975*, Ass. Ind. Chem. Engrs Symp. Series 72, p. 151, Association of Industrial Chemical Engineers, New York, 1976.
5. Dale, J., J. Malcolm, and I. M. Klotz: *Ind. Eng. Chem.*, vol. 38, pt. 1, p. 289, 1946.
6. Hutchins, R. A.: *Chem. Engrs*, vol. 80, pt. 19, p. 133, 1973.
7. Weber, W. J., and B. E. Jones: EPA NTIS PB86-182425/AS, 1983.
8. Berndt, C., and L. Polkowski: "A Pilot Test of Nitrification with PAC," 50th Ann. Control States WPCA, 1977.

CHAPTER 9

ION EXCHANGE

9.1 THEORY OF ION EXCHANGE

Ion exchange can be used for the removal of undesirable anions and cations from a wastewater. Cations are exchanged for hydrogen or sodium and anions for hydroxyl ions.

Ion exchange resins consist of an organic or inorganic network structure with attached functional groups. Most ion exchange resins used in wastewater treatment are synthetic resins made by the polymerization of organic compounds into a porous three-dimensional structure. The degree of crosslinking between organic chains determines the internal pore structure, with higher crosslink density giving smaller pore sizes. From a kinetic viewpoint, a low degree of crosslinking would enhance diffusion of ions through larger pores. However, physical strength decreases and swelling in water increases as crosslink density is lowered. The functional ionic groups are usually introduced by reacting the polymeric matrix with a chemical compound containing the desired group. Exchange capacity is determined by the number of functional groups per unit mass of resin.

Ion exchange resins are called cationic if they exchange positive ions and anionic if they exchange negative ions. Cation exchange resins have acidic functional groups, such as sulfonic, whereas anion exchange resins contain basic functional groups, such as amine. Ion exchange resins are often classified by the nature of the functional group as strong acid, weak acid, strong base, and weak base. The strength of the acidic or basic character depends upon the degree of ionization of the functional groups, similar to the situation with soluble acids or bases. Thus, a resin with sulfonic acid groups would act as a strong acid cation exchange resin.

The most common strong acid ion exchange resin is prepared by copolymerizing styrene and divinylbenzene followed by sulfonation of the copolymer. The degree of crosslinking is controlled by the fraction of divinylbenzene in the initial mixture of monomers.

For the other types of ion exchange resins, the most common functional groups are carboxyl (—COOH) for weak acid, quaternary ammonium ($-R_3N^+OH^-$) for strong base, and amine ($-NH_2$ or $-RNH$) for weak base. The acidic or basic strength of an ion exchange resin can be measured by titration techniques.

The reactions that occur depend upon chemical equilibria situations in which one ion will selectively replace another on the ionized exchange site. Cation exchange on the sodium cycle can be illustrated by the following reaction:

$$Na_2 \cdot R + Ca^{2+} \rightleftharpoons Ca \cdot R + 2Na^+ \tag{9.1}$$

where R represents the exchange resin.

When all the exchange sites have been substantially replaced with calcium, the resin can be regenerated by passing a concentrated solution of sodium ions through the bed. This reverses the equilibrium and replaces the calcium with sodium.

A 5 to 10 percent brine solution is usually used for regeneration:

$$2Na^+ + Ca \cdot R \rightleftharpoons Na_2 \cdot R + Ca^{2+} \tag{9.2}$$

Similar reactions occur for cation exchange on the hydrogen cycle:

$$Ca^{2+} + H_2 \cdot R \rightleftharpoons CaR + 2H^+ \tag{9.3}$$

Regeneration with 2 to 10 percent H_2SO_4 yields

$$Ca \cdot R + 2H^+ \rightleftharpoons H_2 \cdot R + Ca^{2+} \tag{9.4}$$

Anion exchange similarly replaces anions with hydroxyl ions:

$$SO_4^{2-} + R \cdot (OH)_2 \rightleftharpoons R \cdot SO_4 + 2OH^- \tag{9.5}$$

Regeneration with 5 to 10 percent sodium hydroxide will renew the exchange sites:

$$R \cdot SO_4 + 2OH^- \rightleftharpoons R \cdot (OH)_2 + SO_4^{2-} \tag{9.6}$$

The performance and economics of ion exchange are related to the capacity of the resin to exchange ions and to the quantity of regenerant required. Since exchange occurs on an equivalent basis, the capacity of the bed is usually expressed as equivalents per liter of bed volume. In some cases capacity has been expressed as kilograms of $CaCO_3$ per unit of bed volume or as mass of ions per

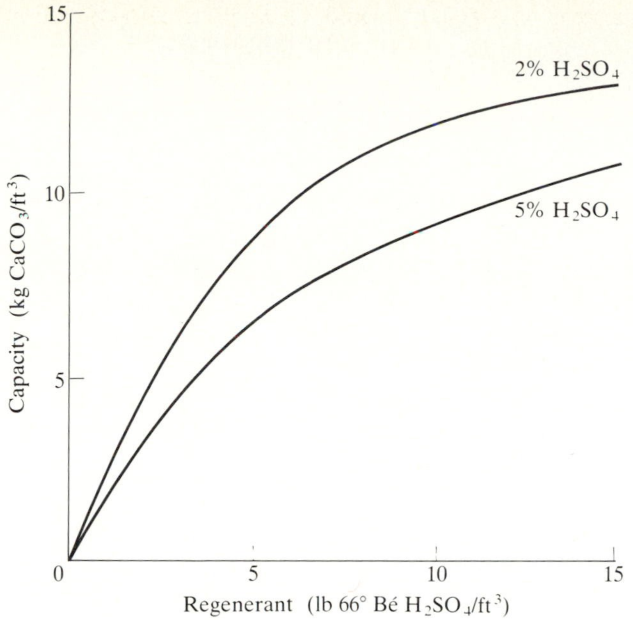

Note:
$ft^3 = 2.83 \times 10^{-2}\ m^3$
$lb/ft^3 = 16\ kg/m^3$

FIGURE 9-1
Performance of a cation exchange resin.

unit volume of bed. In like manner, the quantity of ions to be removed in the wastewater is expressed as equivalents per liter of wastewater to be treated.

In a fixed-bed exchanger, there is a relationship between the operating capacity of the bed and the quantity of regenerant employed. Resin utilization is defined as the ratio of the quantity of ions removed during treatment to the total quantity of ions that could be removed at 100 percent efficiency. The regenerant efficiency is the quantity of ions removed from the resin compared to the quantity of ions present in the volume of regenerant used. The resin utilization will increase as the regenerant efficiency decreases. Typical performance curves for a cation exchange resin are shown in Fig. 9-1. The shape of these curves will vary, depending upon the characteristics of the resin and the concentration of the regenerant used.

Treatment of a wastewater by ion exchange involves a sequence of operating steps. The wastewater is passed through the resin until the available exchange sites are filled and the contaminant appears in the effluent. This process is defined as the breakthrough. At this point treatment is stopped and the bed is backwashed to remove dirt and to regrade the resin. The bed is then regenerated. After regeneration, the bed is rinsed with water to wash out residual regenerant. The bed is then ready for another treatment cycle.

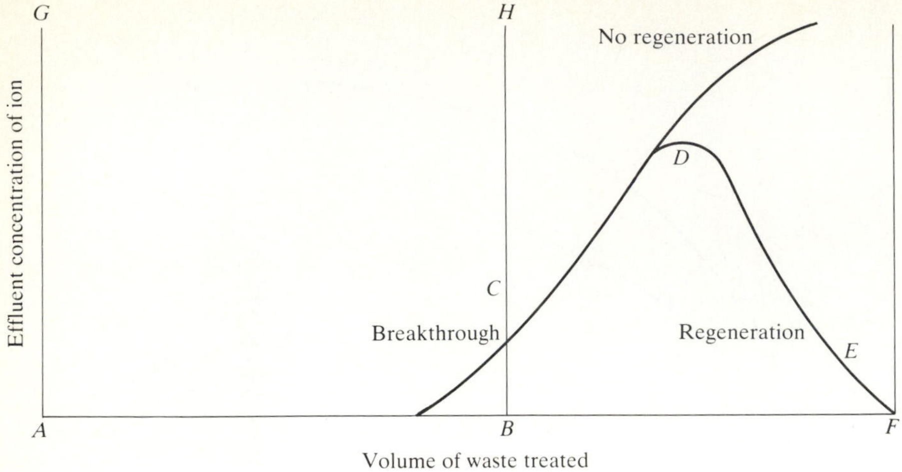

FIGURE 9-2
Treatment and regeneration cycle of an ion exchange resin.

The treatment and regeneration cycle is shown in Fig. 9-2. In this figure, area *ABHG* is the quantity of ions in the volume of solution treated before breakthrough. Area *ABC* is the quantity of ions leaking through the column and area *ACHG* is the quantity of ions removed by the exchange resin. The resin utilization is therefore area *ACHG*/*K*, where *K* is the ultimate capacity of the resin. Area *BCDEF* is the quantity of ions removed from the bed during regeneration. The regeneration efficiency is therefore area *BCDEF*/*R*, where *R* is equal to the concentration of the regenerant times its volume.

To ensure contact of liquid with the resin and to minimize leakage, the minimum bed depth is 24 to 30 in (61 to 76 cm). The treatment flow rate can vary between 2 and 5 gal/(min · ft^3) [0.27 to 0.67 m^3/(min · m^3)], although breakthrough will occur more quickly at the higher flow rates. The regenerant flow rate is 1 to 2 gal/(min · ft^3) [0.13 to 0.27 m^3/(min · m^3)]. A rinsewater volume of 30 to 100 gal/ft^3 (4.0 to 13.4 m^3/m^3), applied at a flow rate of 1 to 1.5 gal/(min · ft^3) [0.13 to 0.20 m^3/(min · m^3)], will usually be sufficient to flush a bed of residual regenerant.

Experimental Procedure

It is frequently necessary to operate a laboratory ion exchange column to develop the necessary design criteria for the removal of ions from complex industrial wastes. A typical laboratory assembly is shown in Fig. 9-3. A suggested procedure based on the apparatus shown in Fig. 9-3 is detailed as follows:

1. Rinse the column for 10 min with deionized water at a flow rate of 50 ml/min.
2. Adjust the flow rate to the column to 50 ml/min of solution containing the waste to be treated.

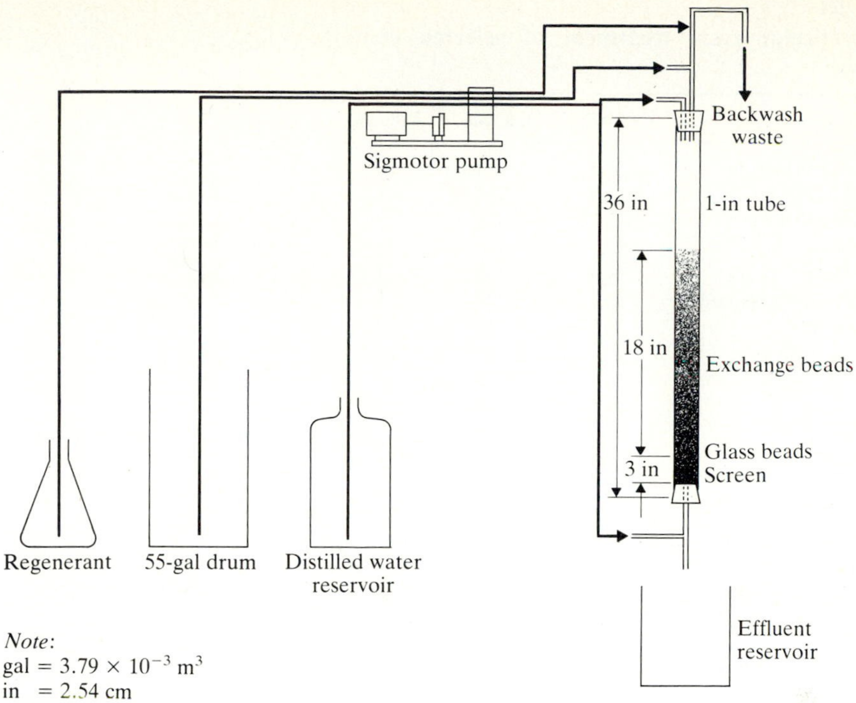

FIGURE 9-3
Laboratory ion exchange column.

3. Measure the initial volume of solution to be treated.
4. Start the treatment cycle. Develop the breakthrough curve until the ion concentration reaches the maximum effluent limit.
5. Backwash to 25 percent bed expansion for 5 to 10 min. (Use distilled water for the backwash operation.)
6. Regenerate at a flow rate of 6 ml/min, using the concentration and volume recommended for the resin. Collect the spent regenerant and measure the recovered ions.
7. Rinse the column with distilled water.

When several runs are made, it is possible to develop a relationship between resin utilization and regenerant efficiency and to select the optimum operating level for the system.

Macroreticular resins are employed for the removal of specific nonpolar organic compounds. These resins are highly specific and can be formulated to remove one compound or a class of compounds. The resins are solvent-regenerated. Treatment results of selected compounds using macroreticular resins are shown in Table 9.1.

TABLE 9.1
Macroreticular resin treatment of selected compounds

Compound	Units	Influent	Effluent
Carbon tetrachloride	μg/l	20,450	490
Hexachloroethane	μg/l	104	0.1
2-Chloronaphthalene	μg/l	18	3
Chloroform	μg/l	1,430	35
Hexachlorobutadiene	μg/l	266	<0.1
Hexachlorocyclopentadiene	μg/l	1,127	1.5
Napthalene	μg/l	529	<3
Tetrachloroethylene	μg/l	34	0.3
Toluene	μg/l	2,360	10
Aldrin	μg/l	84	0.3
Dieldrin	μg/l	28	0.2
Chlordane	μg/l	217	<0.1
Endrin	μg/l	123	1.2
Heptachlor	μg/l	40	0.8
Heptachlor epoxide	μg/l	11	<0.1

9.2 PLATING WASTE TREATMENT

One of the major applications of ion exchange in industrial waste treatment has been in the plating industry, where chrome recovery and water reuse have often resulted in considerable savings.[1–4]

For recovery of the spent chromic acid in plating baths, the chromic acid is passed through a cation exchange resin to remove other ions (Fe, Cr^{3+}, Al, etc.). The effluent can be returned to the plating bath or to storage. Since the maximum concentration of CrO_3 that can be passed through some resins to avoid deterioration is 14 to 16 oz/gal (105 to 120 kg/m^3) as CrO_3, the bath may require dilution and the recovered solution may require makeup to strength.

The rinsewaters are first passed through a cation exchanger to remove metal ions. The effluent from this unit is passed through an anion exchanger to remove chromate and to obtain demineralized makeup water. It is desirable to pass the rinsewater through the cation unit first to avoid precipitation of metal hydroxides on the exchange resin. The anion exchanger is regenerated with sodium hydroxide; this results in a mixture of Na_2CrO_4 and NaOH in the spent regenerant. This mixture is passed through a cation exchanger to recover H_2CrO_4, which is returned to the plating bath. The recovered chromic acid from the spent regenerant will average 4 to 6 percent concentration. The spent regenerant from the cation exchanger will require neutralization and possibly precipitation of metallic ions before it is discharged to the sewer.

Since most of the metal ions are eluted in the first 70 percent of the regenerant volume, neutralization requirements can be reduced by reuse of the last portion of the acid regenerant for the subsequent regeneration.[4] In like manner,

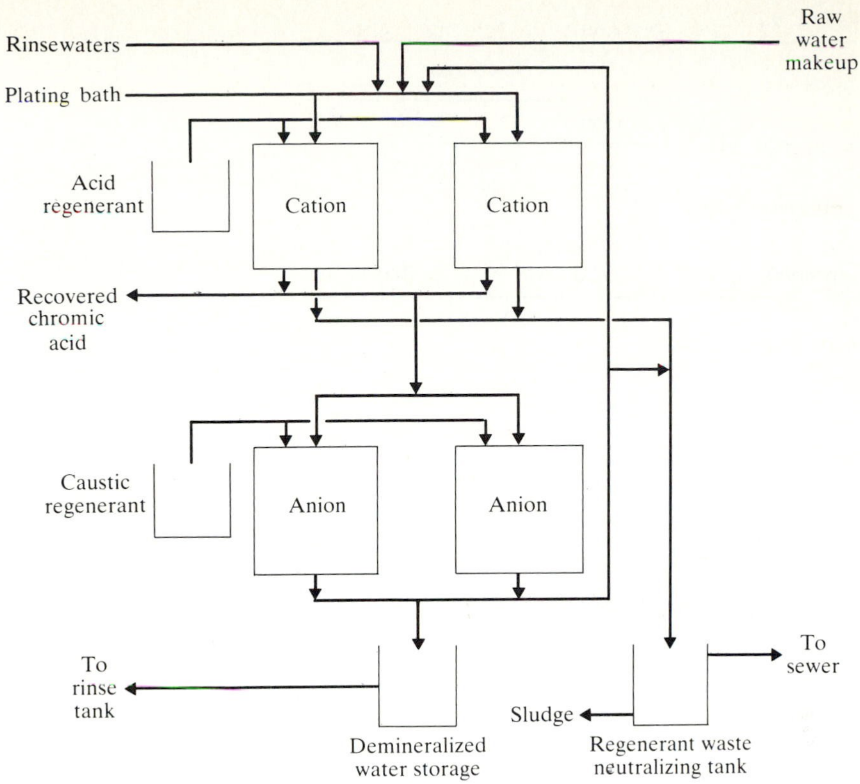

FIGURE 9-4
Ion exchange system for chromate removal and water reuse.

the last portion of the caustic regeneration can be used for neutralization of the spent cation regenerant.

In cation units, the regeneration requirements are higher than in water purification because of the competition with the H^+ present in the waste solution. For water reuse, 4 to 5 lb/ft^3 (64 to 80 kg/m^3) H_2SO_4 may be required for regeneration, but for recovery of H_2CrO_4 as much as 25 lb/ft^3 (400 kg/m^3) H_2SO_4 may be required to reduce leakage of sodium ions. A flow diagram of an ion exchange process for a plating plant is shown in Fig. 9-4.

Example 9.1. A general plating plant operates 16 h/d, 5 d/week. The total discharge of rinsewaters has the following characteristics:

Copper	22 mg/l as Cu
Zinc	10 mg/l as Zn
Nickel	15 mg/l as Ni
Chromium	130 mg/l as CrO_3

The rate of flow is 50 gal/min(0.19 m^3/min), and in-plant separation is not feasible. Design an exchanger system to include water and chromium recovery. The operating characteristics of the cation exchanger are given in Table 9.2.

TABLE 9.2

	Exchanger	
	Cation	**Anion**
Regenerant	H_2SO_4	NaOH
Dosage, lb/ft^3	12	4.8
Concentration, %	5	10
Flow rate	0.5 gal/(min · ft^3)	
Operating capacity	1.5 equiv wt/l	3.8 lb CrO_3/ft^3

Note:
lb/ft^3 = 16.0 kg/m^3
gal/(min · ft^3) = 0.134 m^3/(min · m^3)

Anion exchanger

In the anion exchanger, CrO_3 is exchanged for OH.

$$130 \text{ mg/l} \times 50 \text{ gal/min} \times 60 \text{ min/h} \times 16 \text{ h/d} \times 8.34 \times 10^{-6} \frac{\text{lb/gal}}{\text{mg/l}} = 52 \text{ lb/d}$$

Using a resin capacity of 3.8 lb CrO_3/ft^3 at a regeneration level of 4.8 lb CrO_3/ft^3 and a daily regeneration,

$$\text{Volume of resin} = \frac{52}{3.8} = 13.7 \text{ ft}^3 \ (0.39 \text{ m}^3)$$

Treatment flow rate is 3.6 gal/(min · ft^3) [0.48 m^3/(min · m^3)], using a resin depth of 30 in (0.76 m), 2 units, 2 ft (0.61 m) diameter by 30 in (0.76 m) deep, plus 50 percent for bed expansion.

Regeneration

$$\text{NaOH required} = 4.8 \times 13.7 = 66 \text{ lb/reg (30 kg/reg)}$$

$$\text{Regenerant tank volume} = 66 \text{ lb NaOH} \times \frac{1}{0.10} \times \frac{1}{9.6} \text{ lb reg/gal} = 68 \text{ gal } (0.26 \text{ m}^3)$$

Rinse requirement at 100 gal/ft^3 = 1370 gal (5.2 m^3)

Cation exchanger

The cations to be removed are:

$$\text{Zn} \quad \frac{10 \text{ mg/l}}{32.7 \text{ mg/meq}} = 0.306 \text{ meq/l}$$

$$\text{Cu} \quad \frac{22 \text{ mg/l}}{31.8 \text{ mg/meq}} = 0.693 \text{ meq/l}$$

$$\text{Ni} \quad \frac{15 \text{ mg/l}}{29.4 \text{ mg/meq}} = 0.511 \text{ meq/l}$$

In the cation unit, Cu, Zn, and Ni are exchanged for H^+.

The total daily equivalents are

$$(0.306 + 0.693 + 0.511) \times 10^{-3} \times 50 \times 60 \times 16 \times 3.78 \text{ lb/gal} = 273 \text{ equiv wt/d}$$

Using an operating capacity of 1.5 equiv wt/l at a regeneration level of 12 lb H_2SO_4/ft^3 (5 percent), the resin required for a 2-d regeneration is

$$\frac{273 \times 2}{1.5 \times 28.3 \text{ lb/ft}^3} = 13.0 \text{ ft}^3 \text{ resin } (0.36 \text{ m}^3)$$

The treatment flow rate is 3.8 gal/(min · ft³) [0.51 m³/(min · m³)]. Use 2 units, 2 ft (0.61 m) diameter by 30 in (0.76 m) deep plus 50 percent for bed expansion.

Regeneration
Using 5 percent H_2SO_4 at 12 lb/ft³, H_2SO_4 required is

$$12 \times 13 = 156 \text{ lb } (71 \text{ kg})$$

$$\text{Regenerant tank} = 156 \times \frac{1}{0.05} \times \frac{1}{1.0383 \times 8.34 \text{ lb/gal}} = 360 \text{ gal } (1.36 \text{ m}^3)$$

where 1.0383 is the specific gravity of 5% H_2SO_4.

$$\text{Rinse requirement} = 120 \text{ gal/ft}^3 \times 13 \text{ ft}^3 = 1560 \text{ gal } (5.9 \text{ m}^3)$$

Anion regenerant capacity for chromium recovery:

$$\text{Sodium} = \frac{66 \text{ lb NaOH} \times 453 \text{ g/lb}}{40 \text{ lb/equiv wt}} = 750 \text{ equiv wt}$$

If it is assumed that 70 percent of anion exchanger regenerant will pass through the cation unit and 525 equiv wt must be exchanged, which is compatible with the capacity of the cation units.

PROBLEM

9.1. A metal finishing plant operates 8 h/d, 6 d/week. The total discharge of reuse waters has the following characteristics:

Zn	15 mg/1
Ni	12 mg/l
Cr	190 mg/1 as CrO_3

The flow rate is 100 gal/min (0.38 m³/min). Design an ion exchange system to include water and chromium recovery. The operating characteristics of the resins are given in Example 9.1.

REFERENCES

1. Fadgen, T. J.: *Proc. 7th Ind. Waste Conf.*, 1952, Purdue University.
2. Paulson, C. F.: *Proc. 7th Ind. Waste Conf.*, 1952, Purdue University.
3. Rich, L. G.: *Unit Processes of Sanitary Engineering*, John Wiley, New York, 1963.
4. Keating, R. J., R. Dvorin, and V. J. Calise: *Proc. 9th Ind. Waste Conf.*, 1954, Purdue University.

CHAPTER 10

CHEMICAL OXIDATION

Chemical oxidation of a wastewater may be employed to oxidize pollutants to terminal end products or to intermediate products that are more readily biodegradable or more readily removable by adsorption. Common oxidants are chlorine, ozone, hydrogen peroxide, and potassium permanganate. Chemical oxidation is frequently markedly dependent on pH and the presence of catalysts.

10.1 OZONE

Ozone is a gas at normal temperature and pressure. As with oxygen, the solubility of ozone in water depends on temperature and the partial pressure of ozone in the gas phase and has recently been thought to also be a function of pH. Ozone is unstable and the rate of self-decomposition increases with temperature and pH. The decomposition is catalyzed by the hydroxide ion (OH^-), the radical decomposition products of O_3, the organic solute decomposition products, and by a variety of other substances such as solid alkalis, transition metals, metal oxides, and carbon. Under practical conditions, complete degradation of fairly unreactive compounds such as saturated hydrocarbons and halogenated aliphatic compounds does not occur with O_3 alone, but current research has shown that O_3 with an additional energy source, i.e., sonification or ultraviolet, readily decomposes these refractory compounds.

TABLE 10.1
Ozonation results on secondary clarifier effluent

	Low *F/M* (0.15) time, min†		High *F/M* (0.60) time, min†	
Parameter	**0**	**60**	**0**	**60**
BOD, mg/l	27	22	97	212
COD, mg/l	600	154	1100	802
pH	7.1	8.3	7.1	7.6
Org-N, mg/l	25.2	18.9	40	33
NH_3–N, mg/l	3.0	5.8	23	25
Color (Pt–Co)	3790	30.0	5000	330

† Loading of 155 mg O_3/min.

The mechanisms of ozone oxidation of organics are:

1. Oxidation of alcohols to aldehydes and then to organic acids:

$$RCH_2OH \xrightarrow{O_3} RCOOH$$

2. Substitution of an oxygen atom into an aromatic ring
3. Cleavage of carbon double bonds

Ozone is generated from dry air or oxygen by a high-voltage electric discharge with oxygen yielding twice the O_3 concentration (0.5 to 10 wt %) as air. Theoretically, 1058 g of ozone can be produced per kilowatt-hour of electrical energy but in practical application only a production of 150 g/(kW · h) can be expected.

Ozonation can be employed for the removal of color and residual refractory organics in effluents. In one case, while there was a decrease in TOC in the final filtered effluent, the soluble BOD increased from 10 to 40 mg/l due to the conversion from long-chain biologically refractory organics to biodegradable compounds. Similar results were obtained from the ozonation of a secondary effluent from low- and high-rate activated sludge units treating a tobacco-processing wastewater, as shown in Table 10.1. TOC will not be reduced until the organic carbon has been oxidized to CO_2, while COD will generally be reduced with any oxidation.

Oxidation of unsaturated aliphatic or aromatic compounds causes a reaction with water and oxygen to form acids, ketones, and alcohols. At a pH greater than 9.0 in the presence of redox salts such Fe, Mn, and Cu aromatics may form some hydroxyaromatic structures (phenolic) which may be toxic. Many of the byproducts of ozonation are readily biodegradable.

Organic removal is improved with ultraviolet radiation. It is postulated that the ultraviolet (UV) activates the O_3 molecule and may also activate the

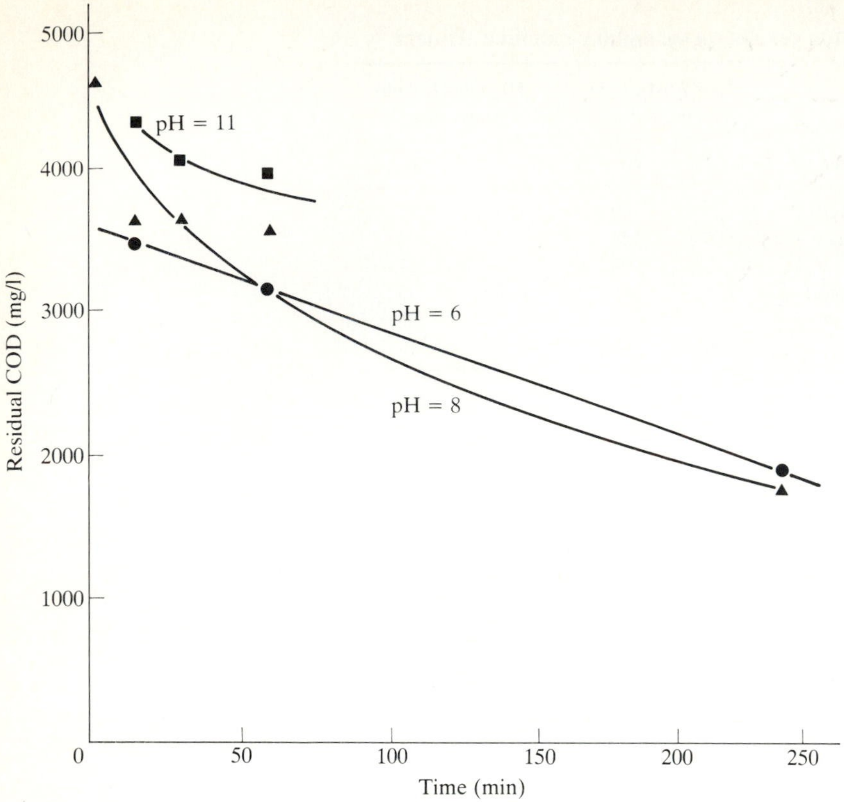

FIGURE 10-1
Effect of pH on COD removal by ozonation-aerated lagoon effluent.

substrate. Ozone-UV is effective for the oxidative destruction of pesticides to terminal end products of CO_2 and H_2O.

For many of the priority pollutants there is a rapid, first-order reduction in COD to some level followed by a slow or even zero removal, indicating the nonreactive nature of the byproducts.

COD reduction by ozonation of an aerated lagoon effluent treating an organic chemicals wastewater is shown in Fig. 10-1.

Phenol can be oxidized with ozone producing as many as 22 intermediate products between phenol and CO_2 and H_2O. The reaction is first order with respect to phenol and proceeds optimally over a pH of 8 to 11. The ozone consumption is 4 to 6 mol O_3 consumed/mol phenol oxidized. This requires in the order of 25 mol O_3/mol phenol to be generated in the gas phase.

10.2 HYDROGEN PEROXIDE

Hydrogen peroxide in the presence of a catalyst, e.g., iron, generates hydroxyl radicals ($^{\cdot}OH$) which react with organics and reduced compounds in a similar

TABLE 10.2
Sulfide oxidation with hydrogen peroxide

Acidic or neutral pH
$H_2O_2 + H_2S \rightarrow 2H_2O + S$
Reaction time: 15–45 min
Catalyst: Fe^{2+}
pH: 6.0–7.5
Reaction time: seconds
Basic pH
$4H_2O_2 + S^{2-} \rightarrow SO_4^{2-} + 4H_2O$
Reaction time: 15 min

manner to ozone:

$$Fe^{2+} + H_2O_2 \rightarrow Fe^{3+} + OH^- + {}^{\bullet}OH$$

$$Fe^{3+} + H_2O_2 \rightarrow Fe^{2+} + {}^{\bullet}HO_2 + H^+$$

$${}^{\bullet}OH + RH \rightarrow R^{\bullet} + H_2O$$

$$R^{\bullet} + H_2O_2 \rightarrow ROH + {}^{\bullet}OH$$

$${}^{\bullet}OH + Fe^{2+} \rightarrow OH^- + Fe^{3+}$$

Hydrogen peroxide is used for the oxidation of sulfide. The typical reaction profile for sulfide oxidation is shown in Table 10.2.

Cyanide is oxidized by hydrogen peroxide:

$$CN^- + H_2O_2 \rightarrow CNO^- + H_2O \qquad \text{pH } 9.5\text{–}10.5$$

$$CNO^- + OH^- + H_2O \rightarrow CO_3^{2-} + NH_3$$

Toxic wastewaters can be treated by hydrogen peroxide to reduce both the toxicity as well as the organic content. Some typical results are shown in Table 10.3.

Cost-effective application of chemical oxidation suggests partial oxidation of refractory organics to render them more biodegradable. H_2O_2 oxidation is most commonly conducted at a pH of 3.5 as corresponding to the maximum rate of free radical formation in the H_2O_2–Fe^{2+} system[1] where Fe^{2+} (Fenton's reagent) is used as a catalyst. TOC and COD indicate the amount of ultimate conversion (TOC) and degree of oxidation (COD). The average oxidation state (OX) of organic carbon in the waste mixture may be expressed as[2]

$$OX = \frac{4(TOC - COD)}{TOC} \tag{10.1}$$

The COD/TOC ratio can be used as a primary parameter to determine the extent of the overall reaction. Figure 10-2 shows a typical case in which the initial COD/TOC ratio before reaction is shown as point *A*. Point *B* identifies the initial

rogen peroxide oxidation of con- streams

Sample	TOC, mg/l	COD, mg/l	BOD, mg/l	LC_{50}, %
Wastewater 1				
Before H_2O_2	92	301	135	16.3
After H_2O_2	52	184	57	29.7
Removal, %	43.4	38.9	57.8	—
Wastewater 2				
Before H_2O_2	2150	2040	300	2.4
After H_2O_2	95	850	42	3.0
Removal, %	95.5	58.3	86.0	—

reaction products and point *C* the ratio after the reaction is substantially complete. An evaluation for toxicity and for biodegradability is made at points *B* and *C* to determine both the H_2O_2 dosage and the time of contact.

The efficiency of treatment is based on the ability of H_2O_2 to produce an acceptable organic byproduct without significant ultimate conversion of organic carbon to carbon dioxide. This can be expressed as

$$f = \frac{H_2O_2}{HPD_L} \tag{10.2}$$

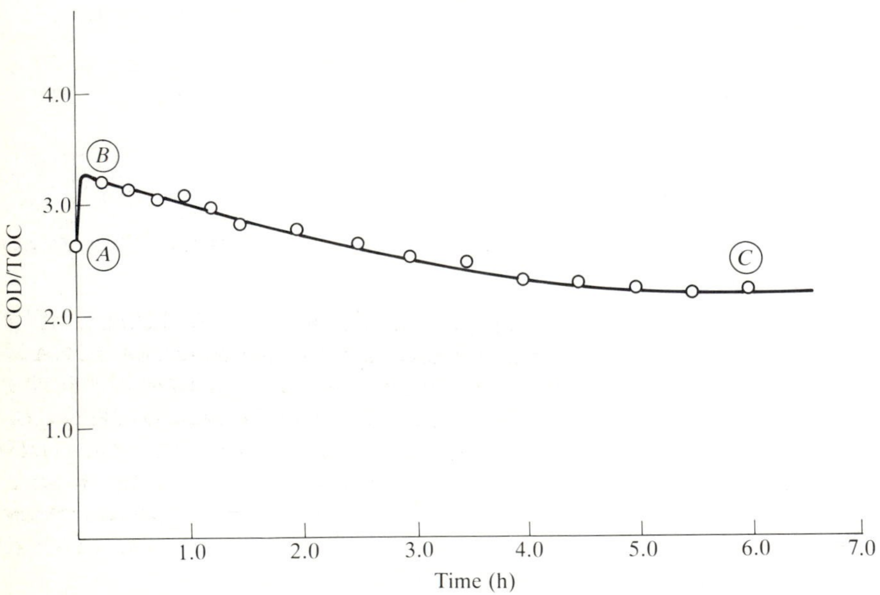

FIGURE 10-2
Change in COD/TOC with addition of H_2O_2.

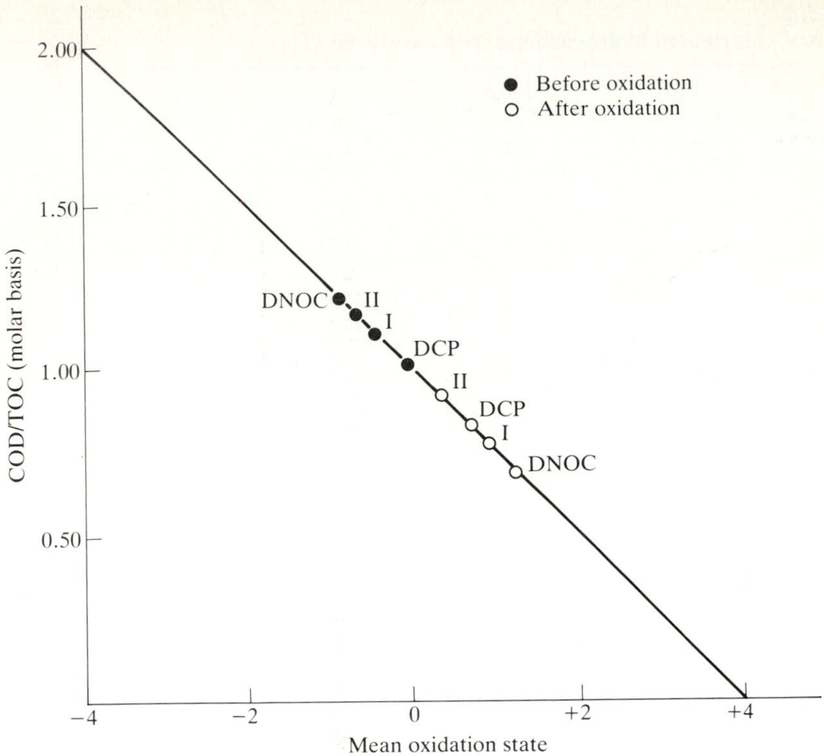

FIGURE 10-3
The COD/TOC ratio versus the mean oxidation state of carbon in the system. Note that a +4 oxidation state corresponds to inorganic carbon (CO_2) and a −4 oxidation state corresponds to methane. I and II refer to different times of oxidation.

in which f = fraction of ultimate H_2O_2 demand
H_2O_2 = hydrogen peroxide consumed, mg/l
HPD_L = 2.13 × initial COD mg/l

Results for the oxidation of 2,4-dichlorophenol (DCP) and 2,4-dinitro-*ortho*-cresol (DNOC) are shown in Fig. 10-3. The results show that the H_2O_2 was not used for ultimate conversion of the organic carbon, but rather the original compounds were drastically altered, leaving byproducts in which the mean oxidation state of the carbon was more highly oxidized (positive instead of negative). DNOC exhibited the greatest change (OX = +2.22) while DCP showed the least (OX = +0.867). The oxidation of chlorinated hydrocarbons is shown in Table 10.4.

The combination of hydrogen peroxide and UV light has been applied for the oxidation of a number of organics including acetates, organic acids, and explosives. The oxidation of trichloroethylene with H_2O_2 alone, UV alone, and

TABLE 10.4
Oxidation of chlorinated hydrocarbons with catalyzed H_2O_2†

Compound	Initial concentration, mg/l	Final concentration, mg/l	COD converted, %	TOC converted, %
Benzene				
Nitrobenzene	615.50	<2	72.40	37.25
Benzoic acid	610.50	<1	75.77	48.36
Phenol	470.55	<2	76.06	44.14
o-Cresol	540.50	<2	74.96	55.64
m-Cresol	540.50	<1	73.27	38.24
p-Cresol	540.50	<2	71.82	39.98
o-Chlorophenol	624.80	<2	75.08	47.87
m-Chlorophenol	624.80	<2	75.00	41.29
p-Chlorophenol	624.80	<2	75.69	21.74
2,3-Dichlorophenol	815.00	<2	70.16	52.63
2,4-Dichlorophenol	815.00	<2	68.89	50.31
2,5-Dichlorophenol	815.00	<2	74.24	41.81
2,6-Dichlorophenol	815.00	<3	61.07	32.51
3,5-Dichlorophenol	815.00	<1	69.09	48.86
2,3-Dinitrophenol	920.55	<1	80.07	50.65
2,4-Dinitrophenol	920.55	<1	72.52	51.00
Aniline	465.50	<1	76.49	43.37

† Sample contacted for 24 h with excess H_2O_2 and 50 mg/l Fe^{2+} catalyst.

H_2O_2 and UV is shown in Fig. 10-4. The oxidation of a variety of halogenated aliphatics at 20°C with UV–H_2O_2 is shown in Fig. 10-5.

10.3 CHLORINE

Chlorine may be used as a chemical oxidant. In reactions with inorganic materials terminal end products usually result while organic oxidations usually produce chlorinated hydrocarbons. Chlorination of a secondary effluent for color removal is shown in Fig. 10-6.

The alkaline chlorination process oxidizes cyanide by the addition of chlorine. The initial reaction forms cyanogen chloride (CNCl); this reaction is instantaneous at all pH levels:

$$NaCN + Cl_2 \rightarrow CNCl + NaCl$$

One part of cyanide requires 2.73 parts of chlorine. The cyanogen chloride formed in this reaction is volatile, toxic, and odorous. In the presence of caustic, sodium cyanate is formed from the cyanogen chloride:

$$CNCl + 2NaOH \rightarrow NaCNO + H_2O + NaCl$$

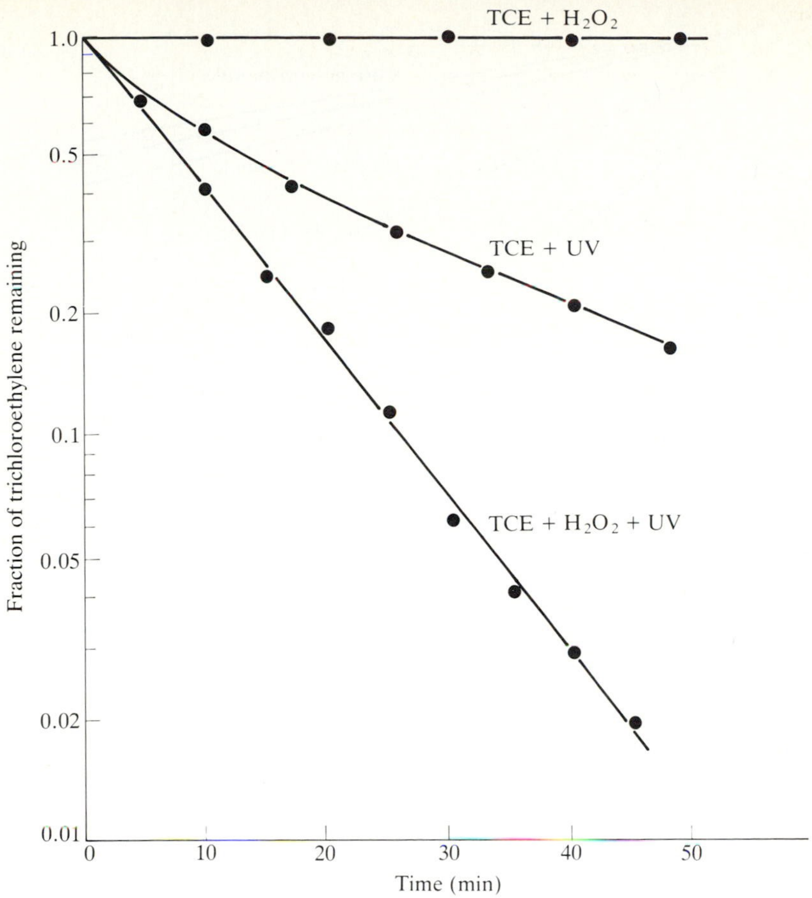

FIGURE 10-4
Effect of H_2O_2 alone, UV alone, and H_2O_2 plus UV on decomposition of trichloroethylene (TCE) at 20°C, pH 6.8. Initial TCE = 58 ppm. Initial H_2O_2/TCE = 4.5 mol/mol.[3]

The rate of this reaction is pH-dependent and is slow below pH 8.0. Above pH 8.5 the reaction goes to completion in 30 min. The reaction is temperature-dependent and can be increased by use of an excess of chlorine. The caustic requirement is 1.13 parts caustic/part Cl_2 applied.

Interferences to this reaction result from the presence of organic compounds and oxidizable metals and complexes. Ferro- and ferricyanide are oxidized at a slow rate.

The cyanate produced by this reaction is only one-thousandth as toxic as cyanide and will hydrolyze under acidic conditions in lakes or streams to ammonia and carbon dioxide. The cyanate will be further oxidized by the addition of chlorine to carbon dioxide and nitrogen:

$$2NaCNO + 4NaOH + 3Cl_2 \rightarrow 2CO_2 + 6NaCl + N_2 + 2H_2O$$

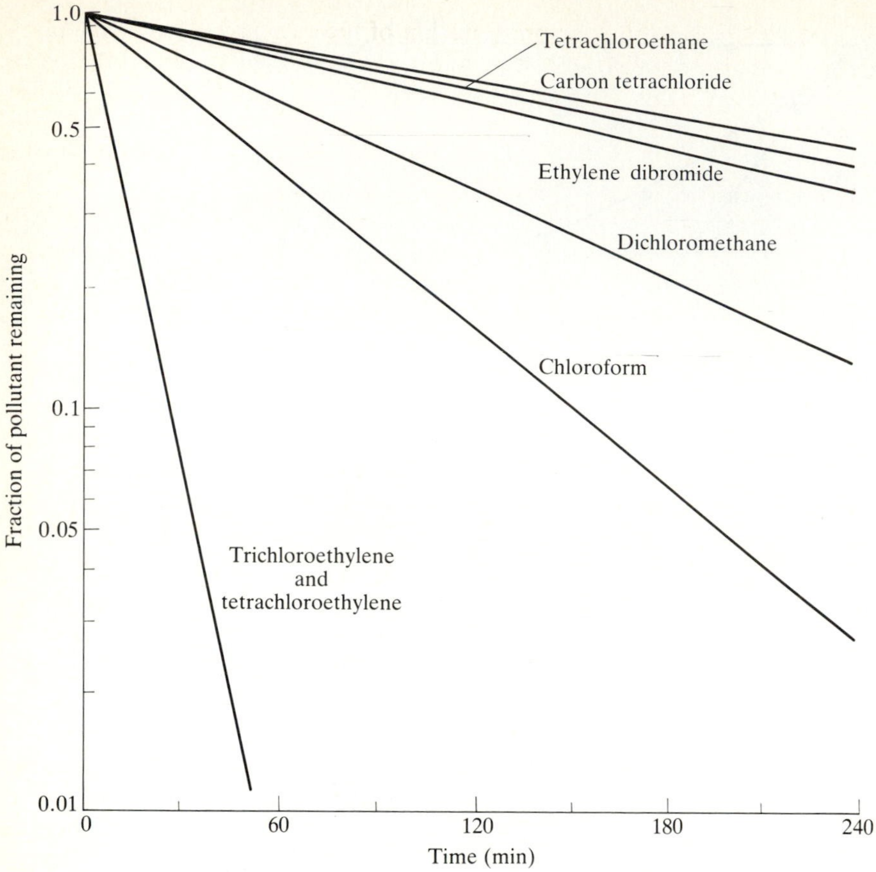

FIGURE 10-5
Comparison of rates of reaction of halogenated aliphatics at 20°C[3] with UV and hydrogen peroxide.

This reaction requires 4.09 parts Cl_2 and 3.08 parts NaOH/part CN^-. Although the reaction is most rapid at pH 6.5 to 6.8, a pH of 8.5 is usually used to simplify operation, to eliminate the formation of nitrogen trichloride, and to permit the last two reactions to occur sequentially in a single treatment unit. At pH 8.4 the reaction is complete in 10 min; at pH 9.6 it is complete in 40 min.

Excess chlorine is used to complete the reaction. Although the theoretical chlorine requirement for the complete oxidation of cyanide is 6.82 parts Cl_2/part CN^-, the reaction is completed with 7.35 parts Cl_2/part CN^-. Oxides of nitrogen that increase the chlorine requirement are formed in the presence of free chlorine. It is essential to avoid an excess temperature buildup, which leads to the formation of chlorates.

Cyanide waste treatment can be accomplished in either batch- or continuous-treatment systems. The batch-treatment system is similar to that described for chromium reduction and precipitation except that only a small volume need be provided for the precipitation of heavy metal sludges. Usually it is necessary to remove sludge at infrequent intervals.

In order to avoid the formation of toxic HCN, strict segregation of cyanide wastes from acidic wastestreams is essential.

The caustic requirements must usually be determined experimentally because of the presence of other contaminants in the wastestream.

In continuous-treatment systems oxidation-reduction potential can be used to control the addition of chlorine, as shown in Fig. 10-7. In the presence of excess chlorine, complete oxidation is assured if the pH and retention time are maintained at their proper levels.

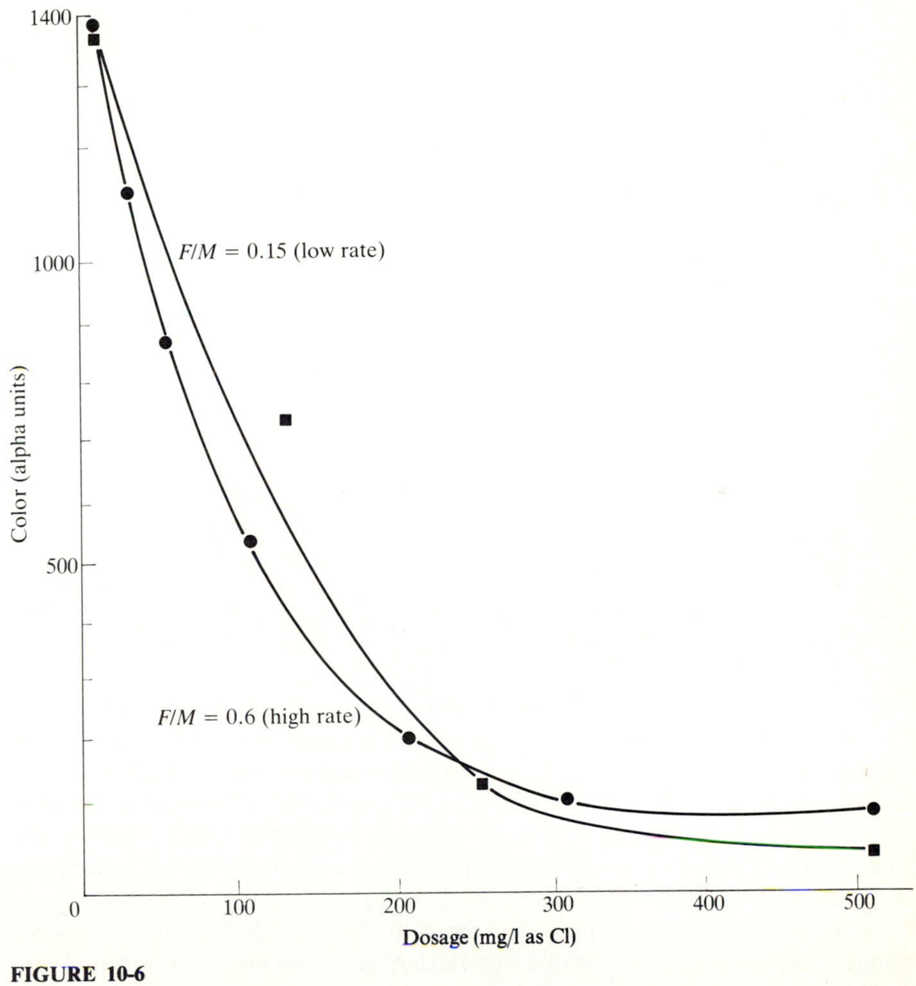

FIGURE 10-6
Color removal by chlorination.

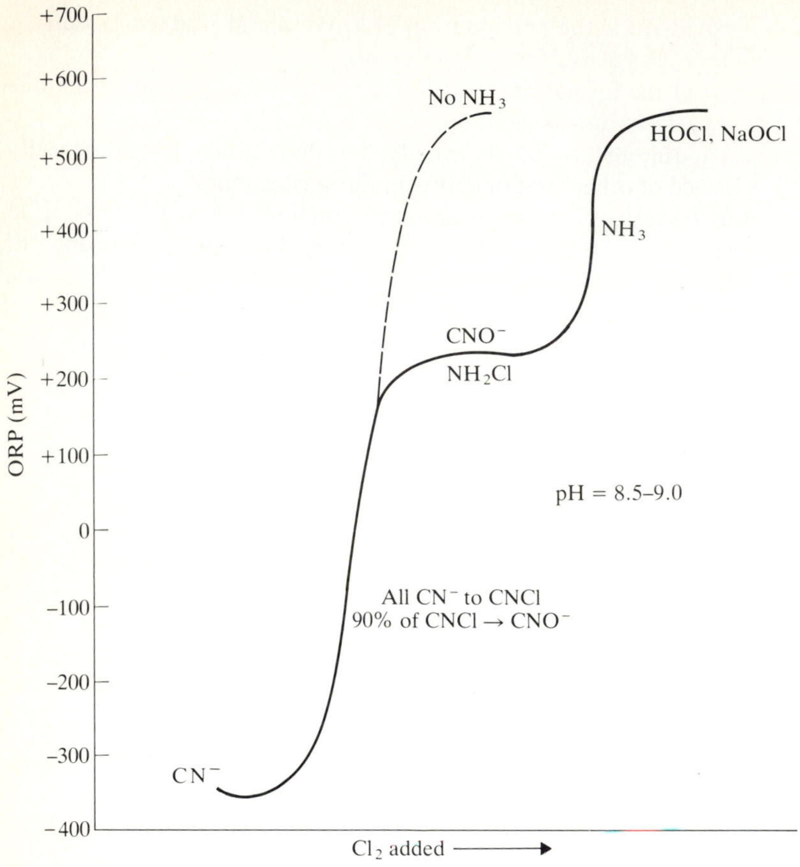

FIGURE 10-7
Oxidation-reduction potential relationships in the alkaline chlorination of cyanide waste.

10.4 WET AIR OXIDATION

Wet air oxidation (WAO) has been successfully applied in a number of applications for organics reduction. WAO is based on a liquid phase reaction between organic material in the wastewater and oxygen supplied by compressed air. The reaction takes place flamelessly in an enclosed vessel, which is pressurized and at high temperatures, typically 2000 lb/in^2 (136 atm) gage and 550°F (288°C). The system temperature, initiated by a startup boiler, is maintained through autothermal combustion of organics once the reaction starts. Typical performance data of WAO on concentrated organics are shown in Table 10.5. Changes in various parameters following WAO for two wastewaters are summarized in Table 10.5.

TABLE 10.5
Wet air oxidation of concentrated organics

	Influent	Effluent	Reduction, %
Total phenols, mg/l	28,700	66	99.77
Organic sulfur, mg/l	3,010	180	94.02
Sulfide sulfur, mg/l	333	1	99.70
Cyanide, mg/l	25,400	82	99.70
Pesticides (malathion, carbaryl, methoxychlor, dinoseb), mg/l	169	0.93	99.45
Total COD, mg/l	76,000	2500	96.70

	Wastewater	
	A	B
Influent COD, g/l	28.8	198.2
Reduction, %	79.1	86.3
Influent BOD, mg/l	228	82,500
Reduction, %	99.5	83.8
Influent color, APHA	1×10^6	—
Reduction, %	98.5	—
LC_{50} Influent†	1.1	3.2
Effluent†	4.1	87
Oxidation temperature 260°C for 60 min		

† 48 h LC_{50} mysid shrimp at 1:25 dilution.

PROBLEM

10.1. A wastewater has a flow of 40,000 gal/day (151 m^3/d) containing 35 mg/l CN^-. Compute the chemical requirements for the alkaline chlorination of this wastewater for CN^- removal.

REFERENCES

1. Bishop, D. F., *et al.*: *Ind. Eng. Chem. Process Des. Dev.*, vol. 7, p. 110, 1968.
2. Stumm, W., and Morgan, J. J.: *Aquatic Chemistry*, John Wiley, New York, 1981.
3. Sundstrom, D. W., *et al.*: Presented at the Ass. Ind. Chem. Engrs Spring Meeting, April 1986.

CHAPTER 11

SLUDGE HANDLING AND DISPOSAL

Most of the treatment processes normally employed in industrial water-pollution control yield a sludge from a solids-liquid separation process (sedimentation, flotation, etc.) or produce a sludge as a result of a chemical coagulation or a biological reaction. These solids usually undergo a series of treatment steps involving thickening, dewatering, and final disposal. Organic sludges may also undergo treatment for reduction of the organic or volatile content prior to final disposal.

In general, gelatinous-type sludges such as alum or activated sludge yield lower concentrations, whereas primary and inorganic sludges yield higher concentrations in each process sequence.

Figure 2-1 shows a substitution flowsheet for the various alternative processes available for sludge dewatering and disposal. The processes selected depend primarily on the nature and characteristics of the sludge and on the final disposal method employed. For example, activated sludge is more effectively concentrated by flotation than by gravity thickening. Final disposal by incineration desires a solids content that supports its own combustion. In some cases, the process sequence is apparent from experience with similar sludges or by geographical or economic constraints. In other cases, an experimental program must be developed to determine the most economical solution to a particular problem.

11.1 CHARACTERISTICS OF SLUDGES FOR DISPOSAL

The physical and chemical characteristics of sludges dictate the most technically and economically effective means of disposal. For thickening, the concentration ratio C_u/C_0 (the concentration of the underflow divided by the concentration of the influent) is related to the mass loading [lb solids/(ft$^2 \cdot$ d) or kg/(m$^2 \cdot$ d)], which indicates the feasibility of gravity thickening.

The dewaterability of a sludge by filtration is related to the specific resistance. While the specific resistance of a sludge can be reduced by the addition of coagulants, economic considerations may dictate alternative dewatering methods.

Ultimate disposal usually considers land disposal or incineration. When considering incineration, the heat value of the sludge and the concentration attainable by dewatering dictates the economics of the operation. Land disposal may use the sludge as a fertilizer or soil conditioner, as in the case of waste-activated sludges or in a confined landfill for hazardous sludges. It is important that if a sludge is to be used for land disposal, heavy metals must be removed by pretreatment.

11.2 AEROBIC DIGESTION

Aerobic digestion when applied to excess biological sludges involves the oxidation of cellular organic matter through endogenous metabolism. The oxidation of cellular organics have been found to follow first-order kinetics when applied to the degradable volatile suspended solids.[1] Under batch or plug-flow conditions,

$$\frac{(X_d)_e}{(X_d)_0} = e^{-k_d t} \tag{11.1}$$

where $(X_d)_e$ = degradable volatile solids after time t
$(X_d)_0$ = initial degradable volatile solids
k_d = reaction rate coefficient, d^{-1}
t = time of aeration, d

If the total volatile suspended solids are considered, Eq. (11.1) becomes

$$\frac{X_e - X_n}{X_0 - X_n} = e^{-k_d t} \tag{11.2}$$

where X_0 = initial VSS
X_e = effluent VSS
X_n = nondegradable VSS

When considering a completely mixed reactor[1] the relationship is modified to

$$\frac{X_e - X_n}{X_0 - X_n} = \frac{1}{1 + k_d t} \tag{11.3}$$

and the required retention time is

$$t = \frac{X_0 - X_e}{k_d(X_e - X_n)} \tag{11.3a}$$

For n multiple mixed reactors in series,

$$\frac{X_e - X_n}{X_0 - X_n} = \frac{1}{(1 + k_d t_n)^n} \tag{11.4}$$

In accordance with the kinetic relationship, multiple reactors in series are more efficient than one mixed reactor.

The oxygen requirements for aerobic digestion can be estimated as 1.4 lb of oxygen consumed for each pound of VSS destroyed (1.4 kg O_2/kg VSS destroyed). Nitrogen and phosphorus will be released by the oxidation process. Under mesophilic digestion conditions, nitrification will usually occur because of the long sludge ages in the reactor. Oxygen and alkalinity must be available for this oxidation. Under thermophilic operation, nitrification will be inhibited.

Temperature will affect the rate coefficient k_d. Conventional aerobic digestion design employs secondary clarifier underflow (0.5–1.5 percent solids) in one or more completely mixed aeration basins. Power levels of 15 to 20 standard $ft^3/(min \cdot thousand\ ft^3)$ [15 to 20 std $m^3/(min \cdot thousand\ m^3)$] using diffused air or 100 hp/million gal (0.02 kW/m^3) using surface mechanical aerators are usually adequate for providing both mixing and oxygen requirements. Prethickening the sludge offers a number of advantages, namely reducing the basin volume requirements and increasing the temperature due to the exothermic heat of reaction. Andrews and Kambhu[2] have estimated the heat combusion as 9000 Btu/lb (2.1×10^7 J/kg) VSS destroyed. Difficulties are encountered with conventional aeration systems to provide adequate mixing and aeration for sludge solids levels above 2 percent. Aerobic digestion designs are illustrated in Example 11.1.

Example 11.1. The data in Table 11.1 were obtained using bench-scale aeration of excess sludge from an activated sludge plant. 8000 gal/d (30 m^3/d) of this excess activated sludge is to be aerobically digested. The sludge has a concentration of 10,000 mg/l of VSS.

TABLE 11.1

Time of aeration, d	VSS, mg/l
0	6434
1	6160
3	5320
6	4300
8	4000
9	3890
14	3550
18	3250
25	3200

(*a*) Design a single-stage and a three-stage digester for 90 percent removal of the degradable solids.
(*b*) Compute the oxygen and power requirements. Assume 1.4 lb $O_2/(\text{hp}\cdot\text{h})$ [0.85 kg/(kW · h)] and a mixing requirement of 100 hp/million gal of basin volume (1.98 kW/m^3).

Solution

1. Determine the non-degradable residue X_n from a plot of data as shown in Fig. 11-1.
2. k_d is determined as the slope from a semilogarithmic plot of the degradable solids (VSS − X_n) versus time, as shown in Fig. 11-2.
3. For 90 percent removal of degradable solids X_e = 10,000 − 0.9 (0.51) × 10,000 = 5410 mg/l VSS.
4. For a single reactor, the retention time is calculated as follows:

$$t = \frac{X_0 - X_e}{k_d(X_e - X_n)} = \frac{10{,}000 - 5410}{0.21(5410 - 3150)} = 9.7 \text{ d}$$

$$\begin{aligned}\text{The reactor volume} &= Qt \\ &= 8000 \text{ gal/d } (9.7 \text{ day}) \\ &= 77{,}600 \text{ gal } (294 \text{ m}^3)\end{aligned}$$

5. For three reactors in series:

$$\frac{5140 - 3150}{10{,}000 - 3150} = \frac{1}{(1 + 0.21t_n)^3}$$

$$t_n = 2.4 \text{ d}$$

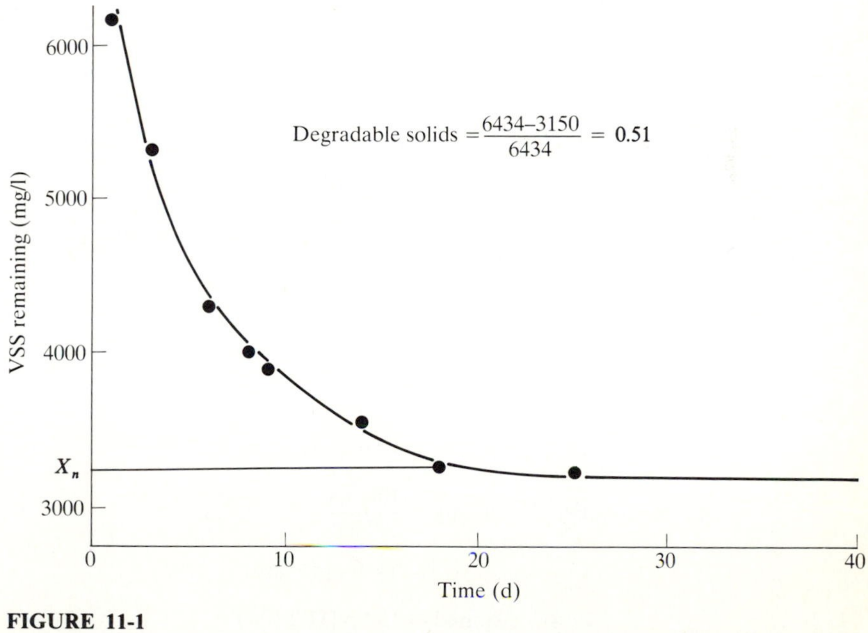

FIGURE 11-1

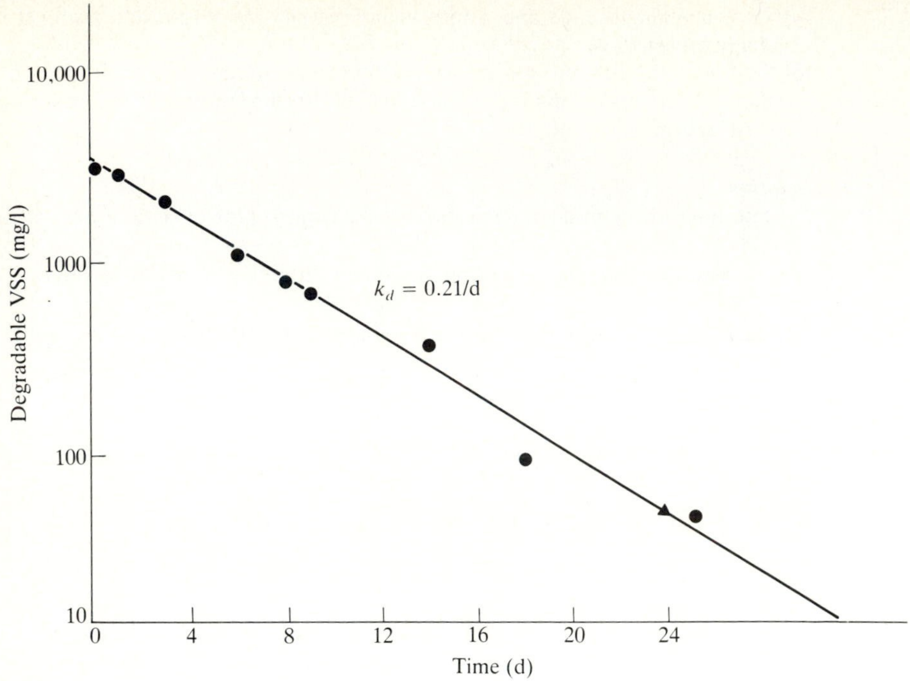

FIGURE 11-2

The volume of each reactor is

$$V = Qt = 8000 \text{ gal/d } (2.4 \text{ d}) = 19{,}200 \text{ gal } (73 \text{ m}^3)$$

6. Calculate oxygen and mixing requirements for each case. Assume 1.4 lb of oxygen is consumed for each pound of VSS destroyed.

Single reactor configuration

$$\text{lb VSS destroyed} = 0.9 \times 0.51 \times 10{,}000 \text{ mg/l} \times 8000 \text{ gal/d} \times 8.34 \text{ lb/gal}/10^6 \text{ mg/l}$$

$$= 306 \text{ lb VSS destroyed/d } (139 \text{ kg/d})$$

$$O_2 \text{ required} = 306 \text{ lb VSS/d} \times 1.4 \text{ lb } O_2/\text{lb VSS}$$

$$= 428 \text{ lb } O_2/\text{d } (194 \text{ kg/d})$$

$$\text{Power required for aeration} = \frac{428 \text{ lb } O_2/\text{d} \times \text{d}/24 \text{ h}}{1.4 \text{ lb } O_2/(\text{hp} \cdot \text{h})}$$

$$= 13 \text{ hp, say } 15 \text{ hp } (11.2 \text{ kW})$$

$$\text{Power required for mixing} = \frac{100 \text{ hp}}{10^6 \text{ gal}} \times 77{,}600 \text{ gal}$$

$$= 7.8 \text{ hp } (5.8 \text{ kW})$$

$$\text{Power required} = 15 \text{ hp } (11.2 \text{ kW})$$

Three reactors in series configuration
Calculate power requirements for each reactor.

For reactor 1, using the following rearrangement of the kinetic equation:

$$X_{e_1} = \frac{X_0 + t_n k_d X_n}{1 + t_n k_d}$$

$$= \frac{10{,}000 + (2.4)(0.21)(3150)}{1 + (2.4)(0.21)}$$

$$= 7700 \text{ mg/l}$$

$$(O_2 \text{ required})_1 = (10{,}000 - 7700)(8000) \times (8.34 \times 10^{-6}) \times 1.4$$

$$= 215 \text{ lb } O_2\text{/d (98 kg/d)}$$

$$\text{Power required for aeration} = \frac{215 \text{ lb } O_2\text{/d} \times \text{d/24 h}}{1.4 \text{ lb } O_2\text{/(hp}\cdot\text{h)}}$$

$$= 6.4 \text{ hp, say 7 hp (5.2 kW)}$$

$$\text{Power required for mixing} = \frac{100 \text{ hp}}{10^6 \text{ gal}} \times 19{,}200 \text{ gal}$$

$$= 1.92 \text{ hp (1.4 kW)}$$

By the same calculations it can be shown that for reactor 2,

$$X_{e_2} = 6175 \text{ mg/l}$$

$$(O_2 \text{ required})_2 = 142 \text{ lb } O_2\text{/d (64 kg/d)}$$

$$(\text{Power})_2 = 4.2 \text{ hp, say 5 hp (3.7 kW)} > 1.92 \text{ hp}$$

$$\text{Thus, (power required)}_2 = 5 \text{ hp (3.7 kW)}$$

For reactor 3:

$$X_{e_3} = 5161 \text{ mg/l}$$

$$(O_2 \text{ required})_3 = 95 \text{ lb } O_2\text{/d (43 kg/d)}$$

$$(\text{Power})_3 = 1.3 \text{ hp, say 1.5 hp (1.1 kW)} < 1.92 \text{ hp}$$

$$\text{Thus, (power required)}_3 = 1.92 \text{ hp, say 2 hp (1.5 kW)}$$

$$\text{Total power required} = 14 \text{ hp (10.4 kW)}$$

11.3 GRAVITY THICKENING

Gravity thickening is accomplished in a tank equipped with a slowly rotating rake mechanism that breaks the bridge between sludge particles, thereby increasing settling and compaction. A typical gravity thickener is shown in Fig. 11-3.

The primary objective of a thickener is to provide a concentrated sludge underflow. The area of thickener for a specified underflow concentration is

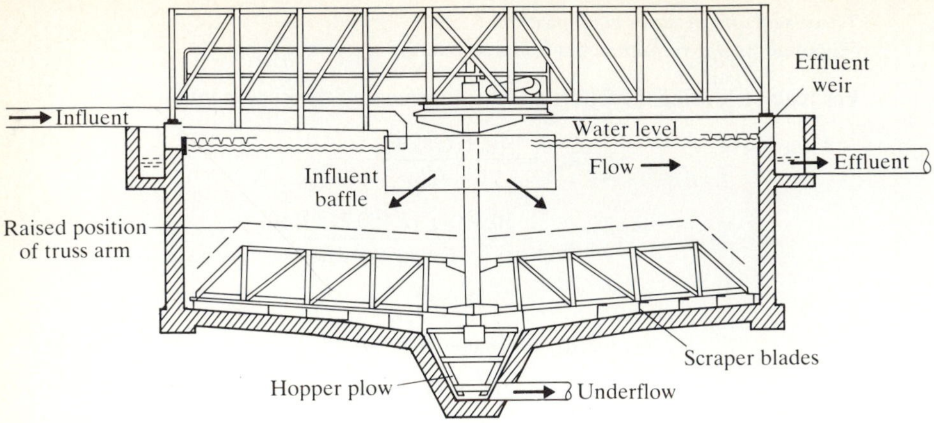

FIGURE 11-3
Gravity thickener. (*Courtesy of Link Belt, FMC Company.*)

related to the mass loading [lb/(ft^2 · d), kg/(m^2 · d)] or to the unit area [ft^2/(lb · d), m^2/(kg · d)].

The mass loading can be computed from a stirred laboratory cylinder test. For municipal sewage, the mass loading might be expected to vary from 4 lb/(ft^2 · d) [19.5 kg/(m^2 · d)] for waste-activated sludge to 22 lb/(ft^2 · d) [107 kg/(m^2 · d)] for primary sludge.

A procedure for the design of gravity thickeners has been developed by Dick.[3] The most important criteria in thickener design and operation is the mass loading or solids flux expressed as pounds of solids fed per square foot per day (or kilograms per square meter per day). The limiting flux that produces the desired underflow for a given area can be defined as

$$G_L = \frac{C_0 Q_0}{A} = \frac{M}{A} \tag{11.5}$$

where Q_0 = influent flow, ft^3/d (m^3/d)
C_0 = influent solids, lb/ft^3 (kg/m^3)
M = solids loading, lb/d (kg/d)
G_L = limiting solids flux, lb/(ft^2 · d) [kg/(m^2 · d)]
A = area, ft^2 (m^2)

The limiting flux can be obtained from the following rationale.

The capacity of a thickener for removing solids under batch conditions is

$$G_B = C_i V_i \tag{11.6}$$

where G_B = batch flux, lb/(ft^2 · d) [kg/(m^2/d)]
C_i = solids concentration, lb/ft^3 (kg/m^3)
V_i = settling velocity at C_i, ft/d (m/d)

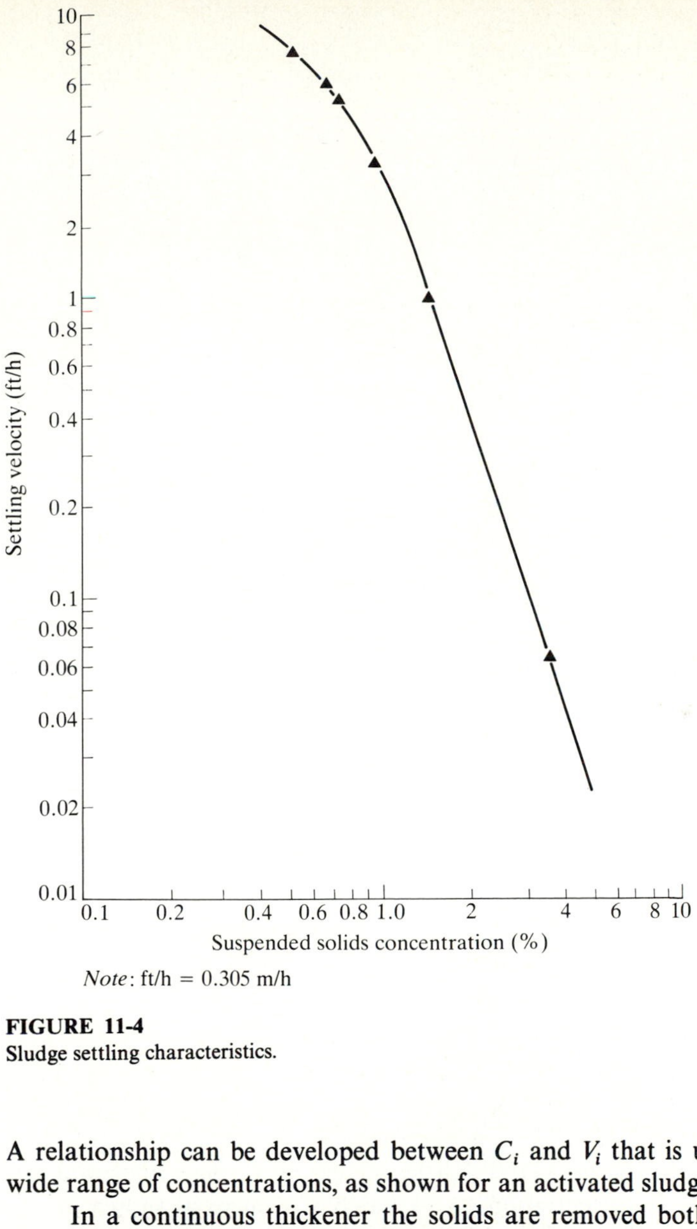

FIGURE 11-4
Sludge settling characteristics.

A relationship can be developed between C_i and V_i that is usually linear over a wide range of concentrations, as shown for an activated sludge in Fig. 11-4.

In a continuous thickener the solids are removed both by gravity and by the velocity resulting from the removal of sludge from the tank bottom:

$$G = C_i V_i + C_i U \tag{11.7}$$

where G = continuous solids flux, lb/(ft$^2 \cdot$ d) [kg/(m^2/d)]

U = downward sludge velocity due to sludge removal, ft/d (m/d)

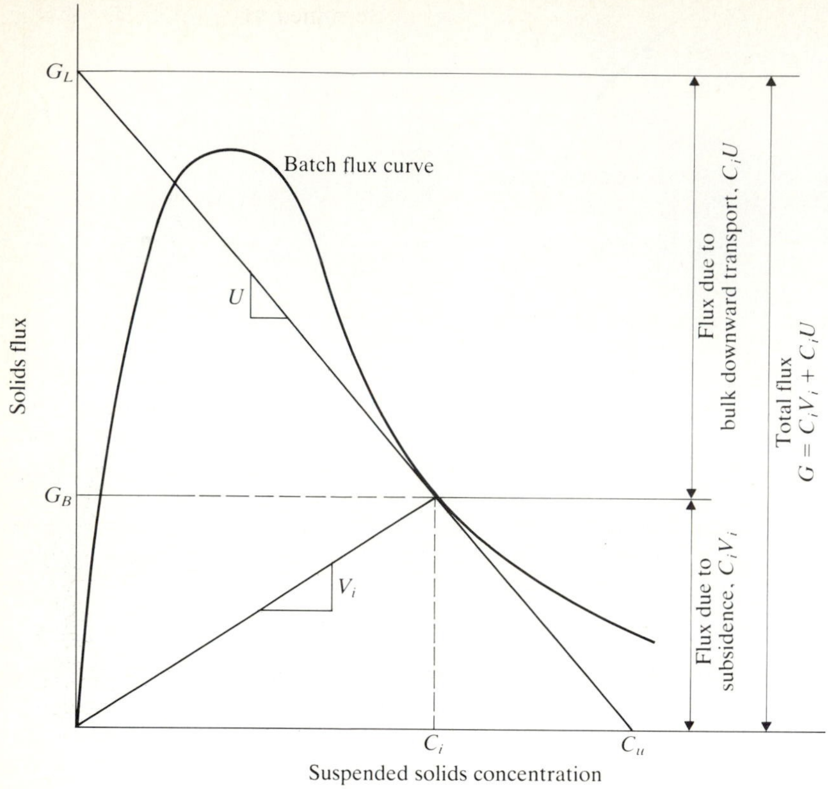

FIGURE 11-5
Batch flux plot illustrating how to determine limiting flux for a continuous thickener.

G can be varied by controlling U since this is determined by the underflow pumping rate. Assuming total solids removal from the bottom:

$$U = \frac{Q_u}{A} = \frac{C_u Q_u}{C_u A} = \frac{M}{C_u A} = \frac{G_L}{C_u} \tag{11.8}$$

where Q_u = underflow, ft³/d (m³/d)
C_u = underflow concentration, lb/ft³ (kg/m²)

It is important to note from Eq. (11.8) that by increasing U, the withdrawal rate decreases the underflow concentration C_u. A batch flux curve as shown in Fig. 11-5 can be employed to determine the limiting flux G_L for a given underflow concentration C_u, since the slope of any line connecting G_L on the y axis with C_u on the x axis on the batch flux curve is the corresponding required underflow velocity U. The batch flux curve as shown in Fig. 11-5 is obtained by plotting G_B as computed from Eq. (11.6) against its corresponding concentration C_i.

The required thickener area, A, is then computed from Eq. (11.5). It should be noted that the selected underflow concentration C_u must be less than the ultimate concentration attainable, C_∞. C_∞ is determined from thickening studies as

$$C_\infty = \frac{C_0 H_0}{H_\infty} \tag{11.9}$$

where C_0 = initial solids concentration
H_0 = initial height
C_∞ = final or ultimate concentration
H_∞ = final height

Example 11.2 illustrates a gravity thickener design.

Example 11.2. Waste sludge from a chemical coagulation process is to be gravity thickened from 0.5 to 4 percent. The average sludge volume is 550,000 gal/d (2082 m^3/d) with a variation of 450,000 to 700,000 gal/d (1703 to 2650 m^3/d). Determine the thickener area required and the underflow solids concentration at minimum flow.

The relationship between zone settling velocity and suspended solids concentration is shown in Table 11.2. The batch flux curve is developed by plotting the flux G versus its corresponding concentration. For example, at 2 percent solids,

$$G = 0.02 \times 62.4 \text{ lb/ft}^3 \times 0.50 \text{ ft/h} \times 24 \text{ h/d} = 15.0 \text{ lb/(ft}^2 \cdot \text{d) [73.3 kg/(m}^2 \cdot \text{d)]}$$

The batch flux curve is shown in Fig. 11-6. For the desired underflow concentration of 4 percent, the limiting flux is found from the batch flux curve by constructing a tangent which extends from 4 percent solids to the limiting flux of 26 lb/($ft^2 \cdot$ d) [127 kg/($m^2 \cdot$ d)]. The required thickener area is calculated:

$$A = \frac{C_0 Q_0}{G}$$

$$= \frac{(0.7 \text{ million gal/d}) (5000 \text{ mg/l}) (8.34) \text{ (lb/million gal)/(mg/l)}}{26 \text{ lb/(ft}^2 \cdot \text{d)}} = 1123 \text{ ft}^2$$

TABLE 11.2
Batch settling data

Solids concentration, %	Settling velocity, ft/h
0.50	7.5
0.75	5.5
1.00	4.2
1.25	3.1
1.50	1.5
2.00	0.50
4.00	0.075
6.00	0.030

Note: ft/h = 0.3048 m/h

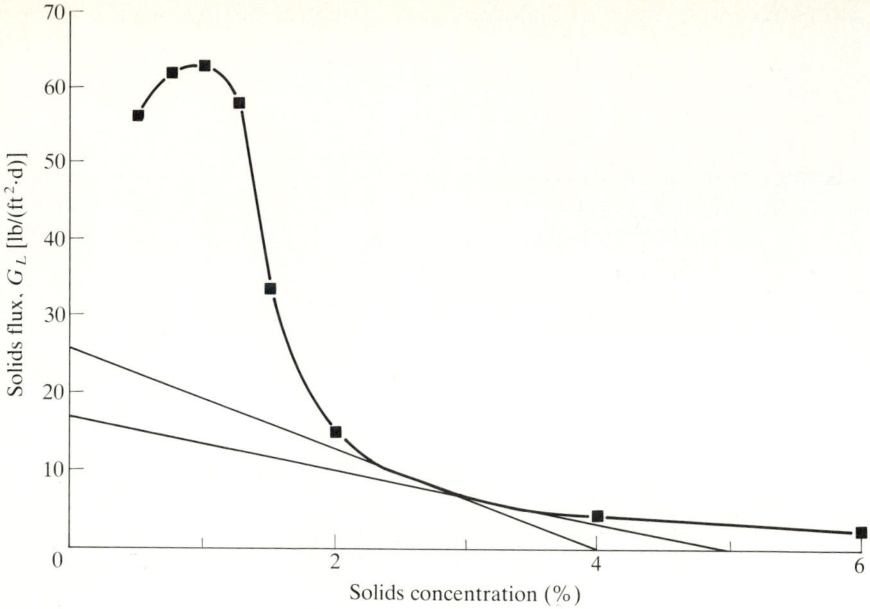

Note: lb/(ft^2·d) = 4.89 kg/(m^2·d)

FIGURE 11-6

When the sludge flow to the thickener is 0.45 million gal/d, the solids flux will be

$$G = \frac{(0.45 \text{ million gal/d})\,(5000 \text{ mg/l})\,(8.34)\,(\text{lb/million gal})/(\text{mg/l})}{1123 \text{ ft}^2}$$

$$= 16.7 \text{ lb/(ft}^2 \cdot \text{d)} \,[81.6 \text{ kg (m}^2 \cdot \text{d)}]$$

From the batch flux curve (Fig. 11-6), the underflow concentration at this loading will be 4.9 percent.

11.4 FLOTATION THICKENING

Thickening through dissolved air flotation is becoming increasingly popular and is particularly applicable to gelatinous sludges such as activated sludge. In flotation thickening, small air bubbles released from solution attach themselves to and become enmeshed in the sludge flocs. The air-solid mixture rises to the surface of the basin, where it concentrates and is removed. The primary variables are recycle ratio, feed solids concentration, air/solids (A/S) ratio, and solids and hydraulic loading rates. Pressures between 50 and 70 lb/in^2 (345 to 483 kPa or 3.4 to 4.8 atm) are commonly employed. Recycle ratio is related to the air/solids ratio and the feed solids concentration. The float solids are related to the A/S ratio, as shown in Fig. 11-7.

Experience has shown that in some cases dilution of the feed sludge to a lower concentration increases the concentration of the floated solids. A flotation

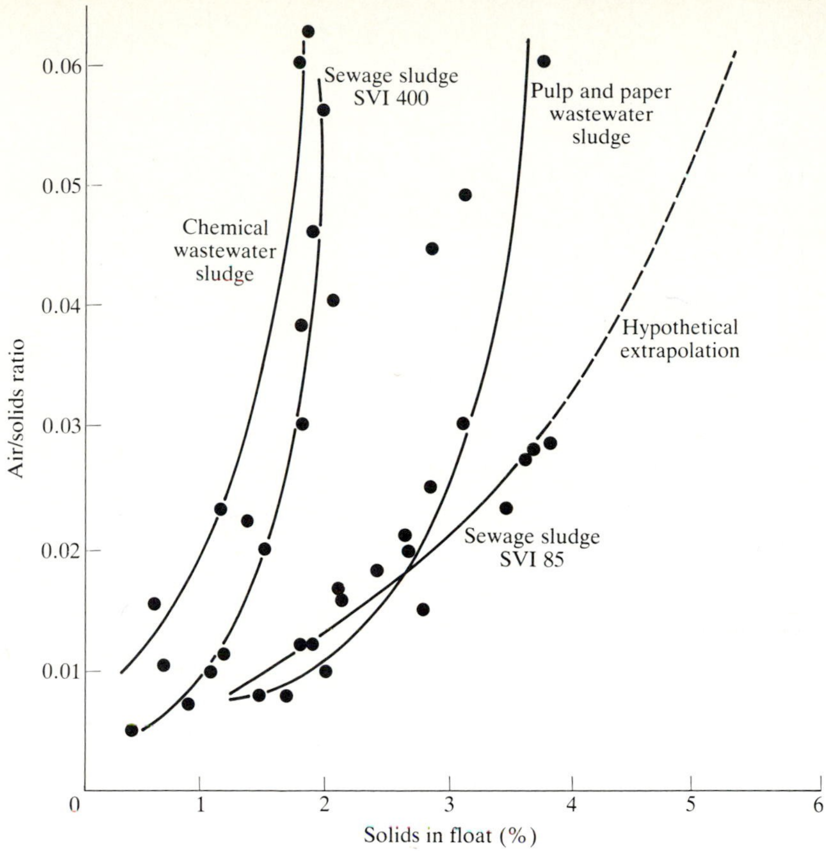

FIGURE 11-7
Influence of air/solids ratio on float solids content.

thickener is shown in Fig. 11-8. Performance data for the thickening of excess activated sludge is shown in Table 11.3. The use of polyelectrolytes will usually increase the solids capture and the thickened sludge concentration.

The quality of the sludge has a significant effect on its ability to thicken by flotation. For example, a filamentous, bulking activated sludge may not achieve 2 percent solids as compared to 4 to 5 percent with a well-flocculated sludge. This in turn will affect all other solids handling operations. Activated sludge can also be thickened using a centrifuge or a gravity belt thickener. Results are shown in Table 11.3.

11.5 CENTRIFUGATION

Centrifugation is employed both for the thickening and the dewatering of sludges. The process of centrifugation is an acceleration of the process of sedimentation

TABLE 11.3
Thickening and dewatering of wastewater sludges

Equipment	Type of sludge	Loading	Cake solids, %	Chemicals, lb/ton polymer	Reference
Thickening					
Gravity	WAS	5–6 lb/(ft^2 · d)	2.5–3.0	None	4
Gravity	Pulp and paper				
	53 percent P, 47 percent WAS	25	4	None	4
	67 percent P, 33 percent WAS	25	6	None	
	100 percent P	25	9	None	
DAF	WAS	2.9–4.5 lb/(ft^2 · h)	4–5.7	Low dosage	4
Solid bowl centrifuge	WAS	75–100 gal/min	5–7	None	4
Basket centrifuge	Citrus, WAS	25–40 gal/min	9–10	10–20	5
Gravity belt	WAS	315 gal/min	5.5	39	—
Solid bowl centrifuge	Paper mill, WAS	100 gal/min	11	10	6
Solid bowl centrifuge	Chemical, WAS	—	7–9	5–10	7
Dewatering					
Basket centrifuge	Citrus, WAS	25–40 gal/min	9–10	10–20	5
Basket centrifuge	Paper mill, WAS	60 gal/min	11	5	6
Belt press	Citrus, WAS	40 gal/(min · m)	18	10–20	5
Belt press	Paper mill, WAS	70 gal/(min · m)	16	6.5	6
Belt press	Chemical, WAS	—	13–15	10–20	7
Belt press	Organic chemical, WAS	190 gal/min	15	25	8
Belt press	Deinking primary	500 l/m	37	4	9
Belt press	Bleached and unbleached kraft; 67 percent P, 33 percent WAS	240 l/m	27	12	9
Belt press	Kraft linerboard WAS	75 l/m	19	25	9

Note:
lb/(ft^2 · d) = 4.88 kg/(m^2 · d)
gal/min = 3.78×10^{-3} m^3/min

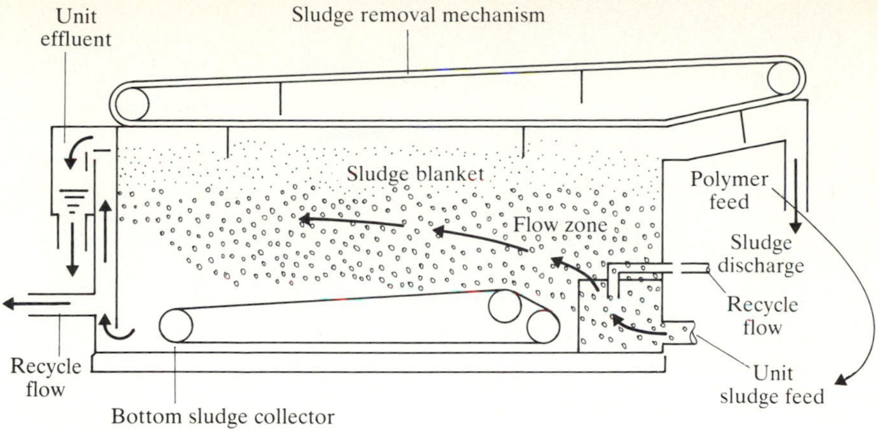

FIGURE 11-8
Flow diagram of a flotation unit. (*Courtesy of Komline-Sanderson Engineering Company.*)

by the application of centrifugal forces. There are three types of centrifuges available: the solid bowl decanter, the basket type, and the disk-nozzle separator. The basic difference between the types of centrifuges is the method in which solids are collected and discharged from the bowl.

Centrifuge performance is affected by both machine and process variables. The significant machine variables for the solid bowl decanter are bowl speed, pool volume, and conveyor speed. Disk-nozzle separator variables include bowl speed, recycle mode of operation, disk spacing and nozzle configuration. Basket centrifuge machine variables are bowl speed, cycle feed time, skimmer nozzle travel rate, and skimmer nozzle dwell time. Process variables include the feed rate of solids to the machine, solids characteristics, chemical addition, and temperature.

The solid bowl decanter consists of an imperforated cylindrical-conical bowl with an internal helical conveyor as shown in Fig. 11-9. The feed sludge enters the cylindrical bowl through the conveyor discharge nozzles. Centrifugal forces compact the sludge against the bowl wall, and the internal scroll or conveyor, which rotates slightly slower than the bowl, conveys the compacted sludge along the bowl wall toward the conical section (beach area) and out.

In the basket-type centrifuge, feed is introduced in the bottom of the basket. At equilibrium, solids settle out of the annular moving liquid layer to the cake layer, which builds up on the bowl wall while the centrate overflows the lip ring at the top. When solids have filled the basket, feed is stopped, the basket speed is reduced, and a knife moves into the cake, discharging it from the bottom of the casing. Cycles are automated and cake unloading requires less than 10 percent of the cycle time. Chemical addition is generally not required for high solids recovery. However, the unit operates at low centrifugal forces, has a discontinuous cake discharge, and a fairly low solids handling capacity.

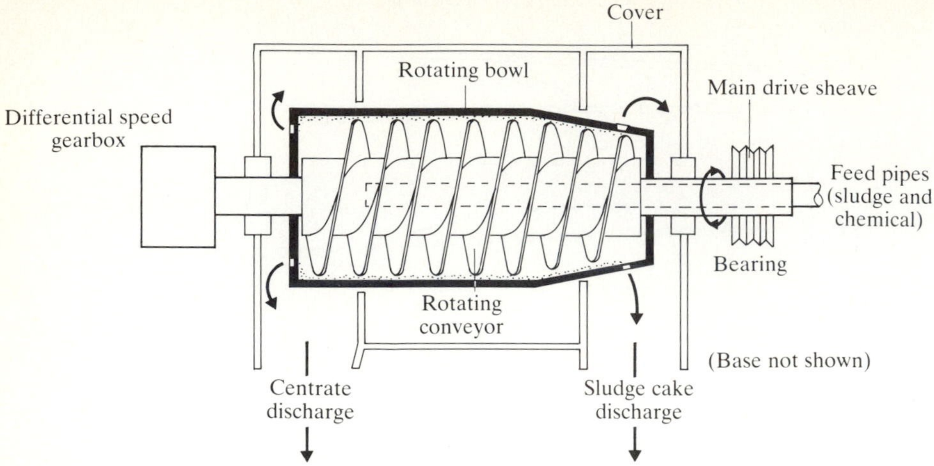

FIGURE 11-9
Continuous countercurrent solid bowl conveyor discharge centrifuge.

Activated sludge can be thickened using up to 2.5 g/kg of cationic polymer to 4.5 to 8.0 percent total solids. A low SVI sludge will yield the higher cake density.

In the disk-nozzle separator, the feed enters at the top and is distributed between a multitude of channels, or spaces between the stacked conical disks. Solid particles settle through the liquid layer, which is flowing in these channels to the underside of the disk and then slide down to a sludge compaction zone. The thickened sludge is flushed out of the bowl with a portion of the wastewater, thus limiting the solids concentration to 10 to 20 times the feed rate. The disk-nozzle separator finds its major application in the thickening of activated and similar sludges. They are very efficient in thickening waste-activated sludge at high feed rates without the addition of polymers. In an industrial waste-treatment plant in Germany excess activated sludge has been thickened from 1 percent to 8 to 10 percent solids.

When the feed rate to a centrifuge is increased, the retention time in the unit is decreased and the recovery decreases. Flow rates are usually limited to 0.5 to 2.0 gal/(min · hp) [3.65 to 14.6 $m^3/(d \cdot kW)$] to obtain satisfactory solids recovery. Since the lower recovery results in only the removal of larger particles, a drier cake is produced. Increasing the feed solids concentration reduces the liquid overflow from the machine, resulting in an increased recovery of solids.

Chemical flocculants (polyelectrolytes) are used to increase recovery. The flocculants both increase the structure strength of the solids and flocculate fine particles. Because of the increased removal of the fine particles, chemical addition usually lowers the cake dryness. Centrifuge performance characteristics are shown in Fig. 11-10.

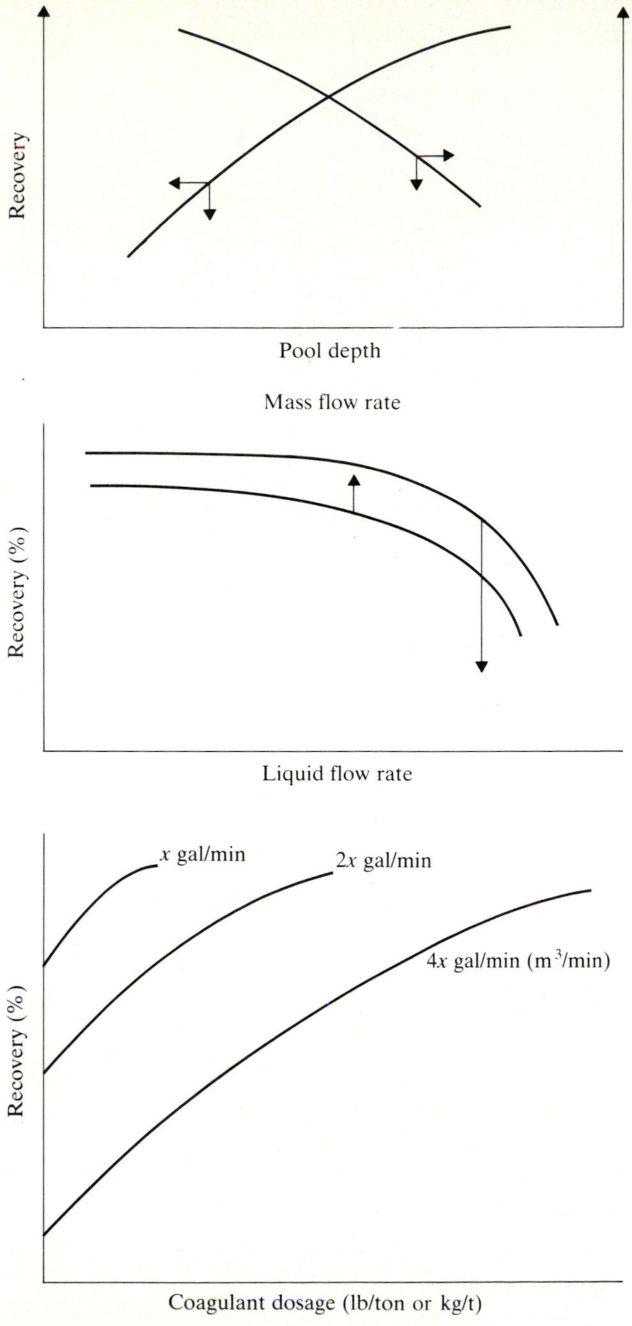

FIGURE 11-10
Centrifuge operating relationships.

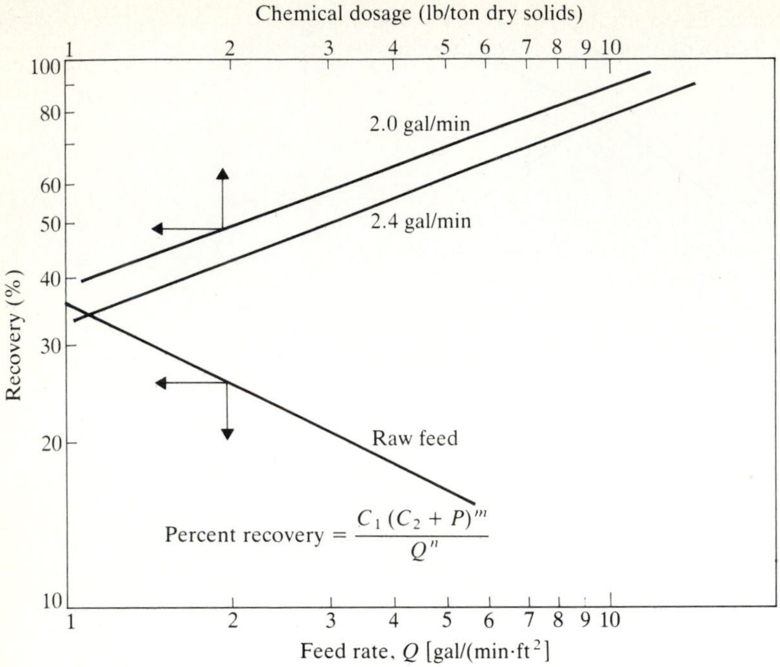

Note:
lb/ton = 0.50 kg/t
gal/min = 3.78×10^{-3} m^3/min
gal/(min·ft^2) = 4.075×10^{-2} $m^3/(min \cdot m^2)$

FIGURE 11-11
Solids recovery for digested activated sludge with cationic polymer.

Bernard and Englande[10] correlated centrifuge performance data in accordance with the relationship

$$R = \frac{C_1(C_2 + P)^m}{Q^n} \tag{11.10}$$

where R = recovery, %
P = polymer dosage, lb/ton dry solids feed or kg/ton
Q = feed rate, gal/(min · ft^2) [$m^3/(min \cdot m^2)$]
C_1, C_2 = constants
m, n = exponents

Centrifuge performance in accordance with Eq. (11.10) is shown in Fig. 11-11. Centrifuge design is shown in Example 11.3.

Example 11.3. Determine the polymer required and size of centrifuge with respect to surface area for dewatering 10,000 lb/d (4536 kg/d) of sludge previously thickened to 4 percent. The centrifuge will operate over 8 hours with a 95 percent solids recovery. The pilot plant data yielded the following relationship:

$$R(\%) = \frac{48(0.47 + P)^{0.37}}{Q^{0.52}}$$

where R = solid recovery efficiency, %
P = polymer dosage, lb/ton of sludge
Q = hydraulic loading, gal/(min · ft²)

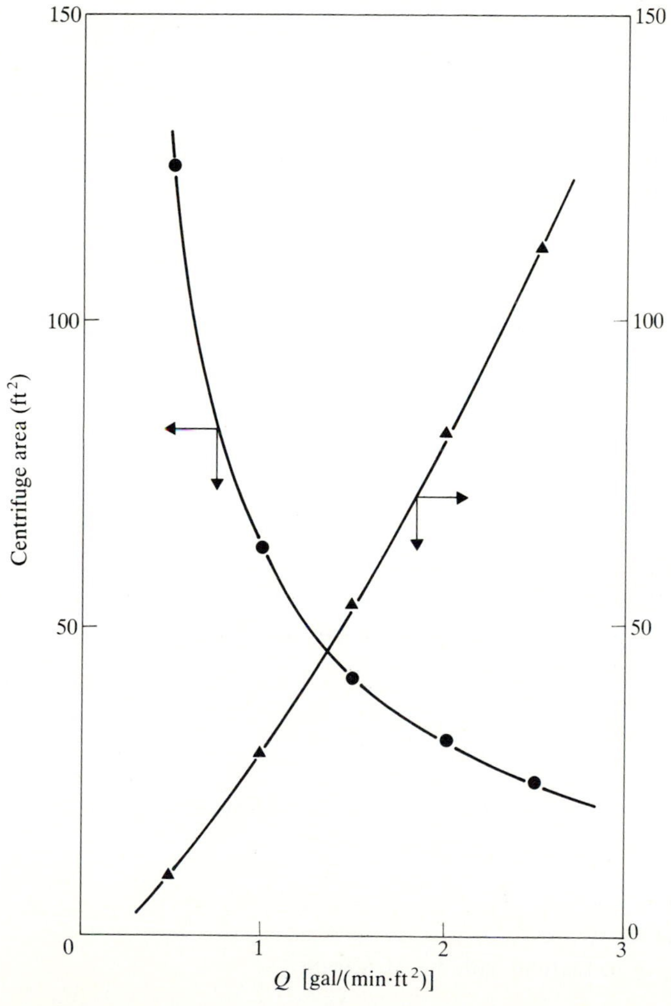

FIGURE 11-12
Required centrifuge area and polymer dosage for various hydraulic loadings.

For 4 percent solids concentration, the sludge flow is

$$Q' = 10{,}000 \times \frac{1}{0.04} \times \frac{1}{8.34}$$

$$= 30{,}000 \text{ gal/d } (114 \text{ m}^3\text{/d})$$

If the centrifuge will operate for 8 hours per day, the total sludge flow to the centrifuge is

$$Q' = 30{,}000 \times \frac{24}{8}$$

$$= 90{,}000 \text{ gal/d or } 62.5 \text{ gal/min } (341 \text{ m}^3\text{/d or } 0.237 \text{ m}^3\text{/min})$$

At 95 percent solids recovery, the relationship between polymer requirement and hydraulic loading can be calculated by

$$95 = \frac{48(0.47 + P)^{0.37}}{Q^{0.52}} \qquad \text{(plot } P \text{ versus } Q \text{ in Fig. 11-12)}$$

The centrifuge size is computed by

$$A = \frac{62.5}{Q} \text{ ft}^2 \qquad \text{(plot } A \text{ versus } Q \text{ in Fig. 11-12)}$$

From Fig. 11-12,

$$A = 46 \text{ ft}^2 \ (4.3 \text{ m}^2)$$

$$P = 46 \text{ lb/d } (21 \text{ kg/d})$$

11.6 SPECIFIC RESISTANCE

Sludge dewaterability can be defined in terms of its specfic resistance.

The rate of filtration of sludges has been formulated according to Poiseuille's and Darcy's laws by Carman and by Coackley and Jones:[11]

$$\frac{dV}{dt} = \frac{PA^2}{\mu(rcV + R_m A)} \tag{11.11}$$

where V = volume of filtrate
t = cycle time (approximate form time in continuous drum filters)
P = vacuum
A = filtration area
μ = filtrate viscosity
r = specific resistance
c = weight of solids/unit volume of filtrate

$$c = \frac{1}{C_i/(100 - C_i) - C_f/(100 - C_f)} \tag{11.12}$$

where C_i = initial moisture content, %
C_f = final moisture content, %

R_m, the initial resistance of the filter media, can usually be neglected, as it is small compared to the resistance developed by the filter cake. The specific resistance r is a measure of the filterability of the sludge and is numerically equal to the pressure difference required to produce a unit rate of filtrate flow of unit viscosity through a unit weight of cake.

Integration and rearrangement of Eq. (11.11) yields

$$\frac{t}{V} = \left(\frac{\mu rc}{2PA^2}\right)V + \frac{\mu R_m}{PA} \tag{11.13}$$

From Eq. (11.13) a linear relationship will result from a plot of t/V against V. The specific resistance can be computed from the slope of this plot:

$$r = \frac{2bPA^2}{\mu c} \tag{11.14}$$

where b = slope of the t/V against V plot.

Although specific resistance has limited value for the design of a sludge dewatering device it provides a valuable tool for the evaluation of the relative filterability of sludges. Typical values are given in Table 11.4.

Most wastewater sludges form compressible cakes in which the filtration rate and the specific resistance is a function of the pressure difference across the cake:

$$r = r_0 P^s \tag{11.15}$$

where s = coefficient of compressibility

The greater the value of s, the more compressible is the sludge. When $s = 0$, the specific resistance is independent of pressure and the sludge is incompressible.

Some generalizations on filtration characteristics can be made. Filterability is influenced by particle size, shape, and density, and by the electrical charge on the particle. Smaller particles exert a greater chemical demand than larger particles. The larger the particle size, the higher is the filter rate and the lower the cake moisture. Municipal and industrial sludges filter very poorly and coagulants must be added; lime and/or ferric salts have been the most common coagulating agents used in the past. Polyelectrolytes have proved effective coagulants in many applications. Frequently, the dual use of anionic and cationic polymers is the most economic and effective procedure. The cationic polymer effects charge neutralization and the anionic effects polymer particle bridging and agglomeration of the particles. Note that excessive coagulant dosages result in a charge reversal and in an increase in specific resistance.

Laboratory Procedures

Filtration characteristics of sludges can be obtained in the laboratory by the Büchner funnel test. The Büchner funnel test can be used to determine specific

TABLE 11.4
Specific resistance of sludges†

Description		Specific resistance, $gr \cdot s^2/g^2 \times 10^{-7}$	Coefficient of compressibility
Domestic activated sludge		2800	
Activated (digested)		800	
Primary (raw)		1310–2110	
Primary (digested)		380–2170	
Primary (digested)		1350	25.00
Primary (digested)			
Detention time	Stage		
7.5 days	1	1590	
10.0 days	1	1540	
15.0 days	1	1230	
20.0 days	1	530	
30.0 days	1	760	
15.0 days	2	400	
20.0 days	2	400	
30.0 days	2	480	
Activated sludge + 13.5% $FeCl_3$		45	
Activated sludge + 10.0% $FeCl_3$		75	
Activated sludge + 125% (by weight) newsprint		15	
Activated digested sludge + 6% $FeCl_3$ + 10% CaO		5	
Activated digested sludge + 125% newsprint + 5% CaO		4.5	
Vegetable-processing sludge		46	7.00
Vegetable tanning		15	20.00
Lime neutralization acid mine drainage		30	10.50
Alum sludge (water works)		530	14.50
Neutralization of sulfuric acid with lime slurry		1–2	
Neutralization of sulfuric acid with dolomitic lime slurry		3	0.77
Aluminum processing		3	0.44
Paper industry		6	
Coal (froth flotation)		80	1.60
Distillery		200	1.30
Mixed chrome and vegetable tannery		300	
Chemical wastes (biological treatment)		300	
Petroleum industry (from gravity separators)			
Refinery A		10–100	0.50
Refinery B		100	0.70

† All values were recorded at 500 gr/cm^2 pressure.

resistance. It is usually possible, however, to determine the sludge compressibility s and the optimum coagulant dosage from a series of Büchner funnel tests.

The procedure for the Büchner funnel test is as follows (see Fig. 11-13):

1. Prepare the Büchner funnel by placing a wire mesh or screen under the filter paper to ensure drainage.

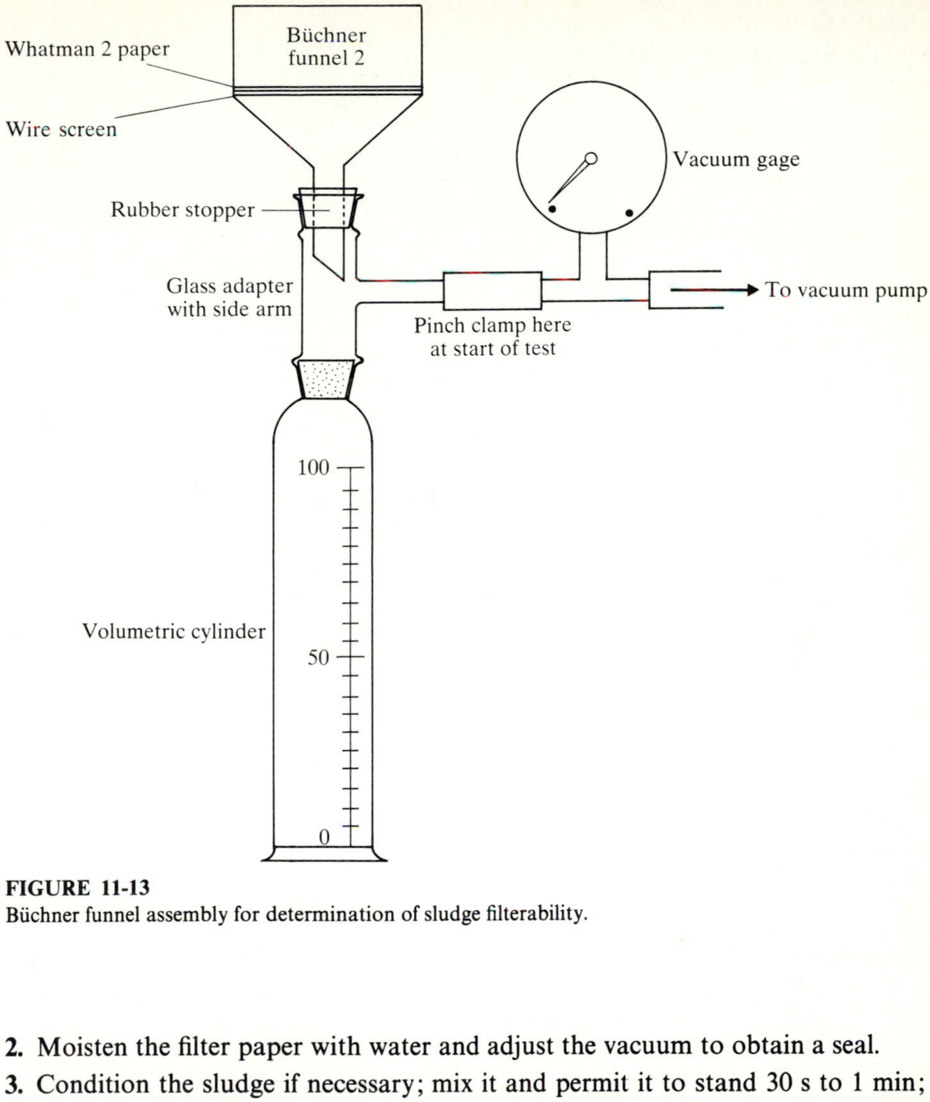

FIGURE 11-13
Büchner funnel assembly for determination of sludge filterability.

2. Moisten the filter paper with water and adjust the vacuum to obtain a seal.
3. Condition the sludge if necessary; mix it and permit it to stand 30 s to 1 min; use 200 ml samples. The relationship between specific resistance and coagulant dosage is shown in Fig. 11-14.
4. Transfer it to the Büchner funnel, allow sufficient time for a cake to form (5 to 10 s), and apply the vacuum.
5. Record the milliliters of filtrate after selected time intervals (usually 5 to 10 s).
6. Continue filtration until the vacuum breaks.
7. Determine the initial and final solids in the feed sludge and cake.
8. Record the data obtained and calculate the specific resistance in accordance with Eq. (11.14).

Calculations for specific resistance are shown in Example 11.4.

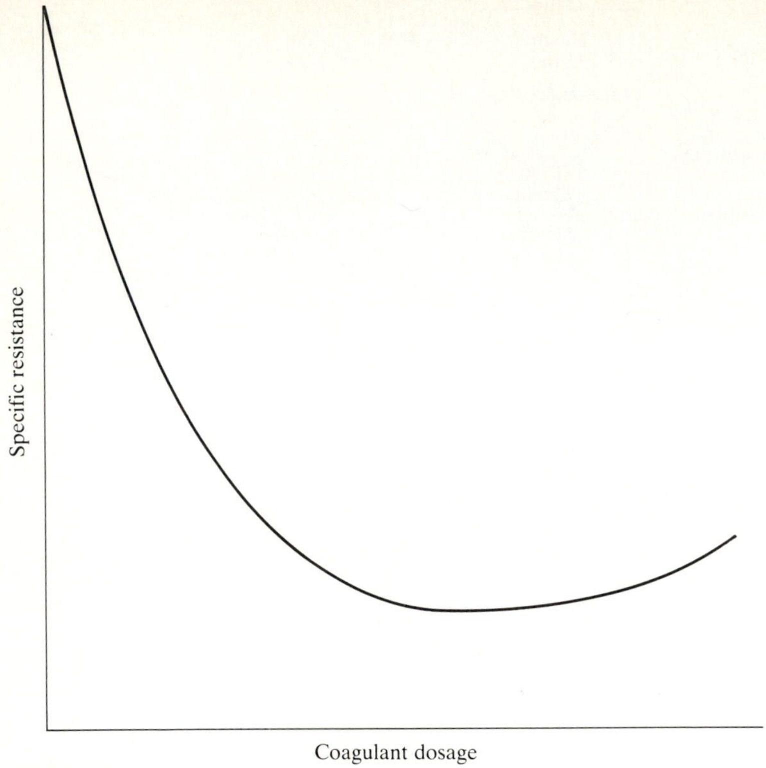

FIGURE 11-14
Relation between specific resistance and coagulant dosage.

Example 11.4. Calculate the specific resistance given the values in Table 11.5 (see Fig. 11-15),
where $A = 176.5\ \text{cm}^2$
$T = 84°\text{F}\ (29°\text{C})$
$P = 20\ \text{inHg} = 704\ \text{gr/cm}^2$
$C_i = 97.6\%$
$C_j = 77.4\%$

TABLE 11.5

Time, s	Volume, ml	t/V
5	78	0.064
10	114	0.088
15	142	0.106
25	178	0.140
35	212	0.165
45	224	0.201
50	228	0.220

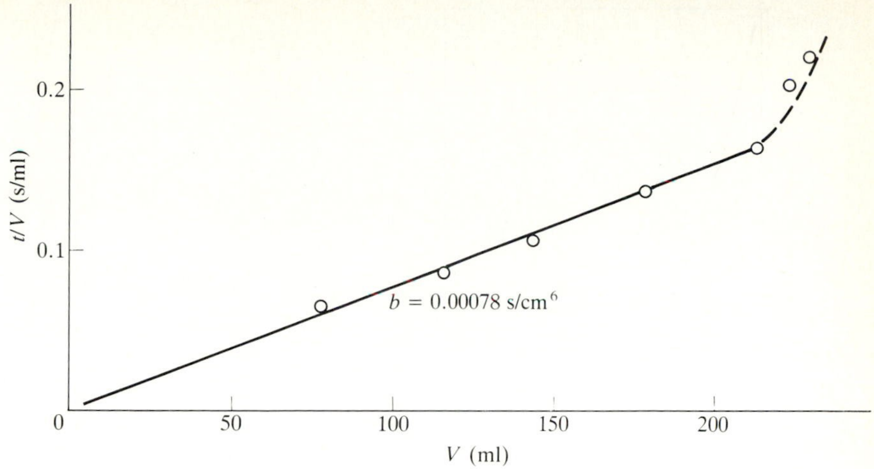

FIGURE 11-15

Solution

$$r = \frac{2PbA^2}{\mu c}$$

$$c = \frac{1}{C_i/(100 - C_i) - C_f/(100 - C_f)}$$

$$= \frac{1}{97.6/2.4 - 77.4/22.6} = 0.0269 \text{ g/ml}$$

$$r = \frac{(2)(0.00078)(704)(176.5)^2}{(0.01)(0.0269)}$$

$$= 12.7 \times 10^7 \text{ gr} \cdot \text{s}^2/\text{g}^2$$

Capillary Suction Time Test

An evaluation technique referred to as the capillary suction time (CST) test has been found to be a rapid, easy, inexpensive, and reproducible method of characterizing the dewaterability of a given sludge. The assembly developed by the Water Research Center (WRC) at Stevenage Laboratory, England, is shown in Fig. 11-16. Filtrate is withdrawn from the sludge sample by capillary action of absorbent filter paper. Filterability is measured by observing the time for a given area of paper to become wetted. The CST can be correlated to specific resistance r. The optimum coagulant and its respective dosage can be determined by a plot of specific resistance or CST versus coagulant dosage.

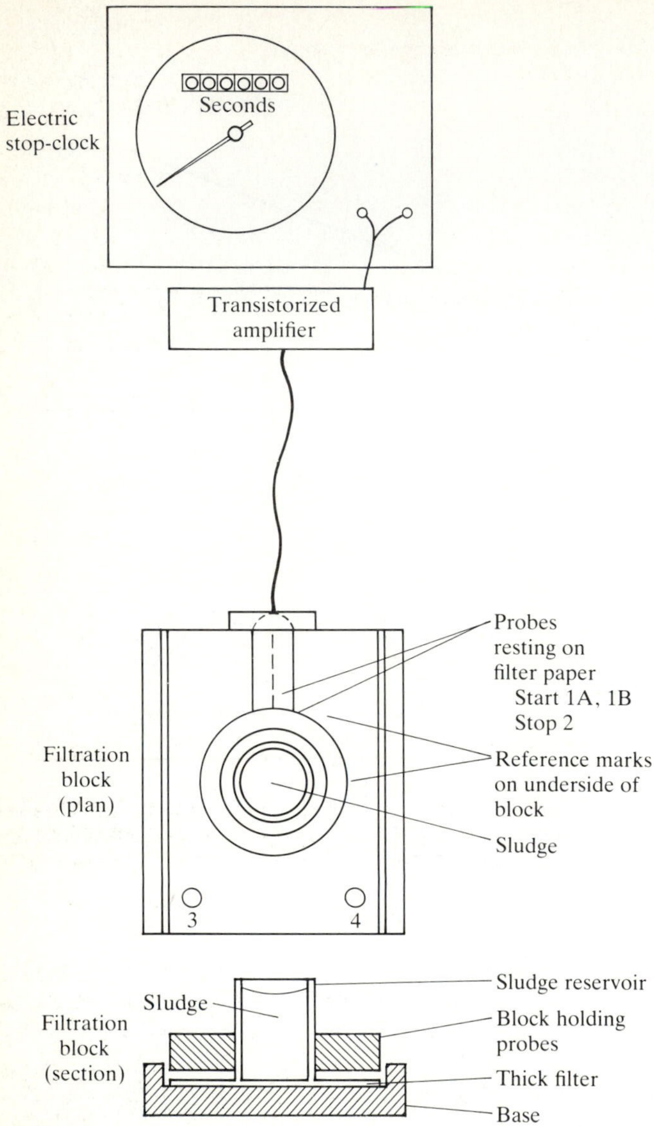

FIGURE 11-16
Capillary suction time apparatus.

11.7 VACUUM FILTRATION

Vacuum filtration is one of the most common methods for dewatering wastewater sludges. Vacuum filtration dewaters a slurry under applied vacuum by means of a porous media, which retains the solids but allows the liquid to pass through. Media used include cloth, steel mesh, or tightly wound coil springs.

In vacuum filter operation a rotary drum passes through a slurry tank in which solids are retained on the drum surface under applied vacuum. The drum submergence can vary from 12 to 60 percent. As the drum passes through the slurry, a cake is built up and water is removed by filtration through the deposited solids and the filter media. The time the drum remains submerged in the slurry is the form time (t_f). As the drum emerges from the slurry tank, the deposited cake is further dried by liquid transfer to air drawn through the cake by the applied vacuum. This period of the drum's cycle is called the dry time (t_d). At the end of the cycle, a knife edge scraps the filter cake from the drum to a conveyor. The filter media is usually washed with water sprays prior to again being immersed in the slurry tank. A vacuum filter is schematically shown in Fig. 11-17.

The variables that influence the dewatering process are solids concentration, sludge and filtrate viscosity, sludge compressibility, chemical composition, and the nature of the sludge particles (size, shape, water content, etc.).

The filter operating variables are vacuum, drum submergence and speed, sludge conditioning, and the type and porosity of the filter media.

Equation (11.13) can be modified to express filter loading (neglecting the initial resistance of the filter media):

$$L_f = 35.7\left(\frac{P^{(1-s)}}{\mu R_0}\right)^{1/2} \frac{c^m}{t_f^n} \tag{11.16}$$

where $R_0 = r_0 \times 10^{-7}$, $\text{gr} \cdot \text{s}^2/\text{g}^2$
P = vacuum, lb/in^2
c = solids deposited per unit volume filtrate, g/ml
μ = filtrate viscosity, centipoise
t_f = form cycle time, min
L_f = filter loading, lb/(ft$^2 \cdot$ h)
m, n = constants related to the sludge characteristics
s = coefficient of compressibility

For routine calculations C_i is used in Eq. (11.16) as c. Equation (11.16) is in terms of form time and is conventionally converted to cycle time by

$$L_c = L_f \frac{\%\ \text{submergence}}{100} \times 0.8 \tag{11.17}$$

The factor 0.8 compensates for the area of the filter drum where the cake is removed and the media washed. The total cycle time on a filter may vary from 1 to 6 min. Submergence of the drum may vary from 10 to 60 percent, resulting in a maximum spread of form time of 0.1 to 3.5 min. This also yields a maximum spread of dry time of 2.5 to 4.5 min. In general, the filter yield from highly compressible cakes is readily unaffected by increases in form vacuum varying from 12 to 17 in (30 to 43 cm) of mercury.

Diatomaceous earth is composed of fossil skeletons of microscopic diatoms and is used as a filter aid in rotary vacuum precoat filtration and pressure filtration. Typically 2 to 6 in (5 to 15 cm) of diatomaceous earth or perlite is

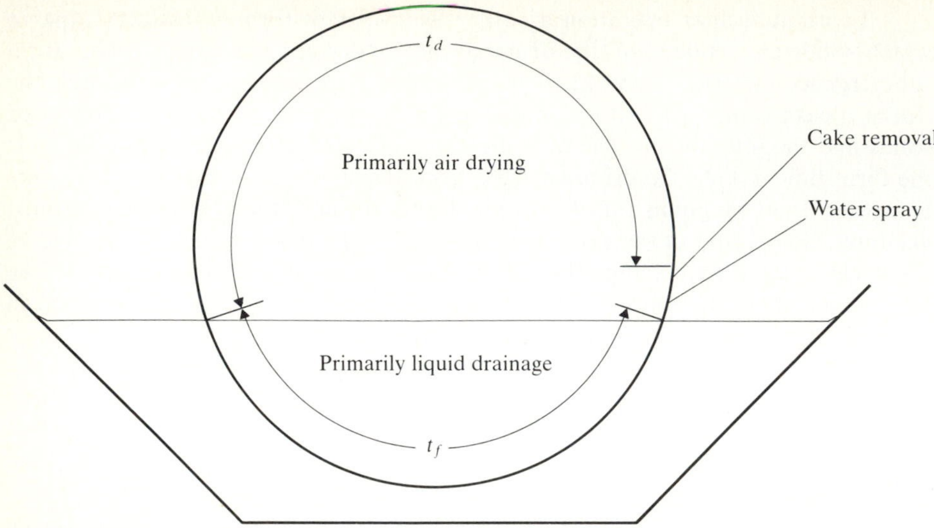

FIGURE 11-17
Mechanism of vacuum filtration.

applied prior to sludge application. To maintain a porous cake, filter aid may also be added continuously to the sludge slurry prior to vacuum filtration. Collected solids plus a few thousandths of an inch of precoat are scraped off by a knife as it advances toward the drum at rates of 0.5 to 5 minutes per revolution. Depending on the waste stream and filter aid, flow rates can vary from 2 to 50 gal/(h · ft^2) [81 to 2035 liters/(h · m^2)]. Operating costs with diatomaceous earth or perlite precoat are high because of the need to replenish expended filter aid. Precoating is especially required for dewatering of gelatinous-type solids such as alum sludge, clarification of oily wastes, and for sludges characterized by a high percentage of fines. Precoating is also generally employed during pressure filtration. As with the vacuum filter operation, it both protects the filter medium against frequent blinding and provides a thin, nonadherent parting plane between the cake and the medium which minimizes cake discharge difficulties. The optimal type and quantity of precoat must be established by laboratory evaluation. Vacuum filtration design is shown in Example 11.5.

Example 11.5. A combined primary and activated sludge from a pulp and paper mill is to be dewatered. The sludge flow is 100 gal/min (0.38 m^3/min) of 6 percent solids. Laboratory and pilot studies have shown that:

1. The coefficient $m = 0.25$.
2. The coefficient $n = 0.65$.
3. An optimum cake solids of 28 percent is obtained at a 3-min dry time.
4. The coefficient of compressibility is 0.85.
5. The specific resistance r_0 is 1.3×10^7 gr · s^2/g^2.

Design a vacuum filter to operate 16 h/d, 7 d/week, using 15 inHg (381 mmHg) vacuum (7.35 lb/in^2) and 30 percent submergence.

Solution

$$L_f = 35.7\left(\frac{P^{(1-s)}}{\mu R_0}\right)^{1/2} \frac{c^m}{t^n}$$

The cycle time is 3/0.7 = 4.3 min.

$$t_f = 4.3 - 3 = 1.3 \text{ min}$$

$$c = \frac{1}{94/6 - 72/28} = 0.077 \text{ g/cm}^3$$

$$L_f = 35.7\left[\frac{7.35^{0.15}}{(1)(1.3)}\right]^{1/2} \frac{0.077^{0.25}}{1.3^{0.65}}$$

$$= 16.3 \text{ lb/(ft}^2\cdot\text{h) [79.7 kg/(m}^2\cdot\text{h)]}$$

Using Eq. (11.17),

$$L_c = 16.3\left(\frac{30}{100}\right) \times 0.8$$

$$= 3.91 \text{ lb/(ft}^2\cdot\text{h)[19.1 kg/(m}^2\cdot\text{h)]}$$

$$\text{lb sludge/h to be filtered} = 100 \text{ gal/min} \times 8.34 \text{ lb/gal} \times 0.06 \times 60 \text{ min/h} \times 24/16$$

$$= 4500 \text{ lb/h (2041 kg/h)}$$

$$\text{Filter area required} = \frac{4500}{3.91} = 1151 \text{ ft}^2 \text{ (107 m}^2\text{)}$$

11.8 PRESSURE FILTRATION

Pressure filtration is applicable to almost all water and wastewater sludges. The sludge is pumped between plates that are covered with a filter cloth. The liquid seeps through the cloth leaving the solids behind between the plates. The filter media may or may not be precoated. When the spaces between the plates are filled, the plates are separated and the solids removed. The pressure exerted on the cake during formation is limited to the pumping force and filter closing system design. Filters are designed at pressures ranging from 50 to 225 lb/in^2 (345 to 1550 kPa). As the final filtration pressure increases, a corresponding increase in dry cake solids is obtained. Most municipal sludges can be dewatered to produce 40 to 50 percent cake solids with 225 lb/in^2 (1550 kPa) filter or 30 to 40 percent solids with 100 lb/in^2 (690 kPa) filters. Filtrate quality will vary from 10 mg/l suspended solids with precoat to 50 to 500 mg/l with unprecoated cloth depending on the media, type of solids, and type of conditioning. Conditioning chemicals are the same as used in vacuum filtration (lime, ferric chloride, or

FIGURE 11-18
Plate and frame pressure filter.

polymers). Materials such as ash have also been used. A pressure filter is shown in Fig. 11-18. Example 11.6 illustrates pressure filter design.

Example 11.6. Size a plate and frame pressure filter to dewater sludge using the following data:

Average loading = 13,300 lb/d (6030 kg/d) dry TSS
Maximum loading = 25,000 lb/d (11,340 kg/d) dry TSS
Average sludge concentration = 3.0 percent
Minimum sludge concentration = 2.0 percent

A series of pilot tests were conducted and the following conditions were found:

Total cycle time = 3.5 h†
Average cake solids = 40 percent
Minimum cake solids = 30 percent
Cake density = 70 lb/ft^3 (1120 kg/m^3)
Conditioning = 100 lb $FeCl_3$/ton (50 kg/t) dry solids
+ 200 lb lime/ton (100 kg/t) dry solids

Size dewatering equipment to treat the average sludge load in 1 shift/d and a 7-d week allows treatment of the maximum load in 2 shifts/d in a 7-day week.

† Allows sufficient time for cloth washing and cake removal.

(*a*) Calculate the volume of sludge to be treated.

$$\text{Average volume} = \frac{13{,}300 \text{ lb dry SS/d}}{0.03 \text{ lb dry SS/lb sludge} \times 8.34 \text{ lb/gal}}$$

$$= 53{,}000 \text{ gal/d } (200 \text{ m}^3\text{/d})$$

$$\text{Maximum volume} = \frac{25{,}000}{0.02 \times 8.34} = 150{,}000 \text{ gal/d } (570 \text{ m}^3\text{/d})$$

(*b*) Calculate the dewatered volume.

$$\text{Average} = \frac{13{,}300 \text{ lb/d} + (300 \text{ lb/ton} \times 5 \times 10^{-4} \text{ ton/lb} \times 13{,}300 \text{ lb/d})}{0.4 \text{ lb TSS/lb cake} \times 70 \text{ lb/ft}^3}$$

$$= 545 \text{ ft}^3\text{/d } (15 \text{ m}^3\text{/d})$$

$$\text{Maximum} = \frac{25{,}000 \text{ lb/day} + (300 \times 5 \times 10^{-4} \times 25{,}000)}{0.30 \times 70} = 1369 \text{ ft}^3\text{/d } (39 \text{ m}^3\text{/d})$$

(*c*) Calculate the number of filter cycles required per day knowing the cycle time of 3.5 h and designing for a single pressure filter.

$$\text{Average number of cycles per day} = \frac{1 \text{ shift/d} \times 8 \text{ h/shift}}{3.5 \text{ h/cycle}} \cong 2 \text{ cycles/d}$$

$$\text{Maximum number of cycles per day} \frac{2 \times 8}{3.5} \simeq 4 \text{ cycles/d}$$

(*d*) Calculate the volume of dewatered sludge or pressure filter volume required per cycle.

$$\text{Average filter volume per cycle} = \frac{545}{2} = 273 \text{ ft}^3 \text{ } (8 \text{ m}^3\text{/d})$$

$$\text{Maximum filter volume per cycle} = \frac{1369}{4} = 342 \text{ ft}^3 \text{ } (10 \text{ m}^3\text{/d})$$

(*e*) Select the size for the filter press. The volume per chamber of the press is 3 ft^3 (0.085 m^3). The maximum filter volume per cycle is 342 ft^3. Therefore, we need a press with a minimum of 114 (342/3) chambers. A standard filter press should then be selected which meets this criteria.

Hydraulic presses have also been applied to further dewater filter cake from paper-mill sludges for incineration. Board-mill sludge has been dewatered to 40 percent solids from 30 percent solids at a pressure of 300 lb/in^2 (2070 kPa) and a pressing time of 5 min.

A belt filter press has recently been developed. Chemically conditioned sludge is fed through two filter belts and is squeezed by force to drive water through these belts, as shown in Fig. 11-19. Variations of this device have been successfully used to dewater municipal and industrial sludges.

A typical belt filter press (Fig. 11-20) employs not only the concept of cake shear with simultaneous application of pressure but also low-pressure filtration

FIGURE 11-19
Belt filter press.

and thickening by gravity drainage. An endless filter belt (*a*) runs over a drive and guide roller at each end (*b* and *c*) like a conveyor belt. The upper side of the filter belt is supported by several rollers (*d*). Above the filter bed a press belt (*e*) runs in the same direction and at the same speed. The drive roller for this belt (*f*) is coupled with the drive roller (*b*) of the filter belt. The press belt can be pressed on the filter belt by means of a pressure roller system whose rollers (*g*) can be

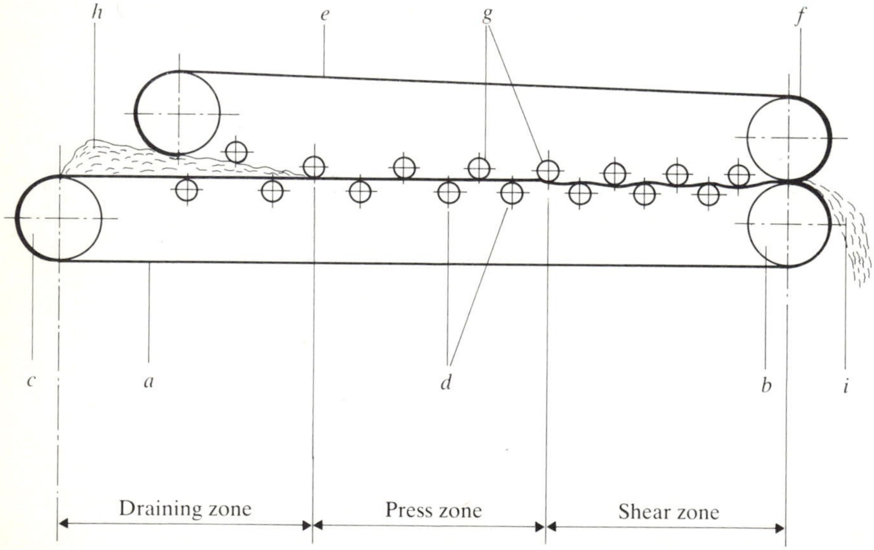

FIGURE 11-20
Schematic of belt filter press.

individually adjusted horizontally and vertically. The sludge to be dewatered (*h*) is fed on the upper face of the filter belt and is continuously dewatered between the filter and press belts. Note how the supporting rollers of the filter belt and the pressure rollers of the pressure belt are adjusted in such a way that the belts and the sludge between them describe an S-shaped curve. This configuration induces parallel displacement of the belts relative to each other due to the difference in radius producing shear in the cake. After dewatering in the shear zone, the sludge is removed by a scraper (*i*). Available data indicate that over a range of 0.5 to 12 percent DS a filter is relatively insensitive to concentration but is very sensitive to rate on flux per unit area. On the average the belt washwater flow approximately equals the sludge application rate. A belt filter design is illustrated in Example 11.7.

Example 11.7. Design a belt filter to dewater 86,600 gal/d (330 m^3/d) of 2 percent thickened sludge. The sludge is produced from a pulp and paper wastewater-treatment facility and is comprised of approximately 23 percent waste activated and 77 percent primary sludge. Pilot scale tests were conducted on a 0.5-m belt width pilot press with the values given in Table 11.6.

The following design criteria have been selected as to fall in the midrange of the operating specification of the full-scale belt press:

Cake total solids	30%
Solids capture	95%
Belt speed	10 ft/min (3 m/min)
Throughput	200 lb/(h · 0.5 m) (181 kg/h/m)
Polymer usage	5 lb/ton dry solids (2.5 kg/t)

TABLE 11.6

Sludge type	77% primary, 23% waste activated				
Run number	1	2	3	4	5
Inlet TS, %	1.76	1.76	2.34	2.34	2.34
Sludge flow, gal/min	15	28	15	20.1	28
Sludge throughput, lb/h	132	247	176	235	328
Cake TS, %	30.2	28.5	31.5	30.9	35.1
Solids capture, %	95.7	94.4	95.2	94.7	94.3
Polymer dosage, lb/ton dry solids	8.3	11.3	5.1	4.6	4.5
Belt speed, ft/min	5	10	5	10	20
Upper belt pressure			5 bar		
Lower belt pressure			5 bar		
Belt tensions			45 bar		

Note:
gal/min = 3.785×10^{-3} m^3/min
lb/h = 0.4536 kg/h
lb/ton = 0.5 kg/t
ft/min = 0.3048 m/min
bar = 100 kPa

Design for two 8-h shifts with a total of 14 h of belt press operation 5 d/week as follows:

$$0.0866 \text{ million gal/d} \times 20{,}000 \text{ mg/l} \times 8.34 = 14{,}445 \text{ lb/d } (6550 \text{ kg/d})$$

$$\frac{14{,}445 \text{ lb/d} \times 7 \text{ d}}{14 \text{ h/d} \times 5 \text{ d}} = 1{,}445 \text{ lb/h } (660 \text{ kg/d})$$

$$\text{Required belt width} = \frac{1{,}445 \text{ lb/h}}{200 \text{ lb/(h} \cdot 0.5 \text{ m)}} = \frac{1{,}445 \text{ lb/h}}{400 \text{ lb/(h} \cdot \text{m)}} = 3.6 \text{ m}$$

11.9 SAND BED DRYING

For small industrial-waste-treatment plants, sludge can be dewatered on open or covered sand beds. Drying of the sludge occurs by percolation and evaporation. The proportion of the water removed by percolation may vary from 20 to 55 percent, depending on the initial solids content of the sludge and on the characteristics of the solids. The design and use of drying beds are affected by climatic conditions (rainfall and evaporation). Sludge drying beds usually consist of 4 to 9 in (10 to 23 cm) of sand over 8 to 18 in (20 to 46 cm) of graded gravel or stone. The sand has an effective size of 0.3 to 1.2 mm and a uniformity coefficient less than 5.0. Gravel is graded from $\frac{1}{8}$ to 1 in (0.32 to 2.54 cm). The beds are provided with underdrains spaced from 9 to 20 ft apart (2.7 to 6.1 m). The underdrain piping may be vitrified clay laid with open joints having a minimum diameter of 4 in (10 cm) and a minimum slope of about 1 percent. The filtrate is returned to the treatment plant.

Wet sludge is usually applied to the drying beds at depths of 8 to 12 in (20 to 30 cm). Removal of the dried sludge in a "liftable state" varies with both individual judgment and final disposal means, but usually involves sludge of 30 to 50 percent solid.

In many cases, the bed turnover can be substantially increased by the use of chemicals. Alum treatment can reduce the sludge drying time by 50 percent. The use of polymers can increase the rate of bed dewatering and also increase the depth of application. Bed yield has been reported to increase linearly with polymer dosage.

A rational method has been developed by Swanwick[12] based on the observed dewatering characteristics of a variety of sewage and industrial waste sludges. In this procedure sludge after drainage (usually 18 to 24 h) is permitted to air dry to the desired consistency. The moisture difference (initial-final) is that which must be evaporated. Depending on the cumulative rainfall and evaporation for the geographical area in question, the time required for various times of the year for evaporation of this moisture is computed. The required bed area may then be determined. Bed yield as related to specific resistance is shown in Fig. 11-21.

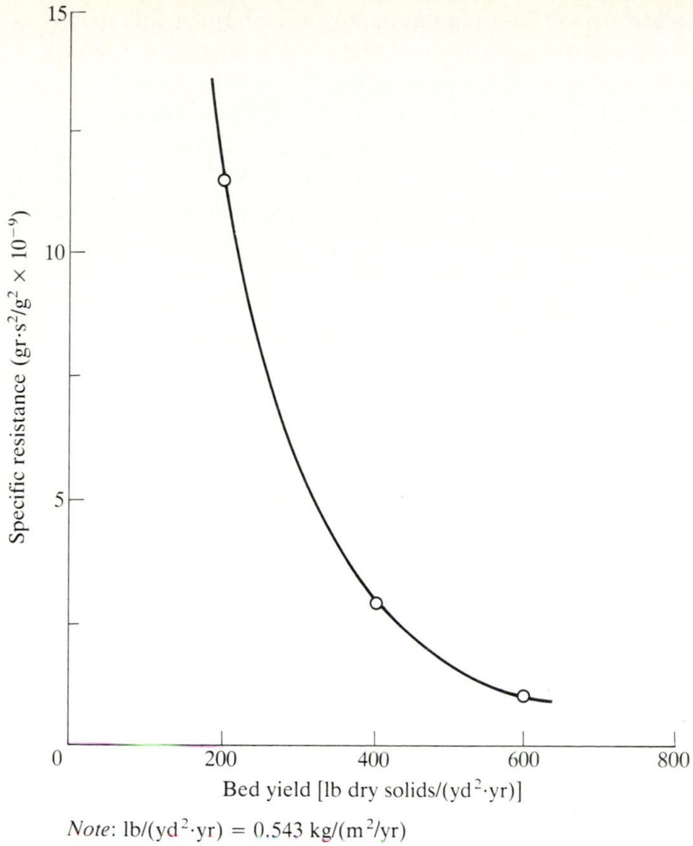

FIGURE 11-21
Relationship between the specific resistance of the sludge and bed yield as liftable sludge. (*After Swanwick.*[12])

11.10 LAND DISPOSAL OF SLUDGES

Land disposal of wet sludges can be accomplished in a number of ways: lagooning or application of liquid sludge to land by truck or spray system or by pipeline to a remote agricultural or lagoon site.

Lagooning is commonly employed for the disposal of inorganic industrial waste sludges. Organic sludges usually receive aerobic and anaerobic digestion prior to lagooning to eliminate odors and insects. Lagoons may be operated as substitutes for drying beds in which the sludge is periodically removed and the lagoon refilled. In a permanent lagoon, supernatant liquor is removed and, when filled with solids, the lagoon is abandoned and a new site selected.

In general, lagoons should be considered where large land areas are available and the sludge will not present a nuisance to the surrounding environment.

Benthal stabilization occurs in a lagoon as a result of anaerobic or a combination of aerobic and anaerobic mechanisms. Below the surface aerobic layer, anaerobic conditions prevail in which methane gas is evolved as well as other products of anaerobic decomposition. Ammonia as well as some of the less-reduced products generated in the anaerobic layer (primarily organic acids) diffuse into the aerobic layer in which they are oxidized. Rich[13] has reported an average benthal stabilization rate of 80 g/(m$^2 \cdot$ d) of biodegradable solids at 20°C. Under these conditions as much as 63 percent of total carbon stabilization can occur via methane fermentation. Assuming oxidation of all ammonia and BOD released to the water, the oxygen uptake would be 86 g O_2/(m$^2 \cdot$ d). The sludge should concentrate in the bottom of the lagoon to 2.5 to 3.0 percent solids.

Assuming continuous input to the lagoon and withdrawal once a year, the average annual stabilization rate is estimated as 68 g biomass/(m$^2 \cdot$ d). Example 11.8 illustrates a sludge lagoon design.

Example 11.8. Design a lagoon to stabilize the sludge generated from an activated sludge plant treating 1.0 million gal/d (3785 m^3/d) of a wastewater with a BOD of 425 mg/l. The activated sludge plant operates at a sludge age of 45 d. The mean temperature is 20°C, $a = 0.55$ g, $b = 0.1$/d, $t = 0.71$ d, MLVSS = 3000 mg/l, and 80 percent is volatile, $S = 10$ mg/l.

Solution. X_d is calculated from Eq. (6.6):

$$X_d = \frac{0.8}{1 + (0.1)(45)(0.2)} = 0.42$$

$$\Delta X_v = [0.55(425 - 10) - 0.1 \times 0.42 \times 3000 \times 0.71](8.34) \times 1.0$$

$$= 1158 \text{ lb VSS d } (525 \text{ kg/d}) \text{ or } \frac{1158}{0.8} = 1448 \text{ lb SS/d } (657 \text{ kg/d})$$

Assume 75 percent of the VSS will degrade in the lagoon; the area required is

$$A = \frac{R_a}{B_{av}}$$

in which R_a is the loading rate in lb VSS to degrade/d and B_{av} is the average annual stabilization rate [600 lb VSS/(acre $\cdot$ d) or 0.067 kg/(m$^2 \cdot$ d)].

The required area is

$$A = \frac{1158 \text{ lb/d} \times 0.75}{600 \text{ lb/(acre} \cdot \text{d)}}$$

$$= 1.45 \text{ acres or } 63{,}162 \text{ ft}^2 \ (5868 \text{ m}^2)$$

The maximum sludge depth can be calculated:

$$\left(\frac{1158}{0.8} - 1158 \times 0.75\right) \text{lb residue SS/d} \times 365 \text{ d/year}$$

$$= 211{,}500 \text{ lb residual sludge/year } (96{,}000 \text{ kg/year})$$

At 3 percent solids the volume of sludge is

$$\frac{211{,}500\ \text{lb/year}}{0.03 \times 62.5\ \text{lb/ft}^3} = 112{,}800\ \text{ft}^3\text{/year}\ (31{,}950\ \text{m}^3\text{/year})$$

The depth will be

$$\frac{112{,}800\ \text{ft}^3\text{/year}}{1.45\ \text{acre} \times 43{,}560\ \text{ft}^2\text{/acre}} = 1.79\ \text{ft/year}\ (0.55\ \text{m/year})$$

If the oxygen requirements are 760 lb O_2/(acre · d) [0.086 kg/(m^2 · d)] the total O_2 requirement will be 1102 lb/d (760 × 1.45). Using a 10-ft (3.05-m) water column and an aerator transfer efficiency of 10 percent, the required air flow will be

$$\text{Required air flow} = \frac{1102\ \text{lb/d}}{1440\ \text{min/d} \times 0.1 \times 0.232\ \text{lb O}_2\text{/lb air} \times 0.0746\ \text{lb air/ft}^3}$$

$$= 442\ \text{standard ft}^3\text{/min}\ (12.5\ \text{m}^3\text{/min})$$

If each diffuser operates at 4 standard ft^3/min, the number of diffusers will be 110.

In several cases, biological sludges after aerobic or anaerobic digestion have been sprayed on local land sites from tank wagons or pumped through agricultural pipes. This employs multiple applications at low dosages from 100 dry tons/acre (22.4 kg/m^2) for average conditions to 300 tons/acre (67.3 kg/m^2) in areas of low rainfall.

Excess activated sludge has been disposed of in oxidation ponds in which algal activity maintains aerobic conditions in the overlaying liquids while the sludge undergoes anaerobic digestion. This procedure has been successfully employed for municipal activated sludge at Austin, Texas, and excess activated sludge from a petrochemical plant in Houston, Texas. Lagoon loading rates of 600 lb VSS/(acre · d) [0.0673 kg/(m^2 · d)] have been employed.

Many organic sludges can be incorporated into the soil without mechanical dewatering. Surface application can be accomplished by spreading from a truck or spraying. Sludge may also be injected into the soil 8 to 10 in (20 to 25 cm) below the surface by a mobile unit, as shown in Fig. 11-22. Injection offers the advantage of minimizing surface runoff and odor problems. An important consideration is the heavy metal content of the sludge. At a pH greater than 6.0 heavy metals will exchange for Ca^{2+}, Mg^{2+}, Na^{+}, and K^{+}. This natural ability to exchange heavy metals by the soil is called the cation exchange capacity (CEC) and is expressed in milliequivalents per hundred grams of dry soil. The amount of heavy metals from sludge is influenced by such factors as pH, and aerobic or anaerobic conditions. The CEC of sandy soil may vary from 0 to 5 while clay soils will have a CEC between 15 and 20. The nutrient content of the sludge will support the growth of plants. The organic portion of the soil will also chelate heavy metals.

Prior to incorporation, sludges should receive a minimum degree of stabilization. Chow[14] has recommended aerobic digestion of 15 d to reduce the volatile content to less than 55 percent. The amount of heavy metals that can be applied

FIGURE 11-22
Sludge injection vehicle.

to the land is expressed in metal equivalents (pounds per ton of dry sludge):[15]

$$\text{Metal equivalent of sludge} = \frac{\text{Zn} + 2\text{Cu} + 4\text{Ni}}{500}$$

where the chemical symbol represents the concentration of the metal in milligrams per kilogram of dry soil.

$$\text{Total metal equivalents that can be applied (lb/acre)} = 65 \times \text{CEC}$$

Cadmium is considered separately:

$$\text{Cd} = 10 \text{ lb/acre } (1.12 \text{ g/m}^2) \text{ (lifetime application)}$$

and

$$\text{Cd} = 2 \text{ lb/(acre} \cdot \text{yr) } [0.224 \text{ g/(m}^2 \cdot \text{yr)] maximum}$$

This is based on the assumption that zinc, copper, and nickel are the most toxic to plant growth. The equation also limits metal additions to 10 percent of the CEC. The U.S. Department of Agriculture has listed the maximum amount of heavy metals that can be applied to the land, as shown in Table 11.7.

Each crop has a nutrient requirement (N, P, K, etc.). The annual quantity of sludge that can be incorporated depends on the available nitrogen content of the sludge and the nitrogen uptake of the selected crop. Excess application of sludge can result in oxidation of ammonia to nitrate, which can go back into the groundwater. Since all the applied organic nitrogen is not available to the crops

TABLE 11.7
Suggested maximum heavy metals can be applied[15]

Metal (lb/acre)	Soil cationic exchange capacity, milliequivalents/100 g		
	0–5	5–15	15
Zn	225	450	900
Cu	110	225	450
Ni	45	90	180
Cd	4.5	9	18
Pb	450	900	1800

Note: lb/acre = 0.112 g/m^2

in the same year, there is a sequential removal of organic nitrogen. Normally, about 40 percent of the organic nitrogen applied in the first year is available for crop growth that year. Subsequently, 20, 10, 5, and 2.5 percent of the organic nitrogen is available for the second, third, fourth, and fifth years.

The Illinois EPA[15] recommends a minimum depth of earth cover to the annual water table of 10 ft (3 m) with a permeability rate of 2 to 20 in/h (5 to 51 cm/h). A maximum land slope of 8 percent is recommended. A minimum soil pH of 6.5 should be maintained. The design of a sludge land incorporation is shown in Example 11.9.

Example 11.9. Design a land incorporation system for an excess activated sludge.

Sludge characteristics:

Quantity, gal/d	6560
Amount, lb/d	3500
NH_3–N, mg/l	235
Org-N, mg/l	865
SS, mg/l	63,000
PO_4, mg/l	30

Note:
gal/d = 3.785×10^{-3} m^3/d
lb/d = 0.4536 kg/d

Metal analysis of sludge:

Al	700 mg/kg (dry solids)
Cd	3.0
Ca	105,000
Cr	400
Cu	60
Fe	6000
Pb	30
Ni	150
Zn	120
K	150

Average CEC of the soil = 16.8 milliequivalents/100 g sludge application

$$\text{Metal equivalent} = \frac{\text{Zn} + 2\text{Cu} + 4\text{Ni}}{500}$$

$$= \frac{120 + 2 \times 60 + 4 \times 150}{500}$$

= 1.68 lb/ton sludge (0.84 kg/t)

Total sludge can be applied = 65 × CEC lb/acre

= 65 × 16.8

= 1092 lb/acre (0.0245 kg/m^2)

$$\text{Sludge application} = \frac{1092}{1.68} = 650 \text{ ton/acre (lifetime) } (146 \text{ kg/m}^2)$$

Maximum allowable sludge application = 650 ton/acre (146 kg/m^2)

Bermuda grass = 200 lb/(acre · yr) for nitrogen loading [0.0224 kg/(m^2 · yr)]

Subsurface incorporation = 100% NH_3 availability and 40% org-N availability

Agronomic loading:

Available N for the first year of application = 235 mg NH_3–N/l + 865 mg org-N/l × 0.4

= 581 mg N/l

= 0.00486 lb/gal

$$\text{Sludge loading} = \frac{200}{0.00486}$$

= 41,152 gal/(acre · yr) [0.0385 m^3/(m^2 · yr)]

$$\text{Acres required} = \frac{6560 \times 365}{41{,}152}$$

= 58 acres (234,720 m^2)

Project life of the plot based on CEC:

Sludge concentration = 63,000 mg/l = 0.53 lb/gal

Sludge applied = 0.53 × 41,152 = 21,690 lb/(acre · yr)

= 10.8 tons/(acre · yr) [2.45 kg/(m^2 · yr)]

$$\text{Years of application} = \frac{650 \text{ tons/acre}}{10.8 \text{ tons/(acre} \cdot \text{yr)}} = 60.2 \text{ years}$$

Project life of the plot based on Cd content in the sludge:

$$\frac{3.0 \text{ mg Cd}}{\text{kg dry solids}} \times \frac{21{,}690 \text{ lb dry solids}}{\text{acre} \cdot \text{yr}} \times \frac{10^{-6} \text{ kg}}{\text{mg}}$$

$$= 0.065 \text{ lb/(acre} \cdot \text{yr)}[7.29 \times 10^{-6} \text{ kg/(m}^2 \cdot \text{yr)}]$$

$$\text{Years of application} = \frac{10 \text{ lb/acre}}{0.065 \text{ lb/(acre} \cdot \text{yr)}} = 154 \text{ years}$$

Subsequent years of application should consider additional organic nitrogen conversion.

Oily sludges have successfully been disposed of on land. Recent data published indicate the following:

1. The oil degradation rate is directly related to the percentage of oil in the soil.
2. Fertilization improved the degradation rate.
3. Aeration (tilling) frequencies vary (from 1 week to 2 months).
4. Between 380 and 400 m^3 (2000 and 2500 bbl) of oil per hectare should be degraded in an 8-month growing season.
5. Sludge farming is about one-fifth as expensive as incineration.

11.11 INCINERATION

After dewatering, the sludge cake must be disposed of. This can be accomplished by hauling the cake to a land-disposal site or by incineration.

The variables to be considered in incineration are the moisture and volatile content of the sludge cake and the thermal value of the sludge. The moisture content is of primary significance because it dictates whether the combustion process will be self-supporting or whether supplementary fuel will be required. The thermal values of sludges may vary from 5000 to 10,000 Btu/lb (1.16×10^7 to 2.33×10^7 J/kg).

Incineration involves drying and combustion. Various types of incineration units are available to accomplish these reactions in single or combined units. In the incineration process, the sludge temperature is raised to 212°F (100°C), at which point moisture is evaporated from the sludge. The water vapor and air temperature are increased to the ignition point. Some excess air is required for complete combustion of the sludge. Self-sustaining combustion is often possible with dewatered waste sludges once the burning of auxiliary fuel raises the incinerator temperature to the ignition point. The primary end products of combustion are carbon dioxide, sulfur dioxide, and ash.

Incineration can be accomplished in multiple-hearth furnaces in which the sludge passes vertically through a series of hearths. In the upper hearths, vaporization of moisture occurs and cooling of exhaust gases. In the intermediate hearths, the volatile gases and solids are burned. The total fixed carbon is burned in the lower hearths. Temperatures range from 1000°F (538°C) at the top hearth to 600°F (316°C) at the bottom. The exhaust gases pass through a scrubber to remove fly ash and other volatile products.

In the fluidized bed, sludge particles are fed into a bed of sand fluidized by upward-moving air. A temperature of 1400 to 1500°F (760 to 815°C) is maintained in the bed, resulting in rapid drying and burning of the sludge. Ash is removed from the bed by the upward-flowing combustion gases.

PROBLEMS

11.1. An aerobic digester has a retention time of 10 days at 20°C and the following characteristics:

Reaction rate k_d	= 0.155/deg
Influent SS	= 10,500 mg/l
Volatile	= 85%
Nondegradable VSS	= 32%
X_d	= 0.68
b	= 0.1/day
Flow	= 100,000 gal/d (379 m^3/d)

(*a*) Calculate the composition of the digester effluent, TSS, VSS.
(*b*) Calculate the oxygen requirements including nitrification.

11.2. A thickener is operating under the following conditions:

Solids flux	= 20 lb/($ft^2 \cdot d$) [97.7 kg/($m^2 \cdot d$)]
Maximum underflow concentration	= 45%
Area	= 600 ft^2 (55.7 m^2)

A change in plant operations modifies the characteristics of the sludge to a flow of 180,000 gal/d (681 m^3/d) with a concentration of 1 percent. The thickening characteristics are those of Example 11.2. What will be the new maximum underflow concentration?

11.3. A refinery has a sludge consisting of waste lime from a neutralization process and several oily wastestreams. The composite waste has a solids concentration of 52.6 g/l and an average flow of 29,600 gal/d (112 m^3/d).
Given: The design coefficients and design operating conditions as shown below:

Process	Design coefficients					Design operating conditions		
	μ, cP	$\frac{1-s}{2}$	m	n	R_0	Vacuum or pressure, lb/in² gage	Cake solids, %	Cycle time, min
Vacuum filter	1	0.087	0.548	−0.562	3.5	9.8	34	6
Filter press	1	0.299	0.306	−0.559	7.63	100	40	1200

Determine:

(*a*) The required vacuum filter area based on a 7 d/week operation with 8 h/week for precoating.

(*b*) The required filter press area based on a 7 d/week operation with 4 h of downtime for every cycle.

Use the relationship:

$$L = \frac{35.7\ P^{(1-s)/2}(c^m)(t)^n}{(\mu R_0)^{1/2}}$$

where L = filter loading, lb/(ft$^2 \cdot$ h)
P = vacuum pressure, lb/in^2 gage
c = solids deposited per unit volume of filtrate, g/ml
t = cycle time, min
μ = viscosity, cP
R_0 = cake resistivity

Note 1: $c = \left[\frac{\text{sludge moisture (\%)}}{\text{sludge solids (\%)}} - \frac{\text{cake moisture (\%)}}{\text{cake solids (\%)}}\right]^{-1}$

Note 2: The above relationship can be used to determine the loading on a filter press where

$$P = \text{filter pressure, lb/in}^2 \text{ gage}$$

REFERENCES

1. Stien, R., C. E. Adams, and W. W. Eckenfelder: *Water Res.*, vol. 8, p. 213, 1974.
2. Andrews, J. F., and K. Kambhu: "Thermophilic Aerobic Digestion of Organic Solid Wastes," Final Progress Report, Clenson University, Clemson, S.C., 1970.
3. Dick, R. I.: "Thickening," in *Process Design in Water Quality Engineering* (Eds. E. L. Thackston and W. W. Eckenfelder), Jenkins, Austin, Tex., 1972.
4. Eckenfelder, W. W.: *Principles of Water Quality Management*, CBI, Boston, Mass., 1980.
5. Bassett, P. J., *et al.*: *Proc. 33rd Ind. Waste Conf.*, 1978, Purdue University.
6. Dickey, R. O., and R. C. Ward: *Proc. 36th Ind. Waste Conf.*, 1981, Purdue University.
7. Poduska, R. A., *et al.*: *Proc. 35th Ind. Waste Conf.*, 1980, Purdue University.
8. Leonard, R. J., and J. W. Parrott: *Proc. 33rd Ind. Waste Conf.*, 1978, Purdue University.
9. Miner, R. G.: *J. Water Pollut. Control Fed.*, vol. 52, pt. 9(2), p. 389, 1980.
10. Bernard, J., and A. J. Englande: "Centrifugation," in *Process Design in Water Quality Engineering* (Eds. E. L. Thackston and W. W. Eckenfelder), Jenkins, Austin, Tex., 1972.
11. Coackley, P., and B. R. S. Jones: *Sewage Ind. Waste*, vol. 28, pt. 8, p. 963, 1956.
12. Swanwich, J. D.: *Advances in Water Pollution Research*, vol. II, p. 387, Pergamon Press, New York, 1963.
13. Rich, L.: *Water Res.*, vol. 16, pt. 9(1), p. 399, 1982.
14. Chow, V.: *Sludge Disposal on Land*, 3M Company, St. Paul, Minn., 1979.
15. Illinois EPA: "Design Criteria for Municipal Sludge Utilization on Agricultural Land," Technical Policy WPC-3, 1977.

CHAPTER 12

MISCELLANEOUS TREATMENT PROCESSES

12.1 LAND TREATMENT

A wide variety of food-processing wastewater including meat, poultry, dairy, brewery, and winery wastewaters have been applied successfully to the land. By 1979, there were an estimated 1200 private industrial land-treatment systems. Disposal of industrial wastes by irrigation can be practiced in one of several ways, depending on the topography of the land, the nature of the soil, and the characteristics of the waste:

1. Distribution of waste through spray nozzles over relatively flat terrain.
2. Distribution of waste over sloping land which runs off to a natural water-course.
3. Disposal through ridge and furrow irrigation channels.

Irrigation

Irrigation includes those systems where loading rates are about 2 to 4 in/week (5 to 10 cm/week) and a crop is grown. Methods of application include various sprinkler systems, ridge and furrow, and surface flooding. It is desirable to have spray application periods followed by rest periods in the ratio of approximately 1 : 4 or greater, for example 0.5 h spray : 2 h rest.

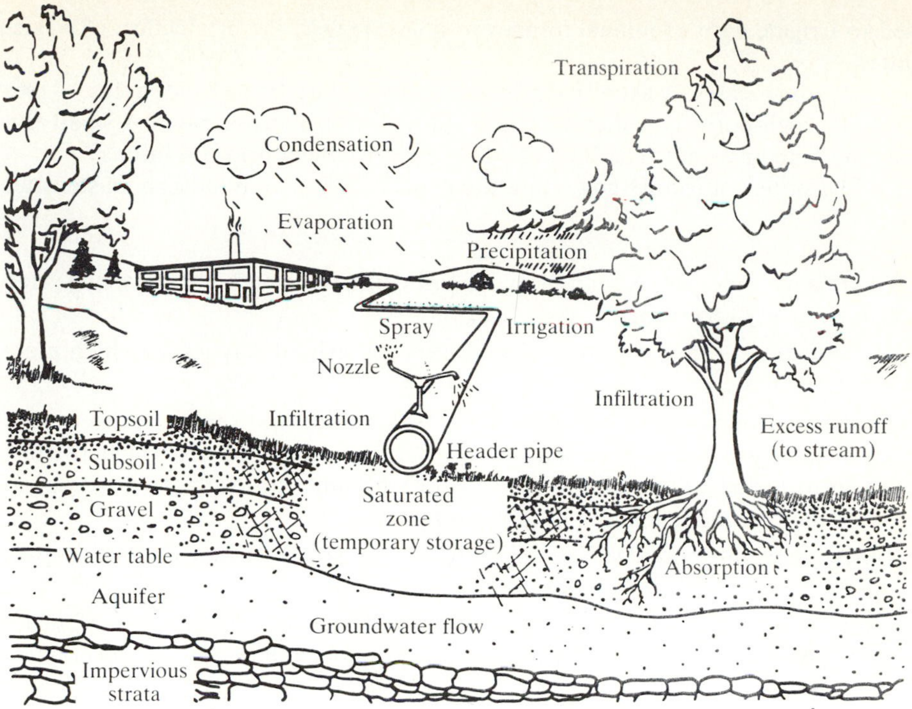

FIGURE 12-1
Spray irrigation system.

Screened waste is pumped through laterals and sprayed through sprinklers located at appropriate intervals, as shown in Fig. 12-1. The waste percolates through the soil, and during this process the organics undergo biological degradation. The liquid is either stored in the soil layer or discharged to the groundwater. Most spray irrigation systems use a cover crop of grass or other vegetation to maintain porosity in the upper soil layers. The most popular cover crop is reed canary grass (*Phalaris arundinacea*). This grass develops an extensive root system, has a relatively large leaf area, and is tolerant to adverse conditions. There is a net waste loss by evapotranspiration (evaporation to the atmosphere and absorption by the roots and leaves of plants). This may amount to as much as 10 percent of the waste flow.

Loamy well-drained soil is most suitable for irrigation systems; however, soil types from clays to sands are acceptable. A minimum depth to groundwater of 5 ft (1.5 m) is preferred to prevent saturation of the root zone. Underdrain systems have been used successfully to adapt to high groundwater or impervious subsoil conditions.

Water-tolerant perennial grasses have been used most commonly because they take up large quantities of nitrogen, are low in maintenance, and keep the soil infiltration rates as high as possible. Seasonal canning wastewaters are often used to irrigate corn or annual forages to coincide with the production of wastewater.

In some cases, wastes have been sprayed into woodland areas. Trees develop a high-porosity soil cover and yield high transpiration rates. A small elm tree may take up as much as 3000 gal/d (11.4 m^3/d) under arid conditions.

The principal factors governing the capacity of a site to adsorb wastewater are:[1]

1. Character of the soil. A sandy type of soil will have a high filtration rate; clay will pass very little water.
2. Stratification of the soil profile. Some soils will exhibit clay lenses which are a barrier to flow.
3. Depth to groundwater. The quantity of wastewater that can be sprayed on a given area will be proportional to the depth of soil through which it must travel to the groundwater. Sufficient soil depth must be available, however, to effect biological degradation of the organics.
4. Initial moisture content. The capacity of the soil to adsorb water is proportional to the initial moisture content of the soil.
5. Terrain and ground cover. A cover crop will increase the quantity of water that can be absorbed by a given area. A sloping site will increase the runoff.

In a spray irrigation field, water absorbed into the ground is held in capillary suspension until approximately 95 percent of saturation. Additional water will flow into the groundwater under a head equal to the distance from the ground surface to the groundwater table. The discharge of wastewater may be at a steady rate or at a short-term rate. The capacity to absorb at a steady rate is proportional to the overall coefficient of permeability for the soil between the ground surface and the groundwater table:[1]

$$Q = 328 \times 10^3 KS \tag{12.1}$$

where Q = gal/(min · acre)
K = overall coefficient of permeability, ft/min
S = degree of saturation of soil (near 1.0 for steady-rate application)

or

$$Q = 1.00KS \tag{12.1a}$$

for Q in $m^3/(\text{min} \cdot m^2)$
K in m/min

The coefficient of permeability K depends on the soil characteristics, as shown in Table 12.1. The overall coefficient will depend upon the variation in soil characteristics for various depths below the surface and can be computed as

$$K = \frac{H}{H_1/K_1 + H_2/K_2 + \cdots + H_n/K_n} \tag{12.2}$$

TABLE 12.1
Variation of permeability with soil characteristics[1]

Description of fine component	K, ft/min
Trace fine sand (0.10%)	1.0–0.2
Trace silt (0–10%)	0.8–0.04
Coarse and fine silt (10–20%)	0.012–0.002
Fissured clay soils (50%) and organic soils (50%)	0.0008–0.0004
Dominating clay soils (up to 100%)	<0.0002

Note: ft/min = 0.3048 m/min

where H is the total depth to the groundwater in feet, H_1, H_2, ..., H_n are the thickness of layers of soil profile, and K_1, K_2, etc., are the average permeability coefficients for each layer. Soil borings are usually necessary to determine the overall coefficient K. Care should be taken to avoid missing clay lenses that may be present. The short-term rate is proportional to the capillarity of the soil and the initial moisture content. It is usually considerably higher than the steady rate.

Rapid Infiltration

Rapid infiltration systems are characterized by percolation of most of the applied wastewater through the soil and into the subsurface. The method is restricted to use with rapidly permeable soils such as sands and sand loams. This type of system is normally thought to be associated with recharge or spreading basins, although in food-processing applications high-rate sprinkler systems have been used to provide distribution of the wastewater.

In rapid infiltration systems, plants play a relatively minor role in terms of treatment of the applied wastewater. Physical, chemical, and biological mechanisms operating within the soil are responsible for treatment. The more permeable the soil, the further the wastewater must travel through the soil to receive treatment. In very sandy soils, this minimum distance is considered to be approximately 15 ft (5 m).

Overland Flow

Overland flow is a fixed-film biological-treatment process. In overland flow, land-treatment wastewater is applied at the upper reaches of the grass-covered slopes and allowed to flow over the vegetated surface to runoff collection ditches. The wastewater is treated by a thin film down the length of the slope. The process is best suited to slowly permeable soils but can also be used on moderately permeable soils that have relatively impermeable subsoils.

Wastewater is usually applied by sprinklers to the upper two-thirds of slopes that are 150 to 200 ft (46 to 61 m) in length. A runoff collection ditch or drain is provided at the bottom of each slope. Treatment is accomplished by

TABLE 12.2
Comparative characteristics of land-treatment systems[2]

Feature	Irrigation	Rapid infiltration	Overland flow
Hydraulic loading rate, cm/d	0.2–1.5	1.5–30	0.6–3.6
Land required, ha†	24–150	1.2–24	10–60
Soil type	Loamy sand to clay	Sands	Clay to clay loam
Soil permeability	Moderately slow to moderately rapid	Rapid	Slow

† Field area in hectares not including buffer area, roads, or ditches for 3785 m^3/d (1 million gal/d) flow.

bacteria on the soil surface and within the vegetative litter as the wastewater flows down the sloped, grass-covered surface to the runoff collection drains. Ideally, the slopes should have a grade of 2 to 4 percent to provide adequate treatment and prevent ponding or erosion. The system may be used on naturally sloped lands or it may be adapted to flat agricultural land by reshaping the surface to provide the necessary slopes.

The characteristics of land-treatment systems are summarized in Table 12.2.

Waste Characteristics

In addition to soil conditions, there are several waste characteristics that require consideration in a spray irrigation system. Suspended solids should be removed from the waste, either by screening or by sedimentation, before it is sprayed. Solids will tend to clog the spray nozzles and may mat the soil surface, rendering it impermeable to further percolation. An excess acid or alkaline pH will be harmful to the cover crop. High salinity will impair the growth of a cover crop and in clay soils will cause sodium to replace calcium and magnesium by ion exchange. This will cause soil dispersion, and as a result drainage and aeration in the soil will be poor. A maximum salinity of 0.15 percent has been suggested to eliminate these problems.[3]

The soil is a highly efficient biological-treatment process and as such the performance of the system is usually governed by the hydraulic capacity of the soil as opposed to the organic loading rate. Oxygen exchange into soils depends on the air-filled pore spaces. In saturated soils oxygen transfer will be similar to oxidation ponds. In well-drained soils oxygen exchange of the surface is rapid because of mass flow. Below the first 4 in (10 cm), however, oxygen exchange is slow due to diffusion.

Research by Jewell[4] showed that organic loading rates onto soil can exceed 16,000 lb/(acre · d) [1.79 kg/(m² · d)] on a COD basis without exceeding bacterial capabilities. Adamczyka[5] reported problems with irrigation systems loaded at 2000 to 5000 lb/(acre · d) [0.22 to 0.56 kg/(m² · d)] of BOD but no problems at 500 lb/(acre · d) [0.056 kg/(m² · d)]. The problems of the very high BOD loadings can include damage to vegetation, odors, and leaching of undegraded organics into the subsurface. Loadings in the range of 535 lb/(acre · d) [0.060 kg/(m² · d)] are generally acceptable for irrigation and rapid infiltration systems. For overland flow the limiting loading rate has not been defined but is probably 134 to 180 lb/(acre · d) [0.017 to 0.020 kg/(m² · d)]. Nutrient addition in the form of nitrogen or phosphates may be necessary for nutrient deficient wastes unless the fields are adequately fertilized.

Design of Irrigation Systems

A model has been developed[6] for irrigation that includes mass balances of water and nitrogen and provides estimates of nitrogen losses to the groundwater. The principal mechanism for nitrogen removal is crop growth.

The land-application system has four components: pretreatment, wastewater transmission to the irrigation site, a lagoon for storage of the wastewater during time periods in which irrigation is infeasible, and the irrigation site. The principal variables are

$$Q_m = \text{wastewater flow to land-application site } (L^3/T)$$

$$A = \text{irrigated land area } (L^2)$$

$$r = \text{average wastewater application rate } (L/T)$$

If Q_m is in million gallons per day, A in acres, and T in inches per week,

$$\frac{Q_m}{A} = \frac{r}{258} \tag{12.3}$$

If Q_m is in cubic meters per day, A in square meters, and r in centimeters per week,

$$\frac{Q_m}{A} = \frac{r}{700}$$

If T is the number of weeks of the irrigation season and P and ET are irrigation season precipitation and evapotranspiration (in inches), respectively, the amount of water entering the groundwater below the irrigation site is $7Q_m T/A + 0.02715\ (P - \text{ET})$, in million gallons per acre. If P and ET are in centimeters,

$$7Q_m(\text{m}^3/\text{d})\,\frac{T}{A(\text{m}^2)} + 0.01(P - \text{ET})$$

Similarly, if n is the nitrogen concentration of the pretreated wastewater (in milligrams per liter) and NC is the nitrogen removal by the growing crop (M/L^2), an estimate of nitrogen entering the groundwater from the irrigation site is $7(8.34)\, nQ_m T/A - \text{NC}$ if NC is in pounds per acre, or

$$7 \times 10^{-3} nQ_m(\text{m}^3/\text{d}) \frac{T}{A(\text{m}^2)} - \text{NC}$$

if NC is in kilograms per square meter.

If groundwater standards require the average nitrogen concentrations in seepage water to be less than the drinking water standard of 10 mg/l,

$$\frac{7nQ_m T/A - \text{NC}/8.34}{7Q_m T/A + 0.02715(P - \text{ET})} < 10 \tag{12.4}$$

which reduces to

$$\frac{Q_m}{A} < \frac{\text{NC}}{58.4T(n-10)} + \frac{0.0388(P - \text{ET})}{T(n-10)} \tag{12.4a}$$

In metric units

$$\frac{7nQ_m T/A - \text{NC} \times 10^3}{7Q_m T/A + 0.01(P - \text{ET})} < 10$$

which reduces to

$$\frac{Q_m}{A} < \frac{143\text{NC}}{T(n-10)} + \frac{(P - \text{ET})}{70T(n-10)}$$

If the nitrogen application in the wastewater is to be equal or less than the crop requirement, NC, then

$$\frac{7(8.34)nQ_m T}{A} < \text{NC} \tag{12.5}$$

or

$$\frac{Q_m}{A} < \frac{\text{NC}}{58.4nT} \tag{12.5a}$$

or in metric units

$$7 \times 10^{-3} nQ_m \frac{T}{A} < \text{NC}$$

or

$$\frac{Q_m}{A} < \frac{143\ \text{NC}}{nT}$$

Equations (12.4) and (12.5) constrain the nitrogen loading rate at the land-application site. The liquid loading rate will be constrained by the drainage

TABLE 12.3
Spray irrigation of industrial wastes

Waste	Acres sprayed	Application rate, gal/(min · acre)	Duration	Average loading lb BOD/(acre · d)	Average loading lb SS/(acre · d)	Reference
Tomatoes	5.63	178	7.5 h/d	413	364	1
	6.40	86	3.7 h/d	155	139	1
Corn	2.28	153.5	10 h/d	864	500	1
Asparagus and beans	0.90	282.0	5.6 h/d	22.5	356	1
Lima beans	6.65	65	16 h/d	65	46	1
Cherries	2.24	96.5	17 h/d	807	654	1
Paperboard	1.30	77	3 h/5 d			8
Hardboard	100	42	12 h/10 d			8
	300	24	18 h/10 d			
Strawboard	1.5	94	6 h/3 d			8
Kraft	70	98	8 h/week			8

Note:
gal/(min · acre) = 9.35×10^{-7} $m^3/(min \cdot m^2)$
lb/(acre · d) = 1.12×10^{-4} $kg/(m^2 \cdot d)$
acre = 4.05×10^3 m^2

capacity of the soil, $\bar{r}$ (in inches per week or centimeters per week):

$$r < \bar{r} \tag{12.6}$$

These equations will define the land area requirements for a specific wastewater. A design is shown in Example 12.1.

Example 12.1. A food-processing wastewater is to be land irrigated. Compute the area requirements for the following conditions:

1 million gal/d (3785 m^3/d)
500 mg/l BOD
25 mg/l N

Regulations limit the application rate to 3 in/week (7.6 cm/week) and 500 lb BOD/(acre · d) [0.056 kg/(m^2 · d)].

$$\frac{Q_m}{A} = \frac{r}{258\ (\text{acre} \cdot \text{in/week}) \times (\text{million gal/d})}$$

$$\frac{1}{A} = \frac{3}{258}$$

$$A = 86 \text{ acres } (35 \text{ ha} = 350{,}000 \text{ m}^2)$$

TABLE 12.4

Spray irrigation of food-processing wastewaters[5]

Design flow, million gal/d	Overall application rate, in/week	Spray field scheduling: Application rate, in/h	Use/rest ratio	Spray time, h	Total spray area, acres	Number of spray fields	Crop cover	Soil	Pretreatment	Product
0.2	2.7	0.25	1 : 3	0.33	23	4	Reed, canarygrass	Coarse and fine sediments	Activated sludge	Dairy-cream butter, cheese, powdered milk
0.189	0.68	0.037	1 : 8	5.25	85.6	6	Reed, canarygrass	Low plastic clay	Lagoons	Beets, cabbage
1.8	5.5	0.25	1 : 5	20	61	5	Forested with heavy undergrowth	Genesee fine sandy loam	Screening and pH adjustment	Beets, corn, peas
1.0	3.15	0.30	1 : 7	12	80	4	Reed, canarygrass, and clover	Ontario loam and Cazeno-via silt loam	Aerated lagoons	Corn, beans, peas, carrots, potatoes
0.64	3.53	0.21	1 : 28	6	73.5	7	Reed, canarygrass	Williamson and Wallington loam	pH adjustment and aerated lagoons	Beans, cherries, beets, apples
0.8	2.7	0.15	1 : 8	8	80	9	Reed, canarygrass	Schoharie silty loam	Lagoons	Sauerkraut, beans
0.26	3.36	0.2	1 : 7	12	20	4	Reed, canarygrass	Gravelly loam	Screening aerated lagoons	Beans, apples
0.025	2.6	0.11	1 : 6	18	9	5	Reed, canarygrass	Chenango gravel	Screening	Cheddar cheese
0.1	1.0	0.1	1 : 17	12	27	9	Japanese millet	Fine, sandy loam and silt loam	Lagoons	Cherries, prunes, apples
0.047	2.02	0.12	1 : 9	12	6	6	Reed, canarygrass	Sandy silt	Screening pH adjustment lagoon	Cider, vinegar

Note:
gal/d $= 3.79 \times 10^{-3}$ m^3/d
in/week = 2.54 cm/week
acre $= 4.05 \times 10^3$ m^2

$$\text{lb BOD/d} = \frac{500 \text{ mg BOD}}{1} \times 1 \text{ million gal/d} \times 8.34 \frac{\text{lb/million gal}}{\text{mg/l}}$$

$$= 4170 \text{ lb/d } (1893 \text{ kg/d})$$

$$A = \frac{4170 \text{ lb/d}}{500 \text{ lb BOD/(acre} \cdot \text{d)}} = 8.34 \text{ acres } (3.40 \text{ ha} = 34{,}000 \text{ m}^2)$$

The nitrogen for the crop requirement with reed canary grass is 200 lb/acre (0.022 kg/m^2). The spraying period for the cannery is 12 weeks.

$$\text{lb N} = \frac{25 \text{ mg N}}{1} \times 1 \text{ million gal/d} \times 8.34 \frac{\text{lb/million gal}}{\text{mg/l}} \times 12 \text{ weeks} \times 7 \frac{\text{days}}{\text{week}}$$

$$= 17{,}514 \text{ lb N } (7951 \text{ kg})$$

$$A = \frac{17{,}514 \text{ lb N}}{100 \text{ lb N/acre}} = 88 \text{ acres } (35.9 \text{ ha} = 359{,}000 \text{ m}^2)$$

or

$$\frac{Q_m}{A} = \frac{\text{NC}}{58.4nT}$$

$$\frac{1}{A} = \frac{200}{58.4 \times 25 \times 12}$$

$$A = 88 \text{ acres } (35.9 \text{ ha} = 359{,}000 \text{ m}^2)$$

The area requirement is 88 acres.

Performance of land-application systems

Wastes that have been successfully disposed of by spray irrigation include cannery,[7] pulp and paper,[8] dairy, and tannery wastes. Spent sulfite liquor[9,10] has been applied up to a rate of 116 lb solids/(d · yd^2) [63 kg/(d · m^2)]; this resulted in up to 95 percent removal of BOD through a 10-ft (3-m) soil layer. Boxboard[8] wastes have been sprayed on a blown silt loam with a gravel underlay and alfalfa cover crop at a short-term rate of 0.7 in/h (1.8 cm/h) and a total daily rate of 0.21 to 0.56 in/d (0.5 to 1.4 cm/d). This is equivalent to 0.2 to 1.0 acres per ton of production per day [0.89 to 4.40 m^2/(kg · d)]. Strawboard cook liquor including beater-washer water and machine water[8] has been sprayed twice a week for 6 h/d at a rate of 0.5 in/d (1.3 cm/d). A rate of 1.2 in/d (3 cm/d) resulted in ponding. Kraft-mill wastes have been found to require 1.5 acres per ton of production (6.7 m^2/kg) or more.[8] Typical data for spray irrigation systems are shown in Table 12.3. Results for food-processing wastewaters are shown in Table 12.4. Overland flow systems are summarized in Table 12.5.

TABLE 12.5
Selected overland flow systems[2]

Name and location	Loading rate, cm/d	Slope length, m	Type of wastewater
Campbell Soup Co., Chestertown, Maryland	1.0	53–76	Poultry
Hunt-Wesson Foods, Davis, California	2.2	51	Tomato
El Paso, Texas	2.5	62	Meat packing
Middlebury, Indiana	1.5	79	Poultry
Campbell Soup Co., Napoleon, Ohio	1.5	53–61	Tomato
Campbell Soup Co., Paris, Texas	1.6–3.6	60–90	Vegetable soup
Sebastopol Co-op, Sebastopol, California	1.5	45	Apple
Nabisco, Inc., Woodbury, Georgia	0.9–1.1	80–84	Peanuts, pimientos
Frito-Lay Inc., Wooster, Ohio	0.5–1.0	67	Vegetable

12.2 DEEP-WELL DISPOSAL

Deep-well disposal involves the injection of liquid wastes into a porous subsurface stratum that contains noncommercial brines.[11] The wastewaters are stored in sealed subsurface strata isolated from groundwater or mineral resources. Disposal wells may vary in depth from a few hundred feet (< 100 m) to 15,000 ft (4570 m) with capacities ranging from less than 10 to more than 2000 gal/min (38 to 7550 l/min). Wastes disposed of in wells are usually highly concentrated, toxic, acidic, or radioactive, or wastes high in inorganic content which are difficult or excessively expensive to treat by some other process.

The disposal system consists of the well and pretreatment equipment necessary to prepare the waste for suitable disposal into the well.

A casing, generally of steel, is cemented in place to seal the disposal stratum from other strata penetrated during drilling of the well. An injection tube transports the waste to the disposal stratum as shown in Fig. 12-2.

Oil or freshwater is used to fill the annular space between the injection tube and the casing and extends to but is sealed from the injection stratum. Leaks in the injection tube or drainage to the casing can be detected by monitoring the pressure of the fluid.

The system includes a basin to level fluctuations in flow, pretreatment equipment, and high-pressure pumps. Pretreatment requirements are determined by the characteristics of the wastewater, compatibility of the wastewater and the formation water, and the characteristics of the receiving stratum.

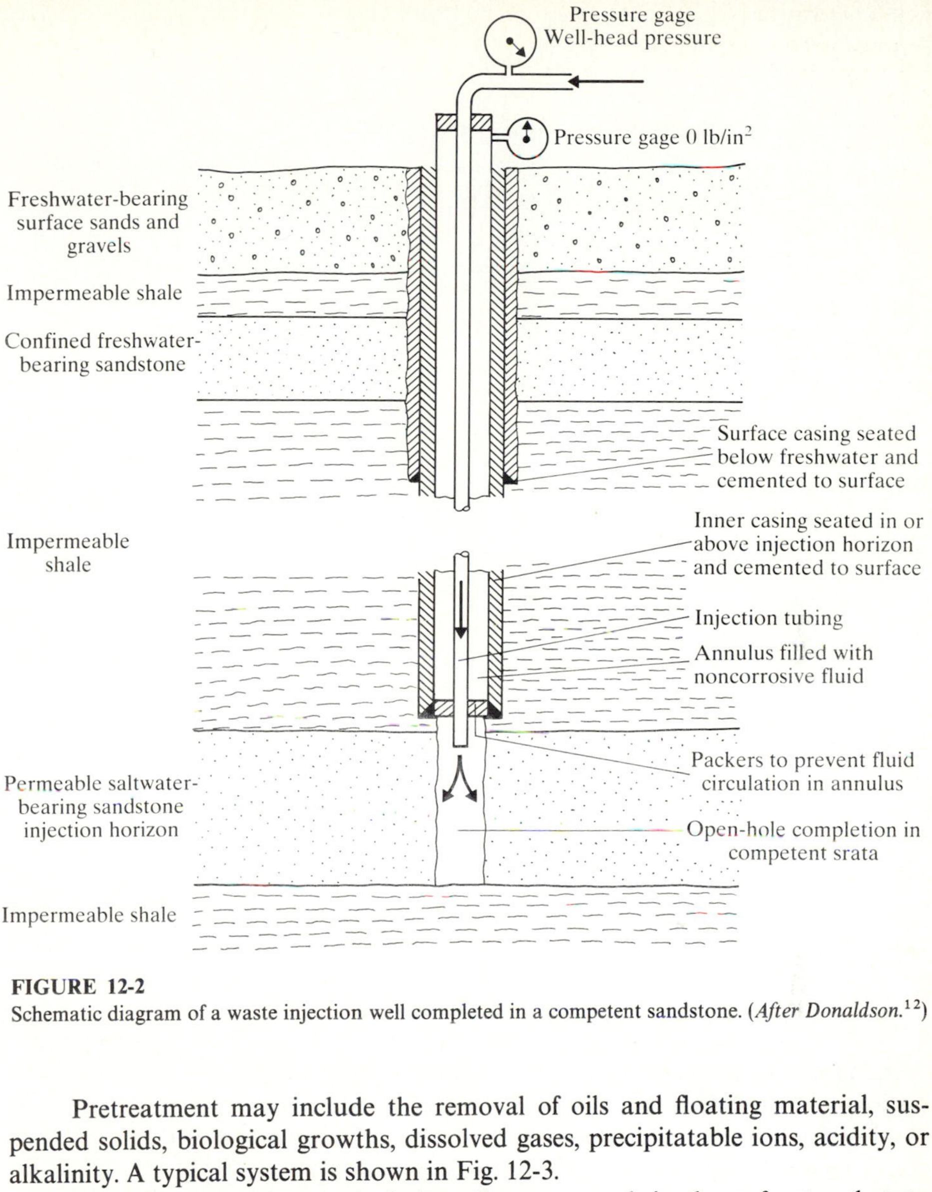

FIGURE 12-2
Schematic diagram of a waste injection well completed in a competent sandstone. (*After Donaldson.*[12])

Pretreatment may include the removal of oils and floating material, suspended solids, biological growths, dissolved gases, precipitatable ions, acidity, or alkalinity. A typical system is shown in Fig. 12-3.

The best disposal areas include sedimentary rock in the unfractured state, including sandstones, limestones, and dolomites and unconsolidated sands. Fractured strata should be avoided because if a vertical fissure exists, groundwater contamination may result.

The well-head pressure is related to the difference between the bottom-hole pressure and the reservoir pressure. Cores of the injection location are needed to evaluate the porosity and permeability of the stratum and any possible reactions between the wastewater and the stratum.

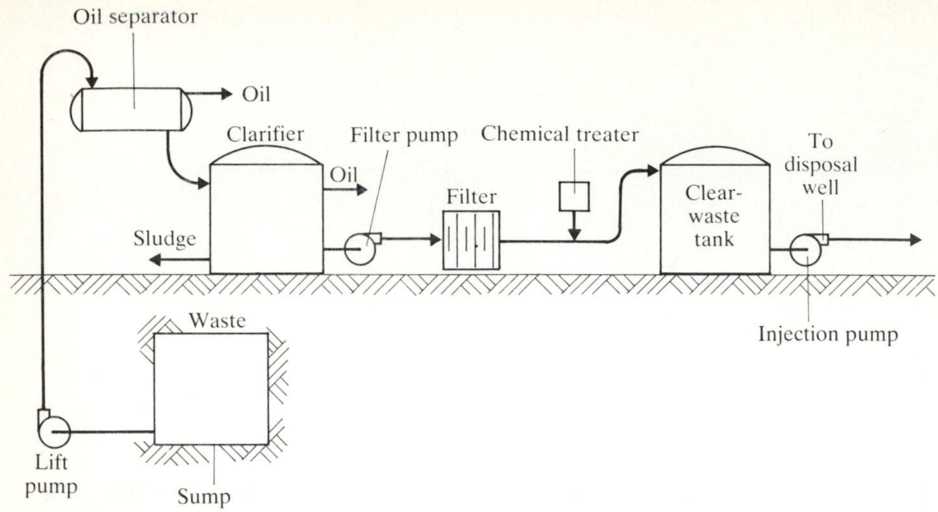

FIGURE 12-3
Typical subsurface waste-disposal system. (*After Donaldson.*[12])

Although wastewaters should be generally free of suspended solids, some vugular formations will accept suspended solids without problems or an increase in injection pressure. In some cases, injection can be increased by well stimulation, which involves the injection of mineral acids to dissolve calcium carbonate and other acid-soluble particulates that tend to plug the stratum. Mechanical procedures involve scratching, swabbing, washing, and underreaming the well bore and shotting the uncased stratum with explosives or hydraulic fracturing.

The cost of a well system will be affected by the depth of the injection well, type of formation, geographic location, waste volume, required pretreatment, and injection pressure.

12.3 MEMBRANE PROCESSES

Membrane filtration includes a broad range of separation processes from filtration and ultrafiltration to reverse osmosis. Generally, those processes defined as filtration refer to systems in which discrete holes or pores exist in the filter media, generally in the order of 10^2 to 10^4 nm or larger. The efficiency of this type of filtration depends entirely on the difference in size between the pore and the particle to be removed.

The various filtration processes relative to molecular size are shown in Table 12.6.

Reverse osmosis employs a semipermeable membrane and a pressure differential to drive freshwater to one side of the cell, concentrating salts on the input or rejection side of the cell. In this process, freshwater is literally squeezed out of the feedwater solution.

TABLE 12.6
Membrane processes

Material to be removed	Approximate size, nm	Process
Ion removal	1–20	Diffusion or reverse osmosis
Removal of organics in true solution	5–200	Diffusion
Removal of organics: subcolloidal—not in true molecular dispersion	200–10,000	Pore flow
Removal of colloidal and particulate matter	75,000	Pore flow

The reverse osmosis process can be described by considering the normal osmosis process. In osmosis, a salt solution is separated from a pure solvent of solution of less concentration by a semipermeable membrane. The semipermeable membrane is permeable to the solvent and impermeable to the solution. Such an arrangement is shown in Fig. 12-4. The chemical potential of the pure solvent is greater than that of the solvent in solution, and therefore drives the system to equilibrium. If an imaginary piston applies an increasing pressure on the solution compartment, the solvent flow through the membrane will continue to decrease. When sufficient pressure has been applied to bring about thermodynamic equilibrium, the solvent flow will stop. The pressure developed in achieving equilibrium is the osmotic pressure of the solution, or the difference in the osmotic pressure between the two solutions if a less concentrated salt solution is used instead of pure solvent in the right chamber of the cell.

If a pressure in excess of the osmotic pressure is now applied to the more concentrated solution chamber, pure solvent is caused to flow from this chamber to the pure solvent side of the membrane, leaving a more concentrated solution behind. This phenomena is the basis of the reverse osmosis process.

The criteria of membrane performance are the degree of impermeability, how well the membrane rejects the flow of the solute and the degree of permeability, or how easily the solvent is allowed to flow through the membrane. Cellulose acetate membranes provide an attractive combination of these criteria. A typical membrane type is shown in Fig. 12-5. Each of these design configurations uses cellulose acetate membranes, except certain of the hollow fine-fiber systems that employ a nylon polymer. The more common membranes are described below:

1. *Tubular*. Manufactured from ceramics, carbon, or any number of porous plastics, these tubes have inside diameters ranging from $\frac{1}{8}$ inch (3.2 mm) up to approximately 1 inch (2.54 cm). The membrane is typically coated on the inside of the tube, and the feed solution flows through the interior from one

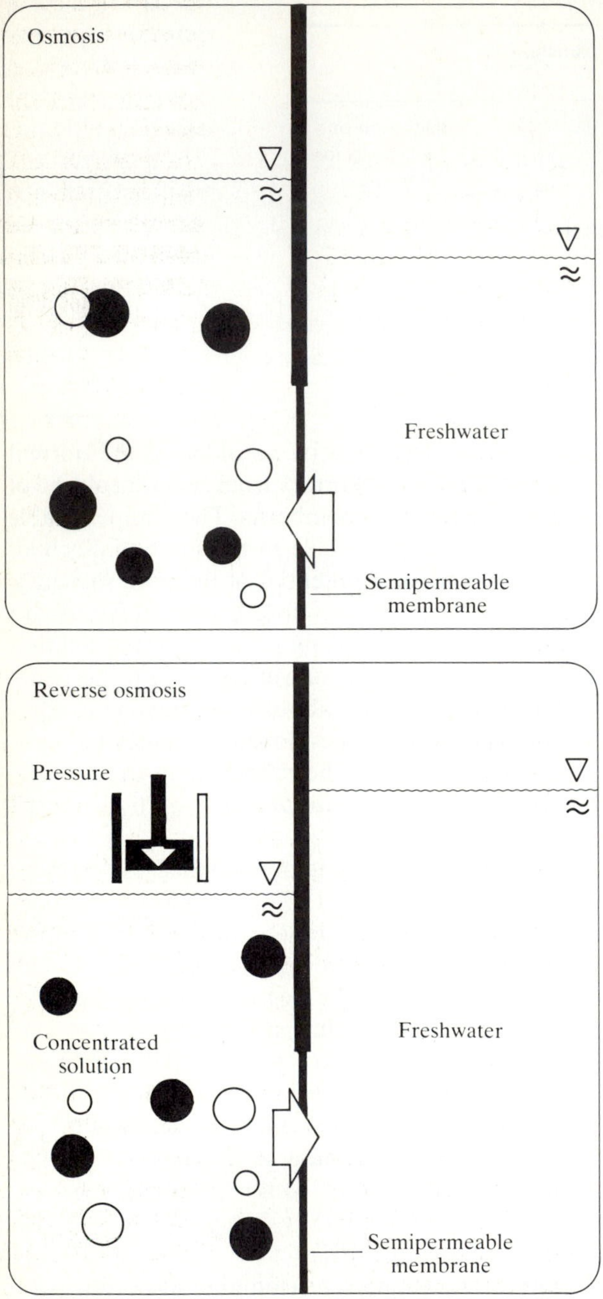

FIGURE 12-4
Osmosis and reverse osmosis.

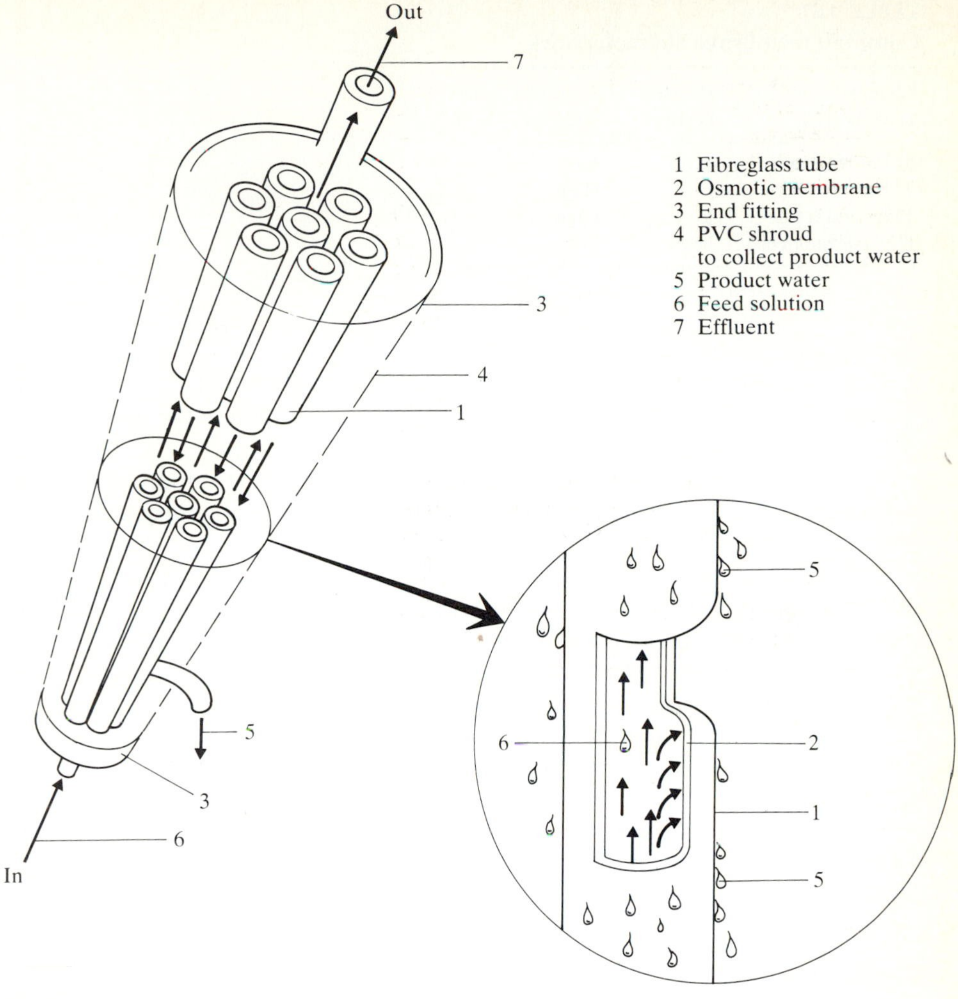

FIGURE 12-5
Membrane filtration systems. Tubular reverse osmosis unit.

end to the other, with the "permeate" or "filtrate" passing through the wall to be collected on the outside of the tube.

2. *Hollow fiber*. Similar to the tubular elements in design, hollow fibers are generally much smaller in diameter and require rigid support such as is obtained from the "potting" of a bundle inside a cylinder. As with tubular elements, feed flow is usually down the interior diameter of the fiber.
3. *Spiral wound*. This device is constructed from an envelope of sheet membrane wound around a permeate tube that is perforated to allow collection of the permeate or filtrate.

TABLE 12.7
Common membrane characteristics

Element configuration	Packing density†	Suspended solids tolerance
Spiral wound	High	Fair
Tubular	Low	High
Plate and frame	Low	High
Hollow fine fiber	Highest	Poor

† Membrane area per unit volume of space required.

4. *Plate and frame*. This device incorporates sheet membrane that is stretched over a frame to separate the layers and facilitate collection of the permeate.

With regard to membrane element configuration, the important physical characteristics of the various membrane element device configurations available today are listed in Table 12.7.

Because of the propensity of suspended or precipitated materials settling out on the membrane surface and plugging the membrane pores, turbulent flow conditions must be maintained (Reynolds numbers in excess of 2000). For high recovery systems, this usually requires recycling a significant percentage of the concentrate back to the feed side of the pump. The addition of this concentrate stream into the feed solution obviously increases the dissolved solids concentration, further increasing osmotic pressure.

In order to minimize membrane fouling, pretreatment is required for the removal of suspended matter, bacteria, and precipitatable ions. A typical reverse osmosis process schematic is shown in Fig. 12-6. The design and operating parameters for a reverse osmosis system are summarized from Agardy.[13]

Pressure

The water flux is a function of the pressure differential between the applied pressure and the osmotic pressure across the membrane. The higher the applied pressure, the greater the flux. However, the pressure capability of the membrane is limited, so the maximum pressure is generally taken to be 1000 lb/in^2 (6895 kPa or 68 atm) gage. Operating experience dictates in the 400 to 600 lb/in^2 (2758 to 4137 kPa or 27 to 41 atm) gage range, with 600 lb/in^2 (4137 kPa or 41 atm) gage normally being the design pressure.

Temperature

The water flux increases with increasing feedwater temperature. A standard of 70°F (21°C) is generally assumed and temperatures of up to 85°F (29°C) are acceptable. However, temperatures in excess of 85°F (29°C) and up to 100°F

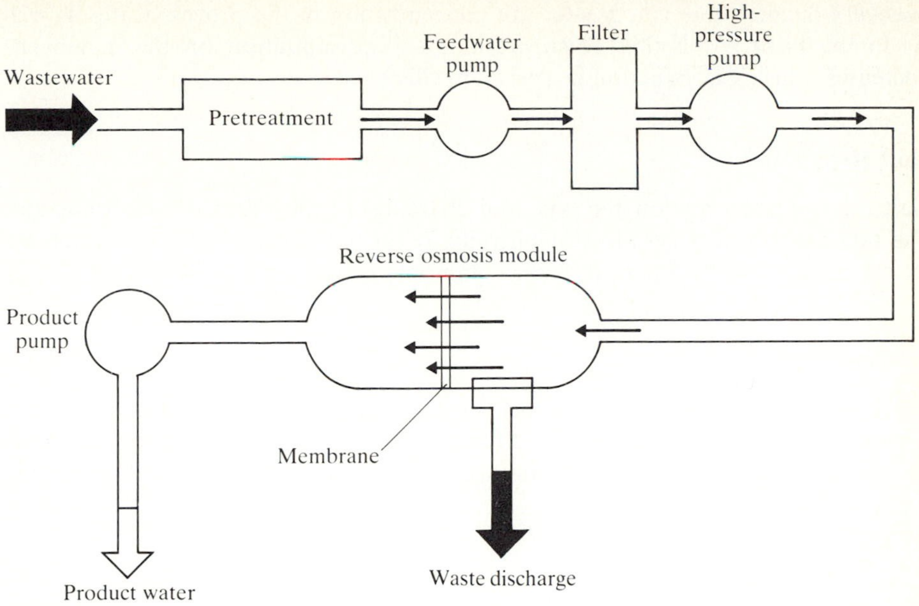

FIGURE 12-6
Basic reverse osmosis process schematic.

(38°C) will accelerate membrane deterioration and cannot be tolerated for long operating periods.

Membrane Packing Density

This is an expression of the unit area of the membrane, which can be placed per unit volume of pressure vessel. The greater this factor the greater will be the overall flow through the system. Typical values range from 50 to 500 ft^2/ft^3 (160 to 1640 m^2/m^3) of pressure vessel.

Flux

Although the flux rates for hollow fibers are between 6.0×10^{-3} to 10.2×10^{-3} $m^3/(d \cdot m^2)$ [0.15 to 0.25 $gal/(d \cdot ft^2)$] versus 6.1×10^{-1} to 10.2×10^{-1} $m^3/(d \cdot m^2)$ [15 to 25 $gal/(d \cdot ft^2)$] for sheet systems, the stacking densities for fibers are about 10 times greater, making the sizes of the two systems competitive. This flux tends to decrease with length of run and over a period of one or two years of operation might be reduced by 10 to 50 percent.

Recovery Factor

This consideration actually represents plant capacity and is generally in the range of 75 to 95 percent with 80 percent being the practical maximum. At high recovery factors there is a greater salt concentration in the process water as well as in the brine. At higher concentrations salt precipitation on the membrane increases, causing a reduction in operation efficiency.

Salt Rejection

Salt rejection depends on the type and character of the selected membrane and the salt concentration gradient. Generally, rejection values of 85 to 99.5 percent are obtainable, with 95 percent being commonly used.

Membrane Life

Membrane life can be drastically shortened by undesired constituents in the feed water, such as phenols, bacteria, and fungi as well as high temperatures and high or low pH values. Generally, membranes will last up to two years with some loss in flux efficiency.

pH

Membranes consisting of cellulose acetate are subject to hydrolysis at high and low pH values. The optimum pH is approximately 4.7 with operating ranges between 4.5 and 5.5.

Turbidity

Reverse osmosis units can be used to remove turbidity from feedwaters. They operate best if little or no turbidity is applied to the membrane. Generally, it is considered that the turbidity should not exceed one Jackson turbidity unit (JTU) and that the feedwater should not contain particles larger than 25 μm.

Feedwater Stream Velocity

The hydraulics of reverse osmosis systems are such that velocities in the range of 0.04 to 2.5 ft/s (1.2 to 76.2 cm/s) are common. Plate and frame systems operate at the higher velocity, while the hollow fine-fiber units operate at the lower velocities. High velocities and turbulent flow are necessary to minimize concentration polarization at the membrane surface.

Power Utilization

Power requirements are generally associated with the system pumping capacity and operational pressures. Values range from 9 to 17 $kW \cdot h$/thousand gal (2.4 to 4.5 $kW \cdot h/m^3$), with the lower figure taking into account some power recovery from the brine stream.

Pretreatment

The present development of membranes limits their direct application to feedwaters having a TDS not exceeding 10,000 mg/l. Further, the presence of scale-forming constituents, such as calcium carbonate, calcium sulfate, oxides and hydroxides of iron, manganese, and silicon, barium and strontium sulfates, zinc sulfide, and calcium phosphate, must be controlled by pretreatment or they will require subsequent removal from the membrane. These constituents can be controlled by pH adjustment, chemical removal, precipitation, inhibition, and filtration. Organic debris and bacteria can be controlled by filtration, carbon, pretreatment, and chlorination. Oil and grease must also be removed to prevent coating and fouling of the membranes.

Cleaning

Recognizing that under continuous use membranes will foul, provision must be made for mechanical and/or chemical cleaning. Methods include periodic depressurizations, high-velocity water flushing, flushing with air-water mixtures, backwashing, cleaning with enzyme detergents, ethylene diamine, tetraacetic acid, and sodium perborate. The control of pH during cleaning operations must be maintained to prevent membrane hydrolysis. Approximately 1 to 1.5 percent of the process water goes to waste as a part of the cleaning operation, with the cleaning cycle being every 24 to 48 hours.

A summary of operational parameters is shown in Table 12.8.

Applications

Reverse osmosis has been applied to the treatment of plating wastewaters for the removal of cadmium, copper, nickel, and chromium at pressures of 200 to 300 lb/in^2 (1378 to 2067 kPa or 13.6 to 20.4 atm). The concentrated stream is

TABLE 12.8
Summary of system operational parameters[13]

Parameter	Range	Typical
Pressure, lb/in^2 gage	400–1000	600
Temperature, °F	60–100	70
Packing density, ft^2/ft^3	50–500	—
Flux, $gal/(d \cdot ft^2)$	10–80	12–35
Recovery factor, %	75–95	80
Rejection factor, %	85–99.5	95
Membrane life,	—	2
pH	3–8	4.5–5.5
Turbidity, JTU	—	1
Feedwater velocity, ft/s	0.04–2.5	—
Power utilization, $kW \cdot h$/thousand gal	9–17	—

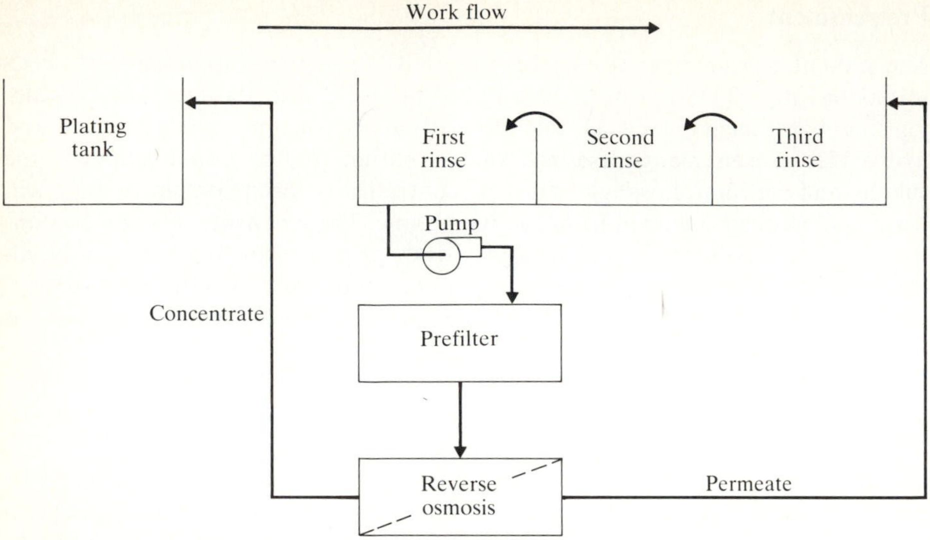

FIGURE 12-7
Treatment of plating wastewaters by reverse osmosis.

returned to the plating bath and the treated water to the next to last rinse tank, as shown in Fig. 12-7.

Pulp-mill effluents have been treated by reverse osmosis at a pressure of 600 lb/in^2 (4137 kPa or 41 atm). Wastestreams were concentrated up to 100,000 mg/l total solids. The flux was found to be a function of total solids level and varied from 2 to 15 $gal/(d \cdot ft^2)$ [0.08 to 0.61 $m^3/(d \cdot m^2)$].[14]

Oily wastes can be treated by ultrafiltration in which the permeate can be recycled as rinsewater and the concentrate can be hauled or incinerated as shown in Fig. 12-8.

An RO design is shown in Example 12.2.

Example 12.2. The following information on a wastewater and reverse osmosis system is to be used for renovation. Design an RO system for this wastewater.

Plant design capacity, Q	10 million gal/d (37,850 m^3/d)
Recovery factor, R	75%
Salt rejection factor, S	95%
Design pressure	600 lb/in^2 gage (40.8 relative atm)
Feedwater temperature	80°F (27°C)
Total dissolved solids	600 mg/l

System sizing. The wastewater (system feedwater) flow, Q_w, is

$$Q_w = \frac{Q}{R} = \frac{10 \text{ million gal/d}}{0.75} = 13.3 \text{ million gal/d } (50{,}341 \text{ m}^3\text{/d})$$

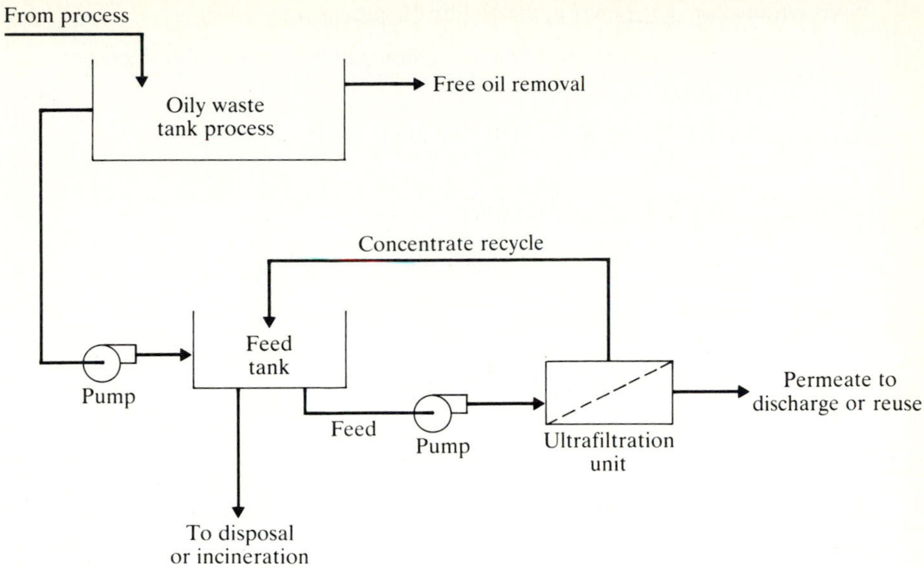

FIGURE 12-8
Treatment of oily wastewaters by ultrafiltration.

The brine flow to be disposed of is then 13.3 − 10, or 3.3 million gal/d (12,491 m^3/d).

Assuming a flux of 20 gal/(d · ft^2) [0.814 m^3/(d · m^2)], the membrane area needed, A, is

$$A = \frac{10 \times 10^6 \text{ gal/d}}{20 \text{ gal/(d} \cdot \text{ft}^2)} = 500{,}000 \text{ ft}^2 (46{,}450 \text{ m}^2)$$

Assuming a packing density of 250 ft^2/ft^3 (820 m^2/m^3), the total module volume, V, is

$$V = \frac{500{,}000 \text{ ft}^2}{250 \text{ ft}^2/\text{ft}^3} = 2000 \text{ ft}^3 (56.6 \text{ m}^3)$$

Assuming 1 ft^3 (0.0283 m^3) of volume per module, 2000 modules would be required, and assuming 10 modules per pressure vessel, 200 pressure vessels would be required.

The approximate total dissolved solids content of the product water, TDS_p, is

$$\text{TDS}_p = \frac{1}{R} (\text{wastewater TDS}) (1 - S)$$

$$= \frac{1}{0.75} (600 \text{ mg/l})(1 - 0.95)$$

$$= 40 \text{ mg/l}$$

Power consumption. The power consumed by the pressurizing pump is

$$\text{Hydraulic hp} = \frac{(\text{wastewater flow in million gal})\,(\text{pressure in lb/in}^2\text{ gage})}{2.47}$$

$$= \frac{(13.3\text{ million gal})\,(600\text{ lb/in}^2\text{ gage})}{2.47}$$

$$= 3230\text{ hp }(2423\text{ kW})$$

The brake horsepower which must be delivered by the pump motor is

$$\text{Brake hp} = \frac{\text{hydraulic hp}}{\text{pump efficiency}}$$

$$= \frac{3230\text{ hp}}{0.95}$$

$$= 3400\text{ hp }(2550\text{ kW})$$

If provision is desired to operate the system at higher pressures, i.e., up to 1000 lb/in gage (68.05 relative atm), the calculated power can be corrected by multiplying by the ratio of the selected pressure to 600 lb/in^2 gage (40.8 relative atm).

Land area required. The land area in acres required for a reverse osmosis system can be estimated by the formula

$$\begin{aligned}\text{Plant area} &= 0.7 + 0.33Q \\ &= 0.7 + 0.33\,(10) \\ &= 4\text{ acres }(16{,}300\text{ m}^2 = 1.6\text{ ha})\end{aligned}$$

Brine disposal. The approximate total dissolved solids content of the rejected brine, TDS_b, is

$$TDS_b = \frac{S}{(1-R)}\,(\text{wastewater TDS})$$

$$= \frac{0.95}{1-0.75}\,(600\text{ mg/l})$$

$$= 2280\text{ mg/l}$$

The total dry weight of salt to be disposed of daily, W, is

$$W = 8.34\,(2280\text{ mg/l})(3.3\text{ million gal/day}) = 62{,}600\text{ lb/d }(28{,}400\text{ kg/d})$$

12.4 PHOSPHORUS REMOVAL

Phosphorus can be removed from wastewater either chemically or biologically.

Chemical Phosphorus Removal

Chemical phosphorus removal involves precipitation with calcium, iron, or aluminum. Phosphorus is precipitated with calcium salts to low residuals depending on the pH. The precipitate is a hydroxyapatite, $Ca_5OH(PO_4)_3$:

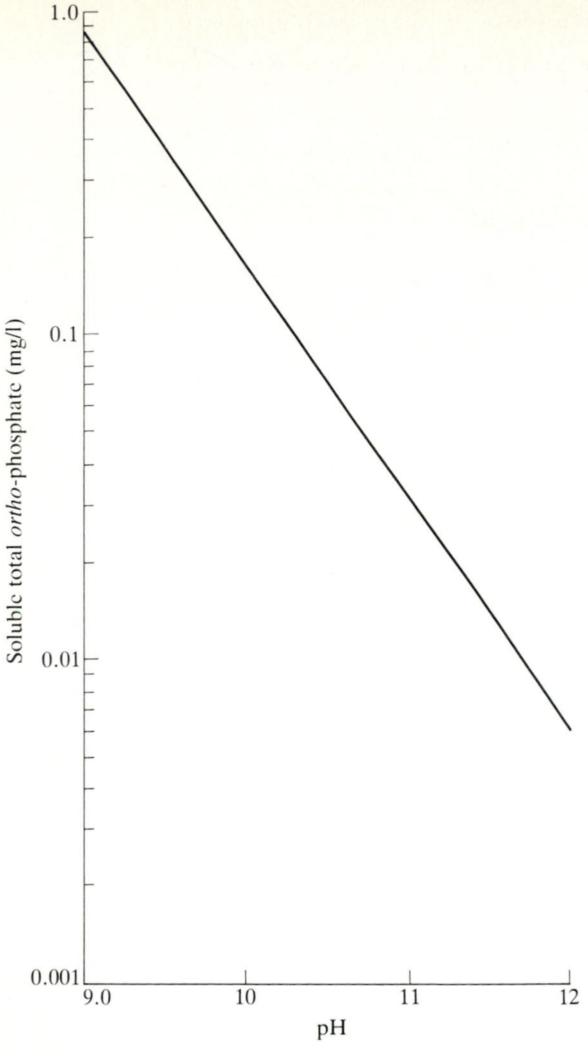

FIGURE 12-9
Ortho-phosphate versus pH for lime precipitation.[15]

$$5Ca^{2+} + 7OH^- + 3H_2PO_4^- \rightarrow Ca_5OH(PO_4)_3 + 6H_2O$$

Between pH 9.0 and pH 10.5 precipitation of calcium carbonate competes with calcium phosphate. Unlike aluminum and iron, calcium phosphate solids nucleate and grow very slowly, especially at neutral pH. The addition of seed enhances the reaction, indicating the advantage of solids recycle. Residual soluble phosphorus with respect to pH is shown in Fig. 12-9. The calcium phosphate precipitate is finely divided so that the presence of $Mg(OH)_2$ floc aids in the removal of the calcium phosphate precipitate.

The lime requirements will be dictated by the hardness and the alkalinity,

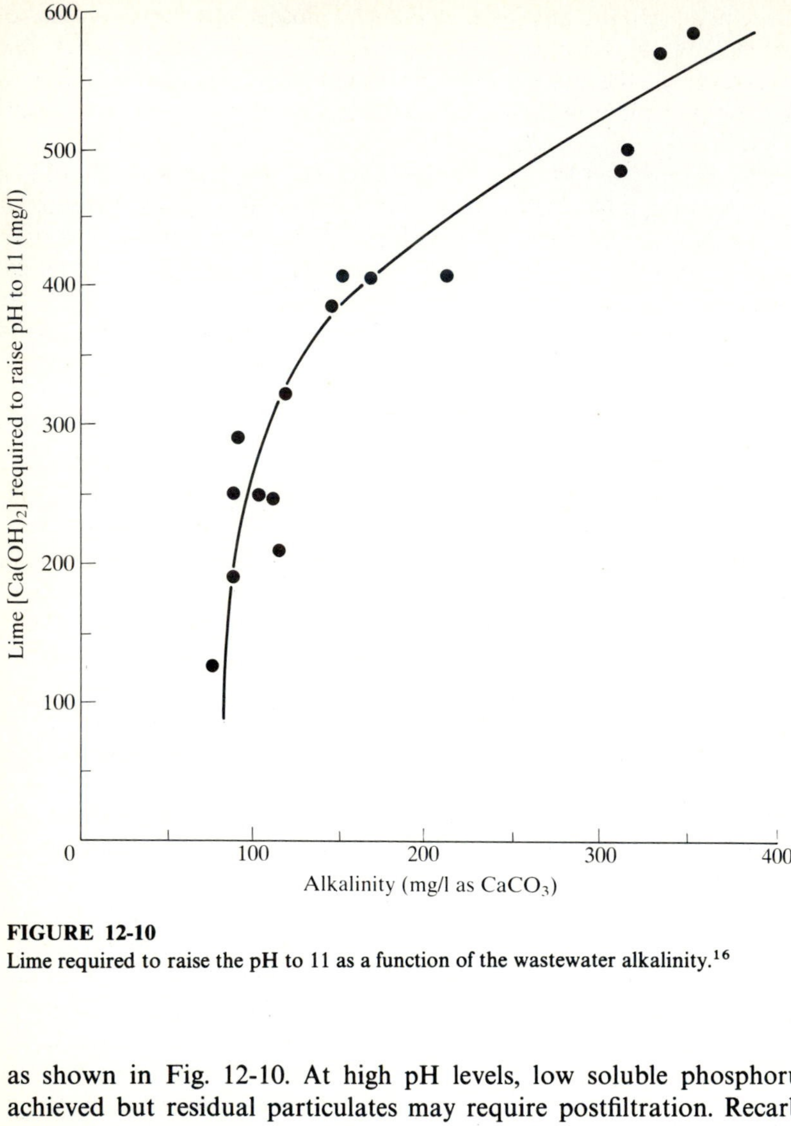

FIGURE 12-10
Lime required to raise the pH to 11 as a function of the wastewater alkalinity.[16]

as shown in Fig. 12-10. At high pH levels, low soluble phosphorus levels are achieved but residual particulates may require postfiltration. Recarbonation for pH adjustment after precipitation may redissolve particulate phosphorus if incomplete removal exists prior to pH adjustment.

Iron or aluminum can be employed as a direct precipitation of the metallic phosphate in the case of inorganic wastewaters or simultaneous precipitation by addition of the coagulating chemicals at the end of the aeration step in the activated sludge process.

Theoretical phosphorus residuals from precipitation with aluminum are a function of pH and the Al/P ratio. The precipitate is usually a mixture of $Al(OH)_3$ and $AlPO_4$, although the $AlPO_4$ precipitation is favored over $Al(OH)_3$. The precipitate tends to be amorphous rather than crystalline.

In order to approach the predicted solubility, a dosage of 1.5 to 3.0 moles of aluminum per mole of phosphorus as P is required over a pH range of 6.0 to 6.5. If the water is alkaline, the pH should be lowered prior to alum addition to minimize $Al(OH)_3$ precipitation. Some turbidity may result from the alum addition.

When the alum is added to the activated sludge process, addition should be immediately prior to the final clarifer in the case of completely mixed systems or at the end of plug-flow aeration basins. This is to avoid phosphorus precipitation in the biological process before microbial utilization and to minimize shear of the chemical flocs in the aeration basin.

In some cases, chemical requirements can be minimized by multiple-point addition; i.e., partial precipitation at the head end of the aeration basin with final precipitation at the end of the basin after microbial assimilation.

Iron can be added as $FeSO_4$ or as $FeCl_3$. The dosage is dependent on the dissolved oxygen level, the pH, biological catalysis, and the presence of sulfur and carbonates. Iron has been employed for phosphorus precipitation in biological-treatment processes but has the disadvantage of some iron in the treated effluent.

The iron dosage will range from 1.5 to 3.0 moles of iron (Fe^{3+}) per mole of phosphorus (as P). The optimum pH is 5.0, which is too low for conventional biological treatment. Precipitation at neutral pH values may produce a colloidal precipitate requiring a polymer to obtain a minimum total phosphorus residual.

The molar ratio of metallic ion to phosphorus increases as the effluent phosphorus concentration decreases, as shown in Fig. 12-11. These ratios are approximate due to side reactions forming byproducts such as hydroxide and carbonates.

For simultaneous precipitation, the optimum pH is in the range of 7 to 8. For direct precipitation, the optimum pH is in the order of 6. For processes with a long precipitation time, such as simultaneous precipitation, the calcium concentration in the water has a beneficial effect with both Al and Fe. Alkalinity has a negative effect on direct precipitation with short detention times and the removal of alkalinity is essential for a low P effluent. This is achieved to a certain extent automatically with the addition of Fe or Al since they behave as acids, although this condition will result in higher chemical consumption and increased sludge production. For low effluent phosphorus concentrations, effluent filtration may be required due to the high phosphorus content of the effluent suspended solids.

Biological Phosphorus Removal

Phosphorus, like N, is assimilated by bacteria and is removed from the water with the excess sludge. The degree of removal is related to the rate of sludge production, as discussed on page 162. It has been found, however, that certain plants can remove more than the usual quantity of phosphorus due to unusually high P content in the sludge mass.

The basic principle of bio-P removal is shown in Fig. 12-12. The bacteria are exposed alternatingly to anaerobic and aerobic conditions. Certain bacteria,

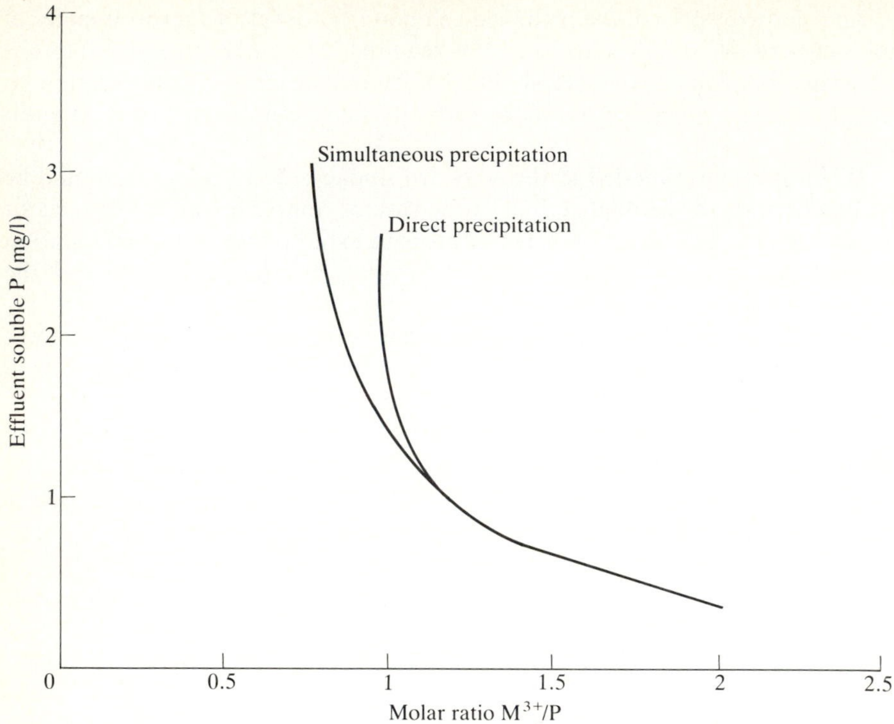

FIGURE 12-11
Phosphorus removal with aluminum and iron.

notably *Acinetobacter*, possess the ability to absorb low-molecular-weight organics (e.g., fatty acids) under anaerobic conditions. The energy required for this is made available by the release of phosphorus bound as polyphosphates in volutin granules in the protoplasm of the bacteria. Under subsequent exposure to aerobic conditions, the organic matter is oxidized and energy is made available for growth and for the reaccumulation of phosphates into polyphosphates in the bacteria. The net effect is an excess content of phosphorus in the bacteria possessing this ability. These bacteria have a competitive edge over other bacteria, since they can hoard the readily available organic matter for their own consumption. Under proper process conditions, they can flourish and dominate the population, resulting in an increased P content in the waste sludge.[18]

In addition to excess biological P uptake, chemical precipitation may also occur with calcium, magnesium, iron, and aluminum present in the wastewater. Recent data indicate that the physical-chemical removal increases with increasing biological P removal. One reason for the biologically mediated P precipitation may be the high phosphate concentration in the anaerobic reactor created by the biological P release.[19]

When nitrification and denitrification is not occurring, the influent soluble

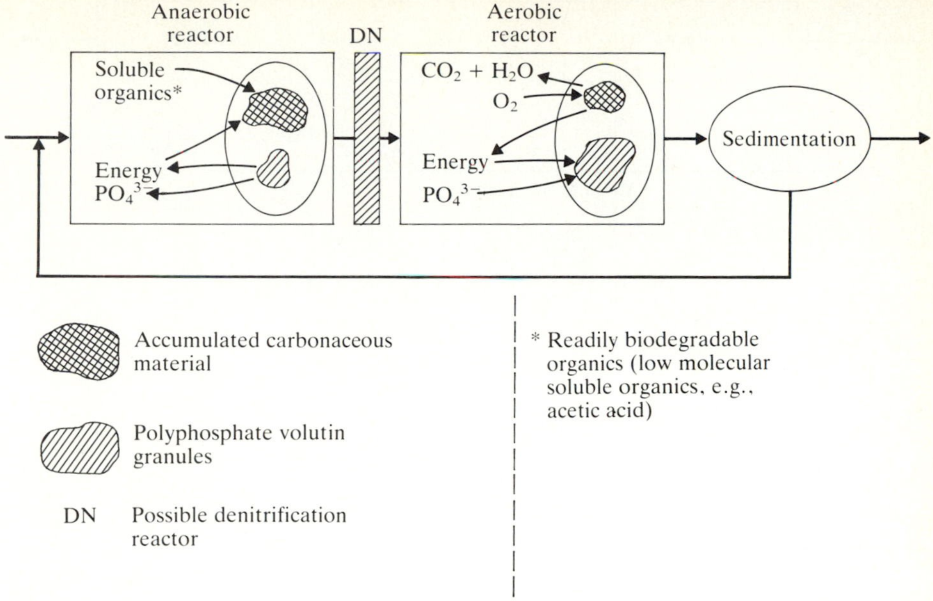

FIGURE 12-12
Biological release and uptake of phosphate while alternating between anaerobic and aerobic conditions.[17]

BOD_5/soluble P ratio should be greater than 10:1 in order to provide substrate for the bio-P organisms.[20] With nitrogen removal the influent COD/TKN must be greater than 12. The P content of the excess sludge is in the order of 5 to 7 percent. When the plant effluent has 10 to 30 mg/l suspended solids, the total content in the effluent will be 0.5 to 2.0 mg/l P. Under optimum conditions, the soluble P in the effluent can be reduced to 0.1 mg/l. A combination of bio-P removal with simultaneous precipitation with small dosages of Al or Fe will achieve low effluent P concentrations.[21]

The principle of a biological-treatment plant with biological removal of phosphorus is the exposure of the organisms to alternating anaerobic and aerobic conditions. This can be implemented with or without nitrogen removal. The alternating exposure to anaerobic and to aerobic conditions can be arranged by recirculation of the biomass through anaerobic and aerobic stages, and an anoxic stage if nitrogen removal is also desired. Generalized plant flowsheets are shown in Fig. 12-13.

12.5 FILTRATION

Filtration is employed for the removal of suspended solids as a pretreatment for low suspended solids wastewaters, following coagulation in physical chemical

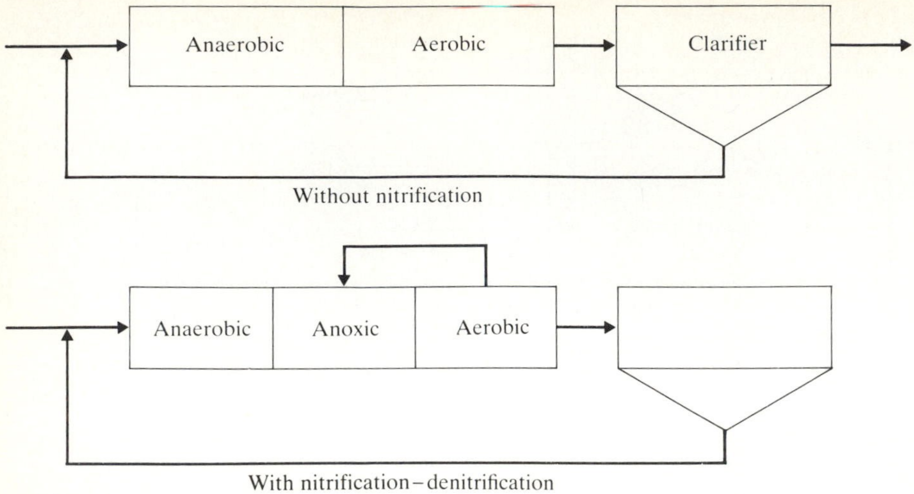

FIGURE 12-13
Biological phosphorus removal with and without nitrification-denitrification.

treatment or as a tertiary treatment following a biological wastewater-treatment process.

Suspended solids are removed on the surface of a filter by straining, and through the depth of a filter by both straining and adsorption. Adsorption is related to the zeta potential on the suspended solids and the filter media. Particles normally encountered in a wastewater vary in size and particle charge and some will pass the filter continuously. The efficiency of the filtration process is therefore a function of:

1. The concentration and characteristics of the solids in suspension
2. The characteristics of the filter media and other filtration aids
3. The method of filter operation

Granular media filters may be either gravity or pressure. Gravity filters may be operated at a constant rate with influent flow control and flow splitting or at a declining rate with four or more units fed through a common header. To achieve constant flow an artificial head loss (flow regulator) is used. As suspended solids are removed and the head loss increases, the artificial head loss is reduced so the total head loss remains constant. In a declining rate filter design, the decrease in flow rate through one filter as the head loss increases raises the head and rate through the other filters. A maximum filtration rate of 6 gal/(min · ft^2) [0.24 m^3/(min · m^2)] is used when one unit is out of service. The filter run terminates when the total head loss reaches the available driving force or when excess suspended solids or turbidity appear in the effluent.

Media size is an important consideration in filter design. The sand size is chosen on the basis that it provides slightly better removal than is required. In

TABLE 12.9
Media options

Type	Material	Size, mm	Depth, in
1. Monomedia			
(*a*) Fine	Sand	0.35–0.60	10–20
(*b*) Coarse	Anthracite coal	1.3–1.7	36–60
2. Dual	Sand,	0.45–0.6	10–12
	anthracite coal	1.0–1.1	20–30
3. Multimedia	Garnet,†	0.25–0.4	2–4
	sand,	0.45–0.55	8–12
	anthracite coal	1.0–1.1	18–24

† Other types of material such as metal oxides are also used.

Note: in = 2.54 cm

dual media filters, the coal size is selected to provide 75 to 90 percent suspended solids removal across 1.5 to 2.0 ft (0.46 to 0.6 m) of media. For example, if 90 percent suspended solids removal is desired across a filter bed, 68 to 80 percent should be removed through the coal layer and the remaining 10 to 25 percent through the sand layer. If the feed suspended solids particle size is larger than 5 percent of the granular media particles, mechanical straining will occur.

A 25-μm particle will be mechanically strained by a 0.5-mm filter media. If the feed solids particles have a density of 2 to 3 times that of the suspending medium, then particles as small as 0.5 percent of the filter media particle size can be effectively removed by in-depth granular media filtration.

Table 12.9 presents available media options. Two general types of monomedias are available, fine and coarse. The fine medias are usually found in propriety-type filters such as the automatic backwash filter or the pulse-bed-type filter and rely on a straining mechanism for removal. Frequent backwashes (or pulsing) are required, especially during plant upset and/or high influent turbidity conditions. The coarse monomedia is usually much deeper than larger medias requiring scour for cost-effective backwash. Operation of coarse media filters is characterized by longer filter runs and the ability to respond to plant upset conditions. Dual media and multimedia have traditionally been used in potable water applications and their use has carried over into the tertiary filtration application wastewater treatment.

Filtration rate will affect the buildup of head loss and the effluent quality attainable. The optimum filtration rate is defined as the filtration rate, which results in the maximum volume of filtrate per unit filter area while achieving an acceptable effluent quality.

Too high a filtration rate will permit solids to penetrate the coarse media and accumulate on the fine media. Too low a filtration rate is insufficient to achieve good solids penetration of the coarse media, resulting in head loss buildup at the top of the coarse media. Filtration rate will also influence effluent quality depending on the nature of the particles to be removed.

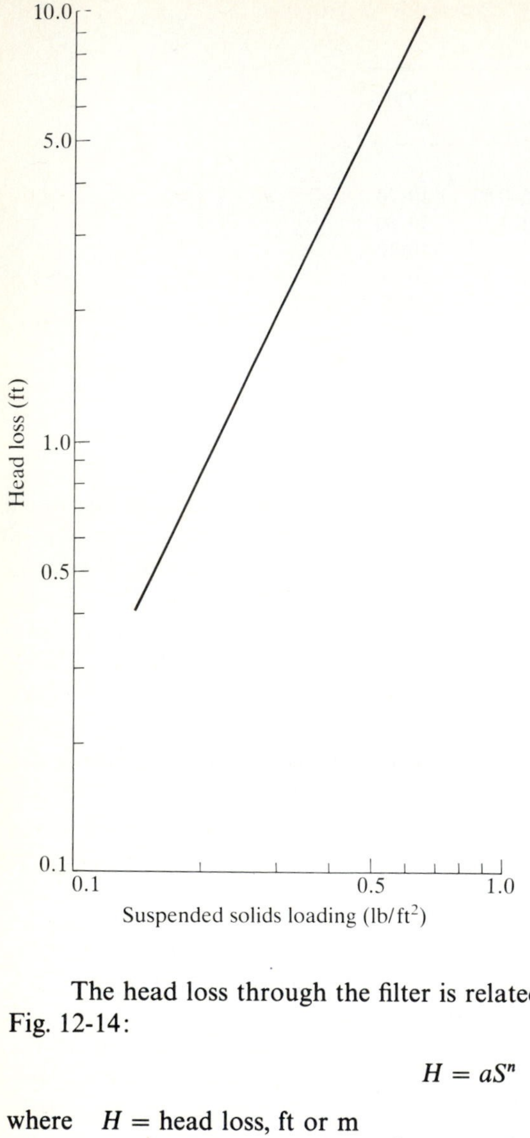

FIGURE 12-14
Head loss versus solids retention in a filter.

The head loss through the filter is related to the solids loading, as shown in Fig. 12-14:

$$H = aS^n \tag{12.7}$$

where H = head loss, ft or m
S = solids captured, lb/ft^2 or kg/m^2
a, n = constants

For a given head loss, the filtration cycle depends upon the influent suspended solids and the hydraulic flow rate. The type of coagulant used may also influence the head loss, as shown in Fig. 12-15.

Improved suspended solids removals can be achieved by the addition of coagulants to the wastewater prior to filtration. The use of alum also results in the precipitation and removal of phosphorus through the filter. Flocculation is not needed since the filter serves as a flocculator. Effective mixing is required to

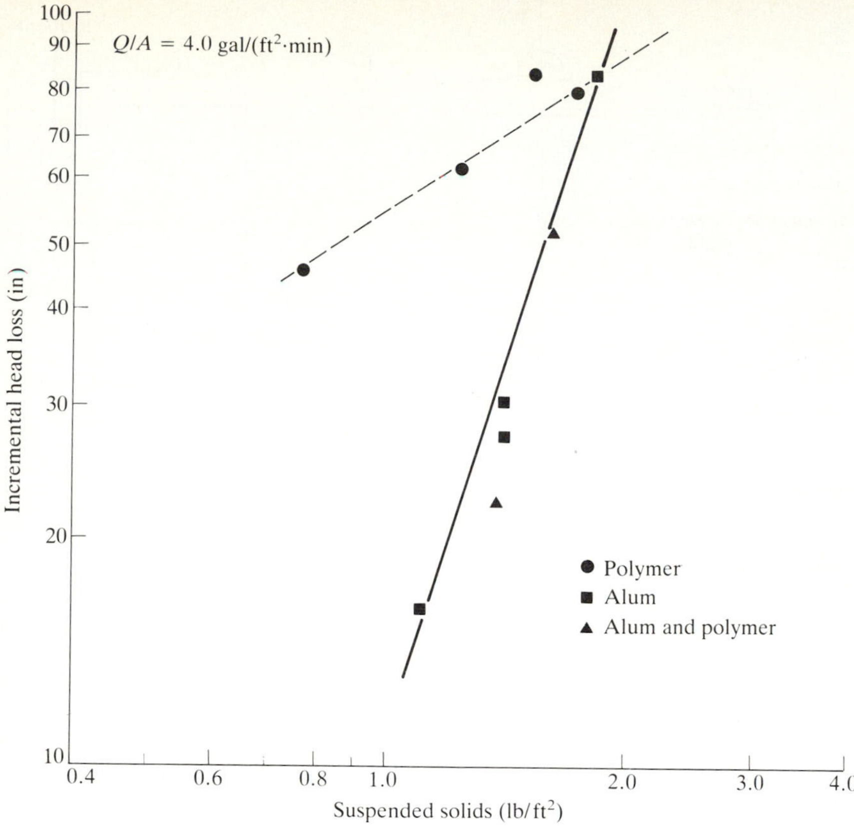

Note:
in = 2.54 cm
lb/ft^2 = 4.89 kg/m^2
gal/(min·ft^2) = 4.07×10^{-2} m^3/(min·m^2)

FIGURE 12-15
Head loss as related to solids deposit in a filter.

disperse the chemicals and initiate the reaction. Since the suspended solids are removed by filtration rather than by sedimentation, 25 to 50 percent less chemicals are required in many cases. For most applications, a maximum of 100 mg/l suspended solids removal is used in order to avoid excessive backwash volumes.

Backwash systems used for the filtration of secondary effluent require an auxiliary scour system to effectively dislodge particulates. The two most frequently used backwash systems are: (1) water with auxiliary surface agitation and (2) water with air scour. Surface systems may be a fixed grid or rotary wash. The filter backwash rate is a function of media size (both effective size and uniformity coefficient), media type (specific gravity), and water temperature. Air scour rates are in the range of 1 to 5 standard ft^3/(min · ft^2) [0.305 to 1.53 std m^3/(min · m^2)] with auxiliary surface wash rates in the range of 2 to 5 gal/(min · ft^2) [0.081 to 0.204 m^3/(min · m^2)].

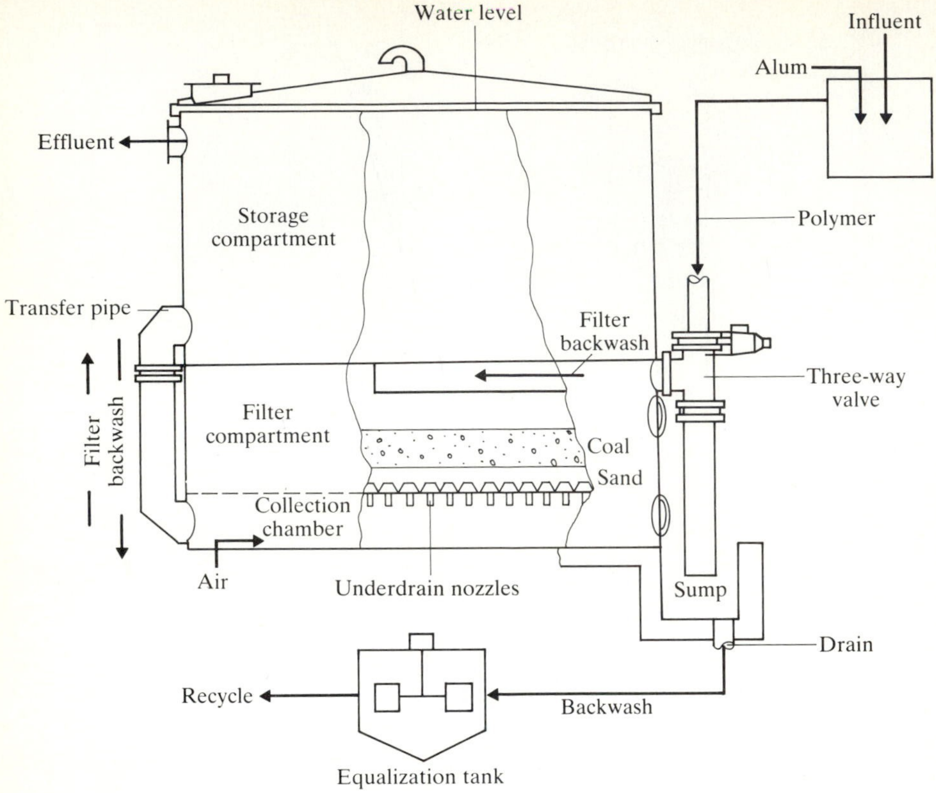

FIGURE 12-16
Typical automatic dual media filter.

Granular media filters can be fully automated by the use of pressure sensors that monitor the head loss across the filter. When the terminal head loss is achieved, the filter would automatically backwash. Turbidimeters serve as a secondary control such that the backwash sequence would be initiated when the turbidity reached the allowable level in the effluent.

While considerable data are available for the design of filters treating domestic secondary effluents, industrial wastewaters require pilot plant studies to define the type of media, filter flow rate, coagulant requirements, head loss relationships, and backwash requirements.

There are several types of filters available today. Three of the more common are the dual media filter consisting of anthrafilt (coal) and sand, the Hydroclear filter, and the continuous backwash filter.[2]

A typical dual media filter is shown in Fig. 12-16. The Hydroclear filter employs a single sand media with an air mix for solids suspension and regeneration of the filter surface. Filter operation enables periodic regeneration of the media surface without backwashing. Typical operating parameters are shown opposite:

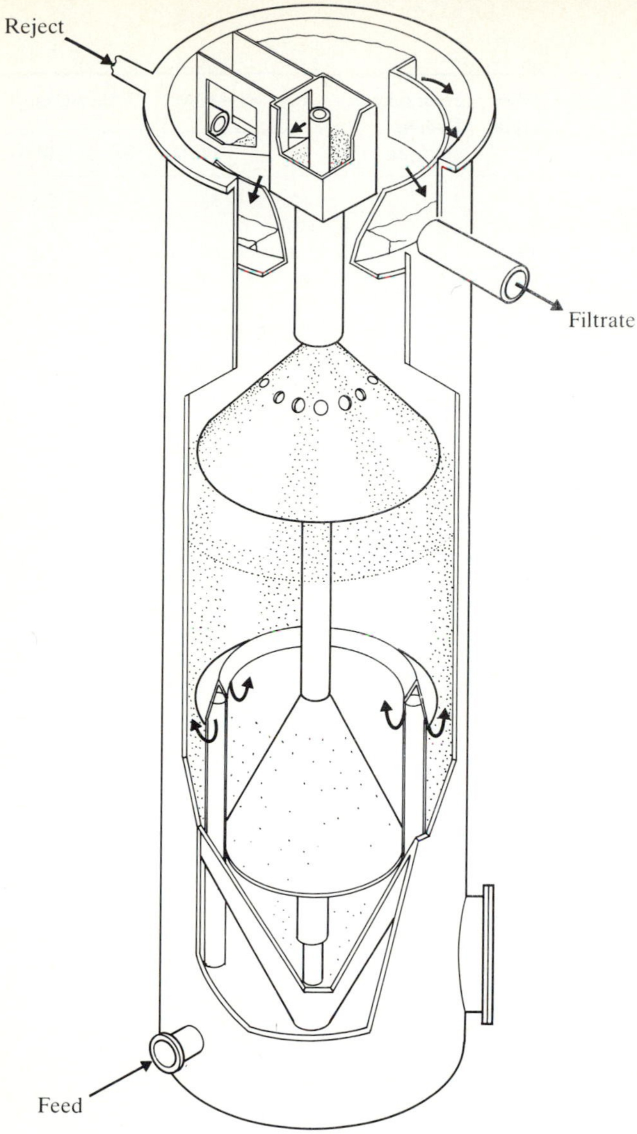

FIGURE 12-17
Dynasand filter (DSF). (*Courtesy of Parkson Corporation.*)

Filtration rate	2–5 gal/(min · ft^2) [0.081–0.204 m^3/(min · m^2)]
Media size	0.35–0.45 mm sand
Bed depth	10–12 in (25.4–30.4 cm)
Backwash rate	12 gal/(min · ft^2)[0.5 m^3/(min · m^2)]
Air mix	0.25 standard ft^3/(min · ft^2) [0.076 std m^3/(min · m^2)]
Terminal head loss	3.5 ft (1.07 m)
Backwash filtrate ratio	0.10

TABLE 12.10
Filtration performance

Filter type	Wastewater	Filter depth, ft	Hydraulic loading, gal/(min · ft²)	Percent removal		Effluent, mg/l	
				SS	BOD	SS	BOD
Gravity downflow	TF effluent	2–3	3	67	58	—	2.5
Pressure upflow	AS effluent	5	2.2	50	62	7.0	6.4
Dual media	AS effluent	2.5	5.0	74	88	4.6	2.5
Gravity downflow	AS effluent	1.0	5.3	62	78	5	4
Dynasand	Metal finishing	3.3	4–6	90	—	2–5	—
	AS effluent	3.3	3–10	75–90	—	5–10	—
	Oily wastewater	3.3	2–6	80–90†	—	5–10†	—
Hydroclear	Poultry	1	2–5	88	—	19	—
	Oil refinery	1	2–5	68	—	11	—
	Unbleached kraft	1	2–5	74	—	17	—

† Free oil.
Note:
ft = 0.305 m
gal/(min · ft²) = 4.07×10^{-2} m³/(min · m²)

The Dynasand (DSF) continuous backwash filter is a continuous self-cleaning upflow deep-bed granular media filter. The filter media is cleaned continuously by recycling the sand internally through an airlift pipe and sand washer, as shown in Fig. 12-17. The regenerated sand is redistributed on top of the bed, allowing for a continuous uninterrupted flow of filtered water and reject water. Filtration performance is shown in Table 12.10.

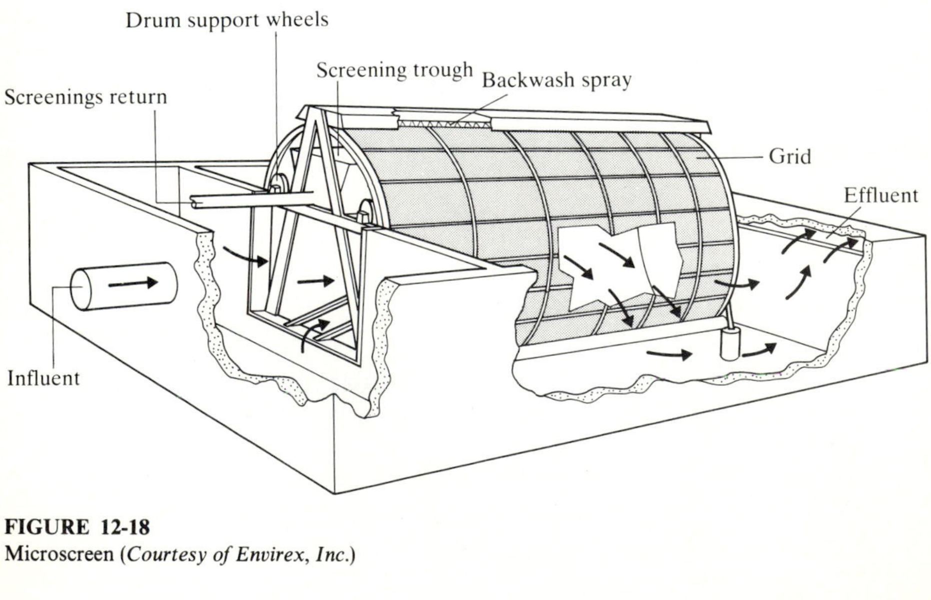

FIGURE 12-18
Microscreen (*Courtesy of Envirex, Inc.*)

TABLE 12.11
Removal efficiencies

Aperture screen, μm	Flow rate, gal/(min · ft²) (submerged)	Percent removal	
		SS	BOD
35	10.0	50–60	40–50
23	6.7	70–80	60–70

Note: gal/(min · ft²) = 4.07 × 10^{-2} m³/(min · m²)

Microscreen

A microscreen is a rotary drum revolving on a horizontal axis covered with stainless steel fabric (Fig. 12-18). The water enters the open end of the drum and is filtered through the fabric with solids being retained on the inside surface of the fabric. As the drum rotates, the solids are transported and continuously removed at the top of the drum by pumping effluent under pressure through a series of spray nozzles that extend the length of the drum. The head loss is less than 12 to 18 inches (30 to 46 cm) of water. The backwash water is 46 percent of the total throughput water. Peripheral drum speeds vary up to 100 ft/min (30.5 m/min) with hydraulic loadings of 2.5 to 10 gal/(d · ft^2) [0.1 to 0.4 m^3/(min · m^2)]. Periodic cleaning of the drum is required for slime control.

For filtration of secondary effluent a maximum solids loading of 0.88 lb/(ft^2 · d) [4.3 kg/(m^2 · d)] at a hydraulic loading of 6.6 gal/(min · ft^2) [0.27 m^3/(min · m^2)] has been reported. Lynam *et al.*[22] reported effluent suspended solids and BOD of 6 to 8 mg/l and 3.5 to 5 mg/l, respectively, with a 20–20 activated sludge effluent on a 23-m microscreen at 3.5 gal/(min · ft^2) [0.14 m^3/(min · m^2)]. For design purposes, the removal efficiencies treating a secondary effluent are given in Table 12.11.

The efficiency of the unit is suspended solids sensitive, as indicated by a decrease in the throughput rate from 60 to 13 gal/min (0.227 to 0.049 m^3/min) with an increase in influent suspended solids from 25 to 200 mg/l.[23]

REFERENCES

1. Eckenfelder, W. W., J. P. Lawler, and J. T. Walsch: "Study of Fruit and Vegetable Processing Waste Disposal Methods in the Eastern Region," U.S. Department of Agriculture, Final Report, 1958.
2. Crites, R. W.: *Proceedings of the Industrial Wastes Symposia*, Water Pollut. Control Fed., 1982.
3. "Diagnosis and Improvement of Saline and Alkali Solids," U.S. Department of Agriculture, Handbook 60, 1954.
4. Jewell, W. J.: *Limitations of Land Treatment of Wastes in the Vegetable Processing Industries*, Cornell University, 1978.
5. Adamczyka, A. F.: *Land as a Waste Management Alternative* (Ed. R. C. Loehr), Ann Arbor Science, Ann Arbor, Mich., 1977.

6. Haith, D. A., and Chapman, D. C.: "Land Application as a Best Practical Treatment Alternative," in *Land as a Waste Management Alternative* (Ed. R. C. Loehr), Ann Arbor Science, Ann Arbor, Mich. 1977.
7. Crites, R. W., *et al.*: *Proc. 5th Natn. Symp. on Food Processing Wastes*, 1974.
8. Gellman, I., and R. O. Blosser: *Proc. 14th Ind. Waste Conf.*, 1959, Purdue University.
9. Wisneiwski, T. F., A. J. Wiley, and B. F. Lueck: *TAPPI*, vol. 89, pt. 2, p. 65, 1956.
10. Billings, R. M.: *Proc. 13th Ind. Waste Conf.*, 1958, Purdue University.
11. Warner, D. L.: "Deep Well Injection of Liquid Waste," Environmental Health Series, U.S. Department of Health, Education, and Welfare, Cincinnati, Ohio, April 1965.
12. Donaldson, E. C.: "Subsurface Disposal of Industrial Wastes in the United States," Bureau of Mines Inform. Circ. 8212, U.S. Department of Interior, Washington, D.C., 1964.
13. Agardy, F. J.: *Membrane Processes, Process Design in Water Quality Engineering* (Eds. E. L. Tackston and W. W. Eckenfelder), Jenkins Publishing Co., Austin, Tex., 1972.
14. Okey, R. W.: *Water Quality Improvement by Physical and Chemical Processes* (Eds. E. F. Gloyna and W. W. Eckenfelder), University of Texas Press, Austin, Tex., 1970.
15. Zoltek, J.: *J. Water Pollut. Control Fed.*, vol. 48, p. 179, 1976.
16. Tshobanoglous, G.: *Proc. 12th Sanitary Engineering Conf.*, 1970, University of Illinois.
17. Arvin, E.: "Biological Removal of Phosphorus from Wastewater," *CRC Critical Rev. J.*, vol. 15, pp. 25–64, 1985.
18. Marais, G. V. R., *et al.*: *Water Sci. Technol.*, vol. 15, nos. 3–4, 1983.
19. Arvin, E.: *Water Sci. Technol.*, vol. 15, nos. 3–4, 1983.
20. Hong, S. N., *et al.*: *EPA Workshop on Biological Phosphorus Removal*, Annapolis, Md., June 1982.
21. Ekema, G. A., *et al.*: *Water Sci. Technol.*, vol. 15, nos. 3–4, 1983.
22. Lynam, B., *et al.*: *J. Water Pollut. Control Fed.*, vol. 41, p. 247, 1969.
23. Convery, J. J.: *FWPCA Symposium on Nutrient Removal and Advanced Waste Treatment*, Tampa, Fla., 1968.

BIBLIOGRAPHY

Adams, C. E., Jr., and W. W. Eckenfelder, Jr.: *Process Design Techniques for Industrial Waste Treatment*, Enviro Process, Nashville, Tenn., 1974.

Adams, C. E., D. L. Ford, and W. W. Eckenfelder: *Development of Design and Operational Criteria for Wastewater Treatment*, Butterworth, Boston, Mass., 1981.

Azad, H. S. (Ed.): *Industrial Wastewater Management Handbook*, McGraw-Hill, New York, 1976.

Benefield, L. D., and C. W. Randall: *Biological Process Design for Wastewater Treatment*, Prentice-Hall, Englewood Cliffs, N.J., 1980.

Besselievre, E. B., and M. Schwartz: *Treatment of Industrial Waste*, McGraw-Hill, New York, 1976.

Busch, A. W.: *Aerobic Biological Treatment of Waste Water and Practices*, Oligodynamics Press, Houston, Tex., 1971.

Callely, A. G., C. F. Forster, and D. A. Stafford (Eds.): *Treatment of Industrial Effluents*, Halsted Press, Somerset, N.J., 1976.

Camp, T. R., and R. L. Meserve: *Water and Its Impurities*, 2d ed., Dowden, Hutchinson, and Ross, Stroudsburg, Pa., 1974.

Eckenfelder, W. W.: *Principles of Water Quality Management*, CBI Publishing, Boston, Mass., 1980.

EPA Treatability Manual EPA-600/2-82-0012, 1982, vol. 1, *Treatability Data*; vol. 2, *Industrial Descriptions*; vol. 3, *Technologies for Control/Removal of Pollutants*; vol. 4, *Cost Estimating.*

Evans, F. L., III (Ed.): *Ozone in Water and Wastewater Treatment*, Ann Arbor Science, Ann Arbor, Mich., 1972.

Fair, G. M., J. C. Geyer, and D. A. Okun: *Water and Wastewater Engineering*, vol. 2, *Water Purification and Wastewater Treatment and Disposal*, Wiley, New York, 1966.

Faust, S. D., and J. V. Hunter (Eds.): *Principles and Applications of Water Chemistry*, Wiley, New York, 1967.

Gehm, H. W., and J. I. Bregman (Eds.): *Handbook of Water Resources and Pollution Control*, Van Nostrand Reinhold, New York, 1976.

Gibbon, D. L. (Ed.): *Aeration of Activated Sludge in Sewage Treatment*, Pergamon Press, Oxford, 1974.

Gloyna, E. F., and W. W. Eckenfelder, Jr. (Eds.): *Advances in Water Quality Improvement*, University of Texas Press, Austin, 1968.

Goransson, B. (Ed.): *Industrial Wastewater and Wastes*, vol. 2, Pergamon Press, Oxford, 1977.

Gurnham, C. F. (Ed.): *Industrial Wastewater Control*, vol. II, *Chemical Technology*, Academic Press, New York, 1965.

Humenick, M. J.: *Water and Wastewater Treatment: Calculations for Chemical and Physical Processes*, Marcel Dekker, New York, 1977.

Jorgenson, S. E., and M. J. Gromiez: *Mathematical Models in Biological Wastewater Treatment*, Elsevier, Amsterdam, 1985.

Loehr, R. C.: *Pollution Control for Agriculture*, Academic Press, New York, 1977.

Mancy, K. H., and W. J. Weber, Jr.: *Analysis of Industrial Wastewater*, Wiley, New York, 1972.

Manual of Methods for Chemical Analysis of Water and Wastes. Technology Transfer, U.S. Environmental Protection Agency, 1974.

Metcalf and Eddy, Inc.: *Wastewater Engineering: Collection, Treatment, Disposal*, McGraw-Hill, New York, 1972.

Middlebrooks, E. J., *et al.*: *Wastewater Stabilization Lagoon Design, Performance and Upgrading*, Macmillan, New York, 1982.

Nemerow, N. L.: *Industrial Water Pollution, Origins, Characteristics, and Treatment*, Addison-Wesley, Reading, Mass., 1978.

Patterson, J. W.: *Industrial Wastewater Treatment Technology*, 2d ed., Butterworth, Boston, Mass., 1985.

Proceedings of Annual Industrial Waste Conferences, Purdue University, West Lafayette, Ind.

Proceedings, A Consolidated Approach to Activated Sludge Process Design, Progress in Water Technology, vol. 7, no. 1, Pergamon Press, Oxford, 1975.

Proceedings of the Third National Conference on Complete Water Reuse, Cincinnati, Ohio, June 1976, American Institute of Chemical Engineers, New York, 1976.

Proceedings of the First Int. Conf. on Fixed Film Biological Processes, Kings Island, Ohio, 1982.

Proceedings of the Second Int. Conf. on Fixed Film Biological Processes, Arlington, Va., 1984.

Process Design Manual for Nitrogen Control, Technology Transfer, U.S. Environmental Protection Agency, October 1975.

Reuse of Water in Industry, Butterworth, London, 1963.

Rice, R., and J. A. Cotruvo: *Ozone/Chlorine Dioxide Oxidation Products of Organic Materials*, Ozone Press International, Cleveland, Ohio, 1978.

Ross, R. D. (Ed.): *Industrial Waste Disposal*, Van Nostrand Reinhold, New York, 1968.

Sanks, R. L., and T. Asano: *Land Treatment and Disposal of Municipal and Industrial Wastewater*, Ann Arbor Science, Ann Arbor, Mich., 1976.

Sarner, E.: *Plastic Packed Trickling Filters*, Ann Arbor Science, Ann Arbor, Mich., 1980.

Sawyer, C. N., and P. L. McCarty: *Chemistry for Environmental Engineering*, 3d ed., McGraw-Hill, New York, 1978.

Schmidtke, N. W., and D. W. Smith, *Scaleup of Water and Wastewater Treatment Processes*, Butterworth, Boston, Mass., 1983.

Speece, R. E., and J. F. Malina, Jr. (Eds.): *Application of Commercial Oxygen to Water and Wastewater Systems*, University of Texas Press, Austin, 1973.

Standard Methods of Water and Wastewater Analysis, 16th ed., American Public Health Association, New York, 1986.

Stumm, W., and J. J. Morgan: *Aquatic Chemistry*, Wiley, New York, 1970.

Schroeder, E. D.: *Water and Wastewater Engineering Treatment*, McGraw-Hill, New York, 1977.

Vesilind, A. P.: *Treatment and Disposal of Wastewater Sludges*, Ann Arbor Science, Ann Arbor, Mich., 1974.

Water-1974. I. Industrial Wastewater Treatment, AIChE Symposium Series, vol. 70, no. 144, American Institute of Chemical Engineers, New York, 1974.

Water-1976. II. Biological Wastewater Treatment, AIChE Symposium Series, vol. 73, no. 167, American Institute of Chemical Engineers, New York, 1977.

Water Technology in the Pulp and Paper Industry, TAPPI Monograph Series, no. 18, New Technical Association of the Pulp and Paper Industry, New York, 1957.

Weber, W. J., Jr.: *Physicochemical Processes for Water Quality Control*, Wiley, New York, 1972.

Verschurenen, K.: *Handbook on Environmental Data on Organic Chemicals*, Van Nostrand Reinhold, New York, 1977.

INDEX